The Tao of Chemistry and Life

The Tao of Chemistry and Life

A Scientific Journey

Eugene H. Cordes

OXFORD
UNIVERSITY PRESS
2009

OXFORD
UNIVERSITY PRESS

Oxford University Press, Inc., publishes works that further
Oxford University's objective of excellence
in research, scholarship, and education.

Oxford   New York
Auckland   Cape Town   Dar es Salaam   Hong Kong   Karachi
Kuala Lumpur   Madrid   Melbourne   Mexico City   Nairobi
New Delhi   Shanghai   Taipei   Toronto

With offices in
Argentina   Austria   Brazil   Chile   Czech Republic   France   Greece
Guatemala   Hungary   Italy   Japan   Poland   Portugal   Singapore
South Korea   Switzerland   Thailand   Turkey   Ukraine   Vietnam

Published by Oxford University Press, Inc.
198 Madison Avenue, New York, New York 10016

www.oup.com

Library of Congress Cataloging-in-Publication Data
Cordes, Eugene H.
The tao of chemistry and life: a scientific journey / Eugene H. Cordes.
    p.   cm.
Includes bibliographical references and index.
ISBN 978-0-19-536963-2
1. Biochemistry—Miscellanea. 2. Molecular structure.
3. Life (Biology). 4. Organisms. I. Title.
QD415.C66 2009
572—dc22   2009005376

9 8 7 6 5 4 3 2

Printed in the United States of America
on acid-free paper

For Shirley, Jennifer, Matthew

Preface

As most writers will testify, writing a book entails a fair amount of hard work. It follows that one needs a reason for undertaking the task in the first place. The easiest motivation to understand for writing a book, and perhaps the most common one, is the desire for monetary reward. I do not know what inspired J. K. Rowling to write her amazingly successful series of books about Harry Potter and his friends at Hogwarts but if monetary reward was it, she has succeeded beyond the wildest dreams of most of us. My stated reason for reading about Harry Potter is the need to keep up with our grandchildren but the basic fact is that I really enjoyed the stories.

Aside from the pull of economic gravity, I think that a lot of people write books for other reasons. That is just as well since most books do not make their authors anywhere near enough money to compensate for the effort involved, let alone to live on. Some writers have insights that they want to share; others have political ends to satisfy; some write to satisfy their desire to make the world better (and some to make it worse); and a few would really like to see a book in print with their name on the cover.

My motivation for taking a word processing program in hand derives largely from a sense of frustration. The frustration derives from seeing all the unnecessary damage that is done out of ignorance of some rather prosaic things in science and, more specifically, in chemistry. Now it is certainly true that a whole lot of stuff in chemistry is profoundly unimportant to anyone but a practicing chemist. It is also true that there are some things in chemistry that are profoundly important to all sentient beings. Those things are the focus of this book. It is written with the primary intent to inform, not entertain.

A lot of people have a substantially negative feeling about chemistry and things chemical. A good bit of this may derive from the way chemistry is taught in both high school and in colleges and universities, specifically for those students who have no intention of becoming chemists. I am more familiar with the introductory courses in universities, having lived a lot in that world, and so will focus on those. The textbooks for general chemistry courses that find wide use in universities are all very carefully done, have loads of multicolored pictures and illustrations, include vast collections of problems, have correlated CDs, cost a great deal of money, and are all basically the same. They generally focus on the stuff exceptionally unimportant for most people. Chemistry majors need to know this stuff but it bores the rest of the world no end (some of it may bore chemistry majors too but they still need to know it). General chemistry courses update their content as new information and insights are developed and take advantage of new technologies as they come along. Beyond that, the teaching of chemistry changes very little over time. Among other things, general chemistry texts focus on quantitative issues, for which it is easy to write problem sets and readily graded exams. Couple the above with the fact that the substantial majority of students in general chemistry courses at the college and university level would just as soon be somewhere else, perhaps anywhere else. Many students see chemistry courses as something to be gotten over with on the way to, for example, medical school or an engineering degree. All of this is a bit frustrating to me since chemistry, particularly as it relates to life and health, is deeply important to most of us and most of us would be better off if we knew more. So that is the rationale for this book: to help the intelligent, interested nonscientist come to grips with some essentials of chemistry and how they relate to life and health.

This book assumes no background in chemistry and very little in biology. If you have one or more chemistry courses in your background and remember some of it, it will help.

There are several key points that form the core message of this book.

- All life is unified: by commonality of the molecules of life, cells, energy interrelationships, and metabolism.
- All life as we know it is based on the chemistry of carbon. Other key elements of life include hydrogen, oxygen, nitrogen, sulfur, and phosphorus.
- Molecular recognition—the fitting together of small molecules with each other, large molecules with each other, and small molecules with large ones—underlies the key phenomena of life.
- Biological outcomes are a sensitive function of molecular structure.

It may be worthwhile to keep these four points in mind as you move forward through the book.

Information is available to help the nonscientist cope more fully and capably with issues that affect your health and well-being and of others who depend on you. It seems to me that there is an obligation to seek out that information, think about it, and use it. In addition to enabling you to make better choices in life and to enjoy the satisfaction of understanding, there is one other very good reason for understanding some of this stuff. The knowledge has been gained in significant part through the work of scientists supported financially by governments. Put bluntly, if you pay taxes, then you have invested in this knowledge. It belongs to you. Take possession.

Acknowledgments

This book covers a fair amount of scientific ground. In an effort to get as many things right as reasonably possible, I have asked a number of people to read one or more chapters in draft form or otherwise make a contribution to the book. Their comments, criticisms, information, and insights have been highly useful in making this a better book and I am grateful for the help.

Thanks go to Paul Anderson, Frank Ascione, Jerome Birnbaum, Lewis Cantley, Jennifer Darnell, Robert Darnell, Bink Garrison, Lori Morton, Mark Murcko, Larry Sternson, and Leslie Vosshall. I am also indebted to Bonnie Bassler, Milos Novotny, and Charles Wysocki who provided reprints and preprints of manuscripts covering important work cited in this book. Aspects of the structure of chapters 3–6 derive in part from a general chemistry textbook that I coauthored with Riley Schaeffer many years ago and thanks are due to Riley for his contributions.

Special thanks go to Mahendra Jain who read more of this book than anyone else and provided a great many useful insights. In addition, Mahendra asked several students in his introductory biochemistry course at the University of Delaware to read and provide comments on several chapters. Their help is also appreciated.

I am indebted to Sandra Geis for her permission to use five illustrations from the elegant work of her father, Irving Geis, and to Donald and Judith Voet for permission to use several figures from their marvelous textbook *Biochemistry*.

Thanks to Samuel Barondes for permission to reprint his clever and useful poem that appears at the beginning of chapter 22.

Finally, I am grateful to Vertex Pharmaceuticals Inc. for supplying the photograph that graces the cover of this book.

Whatever shortcomings in the book remain, they are the sole responsibility of the author.

Contents

The Tao of Chemistry and Life

1

Life

Unity out of diversity

Despite marked diversity in habitat, required resources, size, shape, and structural organization, all life forms are unified by commonality in the molecules of life. These are responsible for inheritance, differentiation, development, cellular organization, and all metabolic events from birth until death.

Life! We celebrate its arrival and bemoan its passing. Between birth and death, we protect life, cling to it, and perhaps prepare for what may come after it.

The birth of a healthy baby is one of the defining events in the course of a marriage. Like most people, I recall the birth of our children, a daughter first and later a son, with joy and pleasure. Without question, these are two of the most memorable events in what is now quite a long life. Later, I derived happiness from the births of our four grandchildren, a granddaughter followed by three grandsons. That happiness was widely shared by family and friends: at each birth, a new life brimming with possibility was brought into the world. My wife and I follow the progress of these lives with love and care. We will continue to do that until our lives have come to their inevitable end.

Each new life brings with it responsibilities: to ensure that this new life is long, happy, and productive. Innumerable hours will be spent in loving, holding, entertaining, nurturing, coaching, correcting, and educating this new life. Cameras, camcorders, and tape recorders record events in the lives of the young for enjoyment throughout life. Talents will be uncovered and developed: perhaps in music or art or

3

athletics or mathematics or cooking. Achievements are greeted with pride and joy; shortcomings with heartache.

From our earliest days, most of us get help in protecting our lives. When young, we are told to eat our fruits and vegetables, look both ways before crossing the street, wash our hands before we eat, and not to talk to strangers. Later, we are counseled to avoid smoking, moderate our consumption of alcohol, avoid drugs of abuse, watch our weight, do our exercise, monitor our blood cholesterol level, not drink and drive, handle guns safely, avoid fried foods and *trans* fats, and fasten our seat belts. We are vaccinated against many childhood diseases when we are young and against others when we age. Our cars must meet safety standards meant to protect us in case of accident. We have sophisticated medical and hospital care designed to nurse us back to health in the event of illness or trauma. Laws intended to protect our lives are passed and enforced. And at the end of life, we usually do cling to it as long as possible, even in the face of disability and suffering. "The last thing that most people want to do is the last thing that they do."

We also protect the quality of our lives. We try to eat a healthy diet though the best diet story evolves over time. Many of us engage in regular physical exercise in an effort to ensure good health. Most of us avoid unnecessary risks. We have a huge multinational pharmaceutical industry to create products to prevent problems, help to diagnose them when they arise, and aid in returning us to a state of good health. We visit our doctors and dentists regularly. We brush and floss our teeth, keep clean, use sunblock, meditate, sleep 7 or 8 hours a night, eat breakfast, and on and on. The point is to be as healthy as possible until the day we die. No one is going to get out of this alive but you can try to be in good shape when you go and to postpone that day as long as reasonably possible.

The mirror image of the joy of a desired and healthy new baby is the sadness occasioned by death of a loved one, family or friend. The potential that life brought into the world has been extinguished. The store of knowledge, experiences, and insights possessed by the deceased is forever lost. Death brings grief and misery and sometimes a compromise to good health in survivors. There is a substantial effort devoted to consoling the grieving: it takes its form in private reflection, family gatherings, churches, funeral homes, florists, condolence cards, and cemeteries. The personal and social cost of debilitation or a life cut short is enormous.

Efforts at self-preservation refer both to the individual life and to the goal of living harmoniously with others who share our habitat. The latter issue reflects the goal of preserving our species and all species. This requires the well-being of progeny on an evolutionary time scale, reason enough to protect our habitat as well as our individual lives.

Reverence for life, in one sense or another, is reflected in societal and political problems. In the United States at this time, vigorous debate, and sometimes violence, is elicited by concerns about birth control, abortion, the cloning of stem cells, and the death penalty.

For many, a substantial part of life is spent in preparing for what may come next. Some elect to pass their lives in service to their God as ministers, priests, rabbis, imams, or missionaries. Others may attend church, prayer meetings, serve as church elders or members of religious lay organizations, read their holy book, and serve their fellow people in a myriad of ways. All this is by way of qualifying for a good outcome following death.

Our love of life is by no means restricted to our own and that of fellow humans. We are biophilic;[1] that is, we love life. Think of the love, attention, and resources lavished on pets: dogs, cats, fish, hamsters, canaries, guinea pigs, and exotica. Many people are avid birders. We go to zoos, hike in wilderness areas in hopes of seeing wild animals or wildflowers, go on safaris, plant and protect trees, and on and on. We love life.

So what is this thing that we know as life? What do we know about the molecular basis of life? How do we decide what is and what is not alive? How have we learned to provide insightful answers to these questions? To provide a meaningful response to the critical questions about life is a central task of this book. We begin the search in this chapter. At the outset, we need to know something about how we know what we know.

Knowledge grows based on observation, experimentation, and rigorous testing

The world works based on a set of mutually agreed understandings. These understandings take many forms: mathematical equations; the laws of physics and chemistry; and the shared experience captured in ideals, stories, parables, and anecdotes. Where we lack such mutual understandings, there is conflict and the world works less well because of them.

These mutual understandings—let me call them "the truth"—change over time as new observations are made and new knowledge is gained. As more refined and more powerful experiments are carried out, new insights emerge: novel elementary particles, new species, unexpected fossils, the discovery of dark matter and dark energy, new chemical compounds having novel properties, identification of neural pathways in the brain, unearthing of human artifacts, discovery of ancient manuscripts, and the like. All of these have the potential to alter the mutual understandings that we think of as "the truth."

I wish to be very clear about one thing: as new knowledge is gained and our understanding of "the truth" is changed a bit, the old knowledge is not lost. Rather, it is refined and expanded and elaborated. Newtonian physics did not become irrelevant when Einstein developed his theories of relativity or when Schrödinger and others developed quantum mechanics. William Harvey's discovery of the circulation of blood is no less important now that we have an enormously better understanding of our cardiovascular system. Certainly the paintings of the old masters are no less valued today in light of impressionism, minimalism, surrealism, or multiple other modern developments in visual arts. Outright mistakes and misunderstandings are eliminated as knowledge builds. But knowledge builds on knowledge rather than displacing it.

So how do we know when we know something? The building of knowledge usually begins with an observation or an idea leading to a hypothesis: a tentative statement of belief. To be any good, a hypothesis must fulfill certain criteria: it must be *testable, falsifiable*, and have *predictive value*. Here are some examples of hypotheses that meet these criteria, though not all of them are true: the Earth moves about the sun; a dietary deficiency of niacin will cause beriberi; vitamin C will prevent scurvy; masturbation causes insanity; *Echinacea* will prevent the common cold; and use of cell phones causes brain cancer. All of these statements are testable, falsifiable, and have predictive value. The first three statements are true while the last three are not but they all meet the basic criteria. Let's see how the vitamin C story meets the criteria.

The hypothesis that vitamin C will prevent scurvy is testable. Divide a population into two groups—one that has adequate vitamin C in the diet and one that has no vitamin C in the diet. Observe what happens over time. If the hypothesis is correct, the first group will be spared scurvy while the second will suffer from it. The hypothesis is falsifiable. If people have abundant vitamin C in their diet and still get scurvy, the hypothesis has been shown to be false. The hypothesis has predictive value: if your diet lacks vitamin C, you will develop scurvy.

There is perhaps no better example of a failed hypothesis than intelligent design, the idea that certain biological structures are so complex that they could not have risen through evolutionary processes and, therefore, must have been designed by some intelligence, generally thought of as God. This hypothesis is not testable, not falsifiable, and has no predictive value.

Formulating, testing, falsifying, and employing the predictive value of hypotheses is how science moves forward. The beauty of science lies in the continuity of thought based on careful observation and experimentation and its relevance to future needs.

Living organisms are both diversified and unified

Life is amazing in two senses that, at first glance, seem contradictory: *life is strikingly diverse; life is strikingly unified*. The diversity reflects variations on a common, unifying biochemical blueprint.

There are many dimensions to the diversity of life. Consider size. The smallest living organisms include a genus of bacteria termed *Mycoplasma* and some members of a great domain of living organisms, the Archaea. These organisms measure only about a thousandth of a millimeter[2] long in their biggest dimension; that is less than the thickness of a human hair. It would take about 25,000 of these organisms laid end-to-end to stretch one inch. These organisms can be visualized only with the aid of a microscope.[3] They are simply too small to be seen by the unaided human eye. The favorite experimental bacterium of scientists is *Escherichia coli*: one *E. coli* bacterium weighs in at one-trillionth (10^{-12} or 0.000000000001) of a gram. It takes about 28.6 grams to make an ounce. So an ounce of *E. coli* would contain about 2.86×10^{13} (28,600,000,000,000) individual bacteria. That is a large number. To get an idea of just how large, let's ask how long that many seconds is. A simple calculation will show that 2.86×10^{13} seconds is about 900,000 years or 450 times as long as the time between the birth of Christ and the present day. Bacteria are really small.

At the other end of the scale are the giant sequoias that tower 200 feet, about 65 meters, above the ground and weigh thousands of tons. A single leaf from such a tree weighs as much as millions of tiny bacteria. Put another way, it would take several hundred million bacteria laid end-to-end to reach from the ground to the height of a sequoia. Life forms occupy most of the included size range.

Moving up the size scale from bacteria, we have the fungus *Saccharomyces cerevisiae,* commonly known as bakers' yeast. It is a single-celled eukaryotic organism significantly larger than a typical bacterium. *Dictyostelium discoideum,* a slime mold and another favorite of biologists, is a multicellular organism. Its life cycle includes a single-celled amoeba stage as well as a multicellular slug stage, 1–2 millimeters (mm) long. The slug contains about 100,000 cells in a volume less than 1 cubic mm. Moving on to bigger stuff, we find *Caenorhabditis elegans,* a nematode worm, roundworm, that we will encounter again later in this book. It is a bit larger than the slug stage of *D. discoideum.* Its simple nervous system serves as a useful model for more complex ones, including our own.

Drosophila melanogaster is the common fruit fly, much beloved by geneticists. A fruit fly is somewhat larger than *C. elegans,* about 3 mm long from head to tip of the wings. A cockroach is bigger than a fruit fly and a butterfly is larger still. We encounter increasingly large and more complex organisms. The range of sizes is truly impressive. However, as noted below, the size range of living organisms is limited.

The diversity of living forms includes variation in shape and form

Structural organization forms another dimension in the diversity of life. Many organisms, including human beings, have bony internal skeletons to maintain shape and protect against physical insult. Our bones provide a substantial measure of rigidity to our bodies, and some, such as the bones of the skull and spinal column, provide important protection against injury as well. Other organisms such as lobsters, crabs, and many insects lack bones but have a hard external skeleton, an exoskeleton. The exoskeleton provides rigidity to the bodies of these organisms. Still other living organisms, such as jellyfish, have neither an internal bony skeleton nor an exoskeleton. The body of the jellyfish lacks any structure that would provide rigidity or offer protection against physical insult. Jellyfish and related organisms survive by spending their life as buoyant organisms in sea water. Finally, trees are quite rigid structures but have neither an internal skeleton nor an exoskeleton. Trees are largely constructed from intertwined strands of a very tough building material—cellulose.

Consider symmetry. One symmetry widely used in living systems is bilateral symmetry. In a bilaterally symmetric organism there is a line or plane through its center that divides it into halves that are mirror images of each other. The left-hand and right-hand sides of our external bodies are mirror images of each other, just as the shoe that fits our left foot is a mirror image of that which fits our right foot. So we exhibit external bilateral symmetry. So do butterflies, bluejays, houseflies, and codfish.

The starfish provides a more complex example of symmetry in a living organism. If we approximate a starfish by a regular five-pointed star, we can recognize that there are five lines or planes through the organism that divide it into mirror images:

Each line in this diagram divides the starfish into two halves that are mirror images of each other. There is one more point here. If we rotate the starfish by $360°/5 = 72°$ around an axis that penetrates the center of the starfish, we will get a structure indistinguishable from the original. We can summarize by saying that the starfish has five mirror planes of symmetry and a fivefold rotation axis through its center.

Life forms occupy many habitats

The diversity of life is also reflected in the diversity of habitat. The cold Arctic seas swarm with marine microorganisms. The ice floes suspended in these seas are home to sea lions, seals, and polar bears. At the other end of the world, penguins huddle together on the landmass of frigid Antarctica, 95% of which is covered by a massive ice cap that averages 1.6 kilometers, about 1 mile, thick.

At the other end of the temperature scale are the habitats of the heat-loving, thermophilic or hyperthermophilic, microorganisms. These organisms live in volcanic vents (black smokers) and hot seeps on the sea floor or in hot springs, steaming geysers, or hot, bubbling mud holes. Many of these organisms can only survive under conditions hostile to most other life forms. For example, several hyperthermophiles reproduce only at temperatures greater than 80°C (176°F). *Pyrodictium* grows optimally in superheated water at 105°C (221°F)!

We find fish, squid, polychaete worms, molluscs, and archaeans in the perpetual dark of the deep sea.[4] Barnacles, limpets, and other animals that adhere tightly to solid supports occupy rocky coasts. Molluscs, polychaete worms, and brittlestars populate the muddy sea bottom. Colonial animals—corals, sponges, bryozoans—wage common cause in support of life on reefs. Perhaps 200 species of orchids beautify rain forests. The arid desert of the Arabian peninsula supports life that includes lizards, insects, and flowering plants. Dark caves are home to fungi, bacteria, beetles, spiders, mites, springtails, and bats. Even the driest place on Earth, the Atacama Desert of Chile and southern Peru, where hundreds or thousands of years may pass without rain, is home to microorganisms that somehow manage to hang onto life.

Life is not confined to the surface of the Earth. Microorganisms are found in rocks well beneath the surface. Some estimates suggest that the mass of life beneath the surface of the Earth may rival that on the surface. An extreme example is provided by the finding of bacteria 1.7 miles below the Earth's surface in the Mponeng gold mine in South Africa. These exotic bacteria ultimately derive energy for maintenance and reproduction from the decay of radioactive isotopes of uranium, thorium, and potassium.

Many living species absolutely depend on oxygen. Others, obligate anaerobes, cannot tolerate oxygen and survive only where they are isolated from it. Still others, facultative anaerobes, live perfectly happily in the absence of oxygen but are capable of tolerating it. Where we search for life, we generally find it.

We have taken samples of the surface of the Moon, by man, and the surface of Mars, by robot, and searched these samples for signs of life: nothing found. Contrast this scenario with one in which some extraterrestrial civilization sampled the surface of the Earth for signs of life. It is difficult to imagine that they could find samples that did not contain signs of life; indeed did not contain an abundance of living organisms. The Earth teems with life.

The diversity of life on Earth as we know it is truly amazing

We do not understand the full range of the diversity of life on Earth. Some estimates suggest that we have recognized and cataloged fewer than 10% of the species living on Earth. In fact, no one knows to within an order of magnitude how many species there are on Earth. Estimates range from as few as ten million or perhaps as many as 100 million. We simply do not know.

About 1.8 million species have been identified. Only a modest fraction of these has been described in detail. Of the known species, about 750,000 are insects and another 250,000 are flowering plants, the angiosperms. Some estimates suggest that there may be 30 million species of arthropods in tropical forests. It is known that there are at least 163 species of beetles that live exclusively on a single species of tree. Should this species of tree become extinct, we shall very probably also lose the 163 species of beetles. In a single gram (about 1/30 of an ounce) of soil or sediment from shallow seawater, 4000–5000 species have been identified. On average, two new species of birds are discovered each year. The fact is that we do not understand the range of living organisms that cohabit the Earth with us well at all. We are far nearer the beginning than the end of that understanding. E. O. Wilson has beautifully described the range of life on Earth in his book: *The Diversity of Life.*[5]

An international effort has been organized to summarize all knowledge of the 1.8 million known species in a publicly available database. It will be known as the *Encyclopedia of Life* and is being pulled together by a consortium, including Harvard University, the Smithsonian Institution, and The Atlas of Living Australia.

Regrettably, we lose species faster than we can identify them. Some estimates suggest that we are losing four to six species an hour, largely through destruction of tropical and subtropical rainforests. As our understanding of the full range of animate nature develops through future research, we will recognize and appreciate an even greater range of diversity. Even if we were to come to know all living organisms on Earth, we would understand only a small part of the whole story of life.

Many life forms have been lost over time

The whole story of life on Earth would include all the life forms that have ever existed. A summary of the temporal development of life on Earth is provided in table 1.1. The number of living species on Earth at present, whether it is 10 million or 100 million, is a small fraction of the total that have existed since the origin of life about 4 billion years ago. Life on Earth first made an appearance as single-celled organisms, similar to the blue-green algae with us currently, during the Precambrian period. Many millions of species have been lost since the beginning of life. We have fossil records of some. Most have left no trace. A minority of the species lost disappeared in one of several periods of mass extinction of species.[6]

The first well-documented episode of extinction came at the time of transition from the Precambrian to Cambrian era, about 600 million years ago. Many species for which we have fossil evidence, the Edicarian animals, simply did not survive this

Table 1.1 A timeline of evolution demonstrates the tremendous expanse of geologic time compared to the period since humans evolved. The indicated times of evolutionary events are subject to change as new information is found.[a]

Millions of years ago	Event
Precambrian era	
4000	Origin of life
3800	Oldest known rocks and fossils
Paleozoic era	
550	First shellfish and corals
500	First fishes
410	First land plants
400	First insects
370	First tetrapods
340	First reptiles
300	First mammal-like reptiles
Mesozoic era	
215	First dinosaurs
210	First mammals
150	First birds
100	First flowering plants
Cenozoic era	
55	First horses
50	First whales
40	First monkeys
30	First apes
20	First hominids
1.8	First modern humans

[a]The indicated dates are derived from a figure in: *Teaching Evolution and the Nature of Science*, National Academy Press, Washington DC, 1998.

transition, never to reappear. A second wave of extinctions occurred in the Ordovician, about 450 million years ago, followed by one in the Devonian, about 360 million years ago.

The Permian extinction,[7] which took place over a few million years about 250 million years ago, dwarfs the loss of species during these earlier extinctions: 75–95% of all marine organisms became extinct during this time, including about half of all marine families. As a kind of compensation, a wealth of new species developed, as they have following each of the major extinctions of species. However, these species are basically variations on themes of structural organization that arose earlier. Other organizational themes were simply lost forever. Although it is difficult to know the precise cause of these extinctions, the most likely explanation is cooling of the surface of the Earth with widespread glaciation. A fourth major extinction occurred during the Triassic, about 210 million years ago.

Finally, we have the extinction which ended the reign of dinosaurs on Earth and which has captured the public imagination in a way that no other can. This is the extinction at the Cretaceous–Tertiary boundary, 65 million years ago.[8] Perhaps the most likely explanation for this extinction is the impact of a large meteorite on the surface of the Earth. This theory suggests that the collision kicked up a tremendous amount of dust that blanketed the planet, ushering in a profound and enduring night. The lack of sunlight caused the loss of plant life and, in turn, loss of those animals that depended on plants as food, most notably perhaps, the plant-eating (herbivorous) dinosaurs. Their demise elicited that of the predator dinosaurs, including the *Velociraptors* and *Tyrannosaurus rex*. The closest relatives of the dinosaurs that survived this extinction and are with us today are believed to be the birds, although there is vigorous debate around this issue.

The extinction of the dinosaurs had one enduring consequence for us mammals: we took over as important players among the living organisms on Earth. While the dinosaurs reigned, the mammals were bit players on the stage of life. If dinosaurs had survived, mammals might have continued to be of minor importance on Earth.

A downside of the rise of mammals, specifically including humans, is that the sixth major extinction of life on Earth is happening now. As noted above, we are losing 4–6 species an hour, 27,000–40,000 species a year, mostly in the tropical and subtropical forests.[9] The tremendous loss of species is the result of habitat destruction, overhunting, introduction of exotic species of animals and plants into new habitats, and the diseases carried by these exotics. One of the most valuable resources on Earth—biodiversity—is being sacrificed, the result of a burgeoning human population and its activities.

Life endures

One point is central: life endures. Despite major changes in the composition of the atmosphere of the Earth, repeated ice ages, changes in the salinity of the oceans, massive movements of the continents and the oceans, and extraterrestrial insults, life endures. Once life had emerged, it proved to be enormously resilient. More than 99% of all the species that ever existed no longer exist, yet we have more species on Earth now than at any time in the past. At the same time, we should realize that the recovery from mass extinctions requires tens of millions of years. Species that are lost do not reappear but are replaced by new ones. For our own welfare, we should protect the biodiversity that we have. E. O. Wilson has laid out supporting arguments in elegant detail in his book: *The Creation: An Appeal to Save Life on Earth.*[10]

There are sound scientific and practical reasons for protecting biodiversity. Here is one example. All prescription drugs that are approved for sale in the United States are collected in a volume known as *The Physician's Desk Reference.*[11] These drugs have been of immeasurable value to the health and well-being of people. About 40% of all entries in *The Physician's Desk Reference* are either natural products or are chemical compounds, molecules, derived from natural products. In short, nature has been a bountiful source of novel molecules that have found important uses in human health. Living organisms create an amazingly diverse collection of molecules. This valuable resource has yet to be fully exploited for good purposes and merits protection.

Living nature is divided into three great domains

The three great domains of life on Earth are the Archaea, the domain of archaeans, the Eubacteria, the domain of bacteria, and the Eukarya, the domain of eukaryotes. The Archaea and Eubacteria, both unicellular organisms, were differentiated from each other largely on the basis of the work of Carl Woese of the University of Illinois in 1977. By determining the structure for a specific class of ribonucleic acid molecules (RNA, see chapter 12), Woese was able to establish that Archaea and Eubacteria diverged from a common progenitor early in the development of life on Earth.[12] The archaeans include many unicellular organisms found in extreme environments: hot springs, black smokers, and the like. However, the Archaea are not restricted to these exotic environments: estimates are that as many as 40% of marine organisms are archaeans, assuring that they are among the most common of Earth's life forms.[13]

There should be no confusion between Eubacteria and Archaea, though both are unicellular and both lack nuclei and subcellular organelles. In addition to differences in the structures of certain RNA molecules, there are a number of other clear distinctions between the two domains. There are distinct sensitivities to antibiotics. For example, antibiotics such as kanamycin and streptomycin that are effective against a broad spectrum of bacteria have no effect on archaeans. Moreover, the genetic complement of Eubacteria and Archaea are distinct: about 30% of all Archaea genes are unique to archaeans. Finally, the lipids that constitute the cell membrane are distinct. There are clear and compelling distinctions between these two great domains of life.

Eukaryotes are differentiated from the Archaea and Eubacteria by the possession of a nucleus in the cell enclosed by a membrane as well as by membrane-enclosed subcellular organelles. The nucleus houses the basic genetic information of these organisms, their genomes, as I will describe in chapter 14. The eukaryotes are a diverse set of species, including but not limited to all plants and animals. Remarkably, the Archaea are more closely related to Eukarya than they are to the Eubacteria. This reflects a striking origin of the eukaryotic cell.

Eukaryotic cells alone possess enclosed subcellular structures, including, for example, the mitochondria. Mitochondria are the powerhouses of eukaryotic cells and I will have much more to say about them in chapter 17. For the present, it has been recognized for some years that the mitochondria are derived from bacteria at some point in the distant past. The basic idea is that an earlier eukaryotic cell captured a bacterium at some point and symbiotic relationships developed. The story may be more complex and more interesting.

The diversity of living forms is the product of evolution over a geological time frame. It is difficult to grasp the events that, over time, have led to the current diversity. In 1951, James Rettie developed a means of thinking about biological evolution through time that makes it more comprehensible. Basically, he constructed the equivalent of a time-lapse motion picture for us. Rettie went back 757 million years, before the Precambrian. Assume that we take one image each year from that time to the present. Now we imagine that we project these images at the normal speed of 24 images per second. So 24 years are collapsed into one second and 2.1 million years into one day of our motion picture. The entire motion picture would require one year to view. Here is what we would see:[14]

> From January through March, not much happens. Unicellular microorganisms appear in April. Small multicellular organisms begin to emerge by the end of that month. Vertebrates appear in May. Land plants have begun to cover the Earth by July. In mid-September, early reptiles appear. The era of the dinosaurs follows and continues through late November. Birds and early mammals have appeared by early November but the dinosaurs dominate until 1 December, when they suddenly disappear. By late December, recognizable ancestors of modern families of mammals appear. But we must wait until noon on New Year's Eve to see our first clear ancestors to human beings. Between 9:30 and 10:00 pm, *Homo sapiens* migrates out of Africa to populate the globe. At 11:54 pm, recorded human history and civilization as we know it began!

Rettie's motion picture needs substantial revision in light of more recent findings concerning the antiquity of life. Rather than going back 757 million years, we should go back perhaps 4 billion years. However, you can get the essential idea from Rettie's original model. Human beings are a very small part of the picture of life on Earth when we look at it over eons.

So far, I have emphasized the amazing diversity of life forms without saying much about the other side of the coin: the unity of life. Since the key point that I wish to make is that the diversity reflects variations on a common theme, it is time to get to the unity side.

The molecules of life bring unity out of diversity

The amazing diversity of animate nature is unified at the level of the molecules of life. Despite the diversity of size, form, and habitat, all living organisms are characterized by and depend upon a set of closely related molecules. Nucleic acids, DNA and RNA, are the universal genetic materials in living systems. That simple fact is proof enough of the unity of all life. Amazingly, intact genes, composed of nucleic acids, can be transferred from one species to another, distantly related in an evolutionary sense, with full retention of function. That seems to me to be another compelling argument for the unity of life. There is a protein known as cytochrome c present in species from wheat to humans with preservation of function across this evolutionary gulf. Nucleic acids and proteins are not found in inanimate nature. They lie at the very heart of the chemistry of life. A single set of metabolic reactions—known as the citric acid cycle—is found in all cells in all forms of life, another compelling argument for the essential unity of life. In subsequent chapters I will cite other metabolic and signaling pathways that are common to basically all living organisms: more evidence—as if more were required—to establish that life as we know it began once and has maintained many of its fundamental chemical themes as it evolved into an amazing diversity of living forms over more than three billion years.

Nucleic acids, proteins, and other molecules characteristic of life are built on a foundation of the element carbon. Life is carbonaceous. Many rocks, in contrast, are built on a foundation of the element silicon and are, therefore, silicaceous, though there are carbonaceous rocks such as limestone and marble. I shall develop a basic understanding of the molecules based on carbon alone in chapters 3–5. There I begin with simple molecules based on carbon and move on to more complex ones. In chapters 6–8, I will elaborate on the theme of carbon chemistry by including that for other elements critical for life: nitrogen, oxygen, sulfur, and phosphorus, among others. That will lead us to molecules of increasing complexity. In chapters 9–14, we will come to know profoundly complex molecules: proteins and nucleic acids. These molecules—small and large, simple and complex—are the molecules of life.

Life on Earth started exactly once

The unity of life as reflected in the molecules of life strongly suggests one dramatic conclusion: all life on Earth, including those species extant and those that have become extinct, started exactly once.[15] That is not the same as saying that life only began once. It is entirely possible that life began several times or even many times but failed to survive the environmental conditions that prevailed at the time. These beginnings have been irretrievably lost. We simply have no way of knowing. However, life as we know it on Earth—the living forms that have survived—almost certainly had a single beginning. This is the simplest way to account for the similarity of the molecules of life extending across the full range of diversity of living organisms, including the molecular fossil record. (The other explanation—more complicated and less parsimonious—is the religious explanation of multiple creations by a supernatural force.)

The origin of life on Earth has challenged scientists and philosophers for many years. Experimental efforts to reconstruct the early events on Earth that may have led to life have been underway for more than 50 years and much of interest has been learned. Some have suggested that life on Earth arrived from some extraterrestrial source. Wherever it originated, it did so once. We are a long way from understanding the sequence of events, and there must have been a very large number of them, which eventually created life. Life is, after all, the result of an experiment of nature that we have not been able to replicate.

The origin of life is a provocative topic. Some have argued that life is too complex to have arisen spontaneously over time. They argue that, for example, the assembly of a protein with a precisely ordered sequence of amino acids is such an incredibly improbable event that it could never have happened. This argument makes little sense. No one believes that creation of a functional protein occurred as a single event. An enormously better way of thinking about the origin of life is as a very long sequence of highly probable events. After all, this is how things actually happen in the world as we know it. The first car was not a Ferrari; the first airplane was not a Boeing 747. These highly sophisticated machines had humble beginnings in machines of very simple design followed by a long and ongoing sequence of modest improvements. If you make small improvements in basic designs long enough, you get something pretty spectacular.

To say that life on Earth began exactly once does not mean that there is a Universal Ancestor from whom all other living forms are derived. Life has no obvious starting point. The key classes of molecules of life were originally created abiotically.

Back in the early nineteenth century and before, there was an argument about whether the molecules found in living organisms possessed a "vital force." The argument suggested that molecules of living systems could not be made from simple inorganic molecules. Friedrich Wöhler laid that hypothesis to rest back in 1828 when he synthesized urea from ammonium cyanate. Urea is an end product of nitrogen metabolism in many species, including humans. Chemists have been making the molecules of life ever since.

A great deal of effort has gone into attempts to mimic the origins of life on Earth in the laboratory. In 1953, Stanley Miller demonstrated the abiotic synthesis of amino acids and other molecules of life in the laboratory.[16] He employed an electrical discharge through an atmosphere believed at that time to mimic that of the early Earth. Many investigators have followed up Miller's pioneering studies, with encouraging results.

The hypothesis is that assembly of these key molecules into simple cellular structures followed their synthesis. These are called progenotes and had genetic information that codes for the molecules of the simple cell. They were characterized by high rates of change (mutations) of the genetic information and fast lateral transfer of genetic information among the progenotes.[17] The high mutation rate and fast transfer of genetic information among these early cells tends to smear out the distinctions among different progenotes. This makes the origins of life vague.

Over time, the progenotes evolved into more complex cellular structures that had a lower mutation rate and a much slower rate of lateral genetic transfer among cells. This was followed by evolution of cellular subsystems, adding a new level of cellular complexity. From these cells came the three great domains of living organisms: the Eubacteria, Archaea, and Eukarya.

The origin of living organisms on Earth has been summed up by Carl Woese, one of the leaders of scientific study and thought in this field, in the following way:[18]

> The universal phylogenetic tree, therefore, is not an organismal tree at its base but gradually becomes one as its peripheral branchings emerge. The Universal Ancestor is not a distinct entity. It is, rather, a diverse community of cells that survives and evolves as

a biological unit. This communal ancestor has a physical history but not a genealogical one. Over time, this ancestor refined into a smaller number of increasingly complex cell types with the ancestor of the three primary groupings of organisms as a result.

The central theme of this introductory chapter has been the unity of life at the molecular level. Despite the diversity of size, shape, symmetry, habitat, or lifestyle, all living organisms share a common set of molecules: proteins and nucleic acids foremost among them. It follows that there are chemical connections between all things biological.

The unity of life testifies that life as we know it began once, perhaps 4000 million years ago, and has evolved since. Now we need to turn to a deeper understanding of life itself: What defines life and how is it reflected in the organization of its components? That is the task of the next chapter.

Key Points

1. Life is strikingly diverse: sizes, shapes, symmetries, structural organization, habitat, life cycle.

2. Life is strikingly unified at the molecular level: one genetic code, limited universal sets of key molecules, common metabolic pathways.

3. About 1.8 million living species are known on Earth. There are many more that we do not yet know, perhaps 10 million in all, perhaps 100 million.

4. Perhaps 99% of all the species that ever existed on Earth are extinct. Many of these were lost in a series of mass extinctions long ago; we continue to compromise our biodiversity today.

5. The three great domains of living organisms are the Eubacteria, Archaea, and Eukarya. Only the Eukarya have a cellular nucleus enclosed by a membrane and intracellular organelles.

6. Life on Earth as we know it today began exactly once. There is no other rational way to account for the unity of life at the molecular level.

2

Life

Central properties

The domain of living organisms is unified by the commonality of cells and in terms of mutual energy dependence. Living organisms are open systems that create order at the expense of disorder in the environment. Life can be defined by a series of characteristics, including evolved programs and chemical properties.

The theme of the last chapter was that, despite all the manifestations of diversity, there is unity of life at the molecular level. That critical conclusion will benefit from some additional elaboration at the levels of cells and energy interdependence. Subsequently, it will be time to take a closer look at life and what it means to be living, as opposed to nonliving.

The cell is the unit of structure and function in living organisms

In the last chapter, I mentioned cells several times but said little about them. The fact is that they, like the molecules from which they are built, bring unity out of diversity for all life forms.

Guy Brown provides a series of revealing insights into cells, including the relative sizes of things, in a paragraph in his book: *The Energy of Life: The Science of What Makes Our Minds and Bodies Work.*[1] Here it is:

> A cell is very small, and of variable size and shape—an average human cell might be 20 microns (0.02 millimeters) across—but it is very large compared to the size of the molecules it contains. If we increased the scale of everything 100 million times, then we could see an atom; it would be one centimeter across—about the size of a pea. Small molecules like sugars, amino acids, and ATP would be 5 to 10 centimeters—the size of apples and light bulbs. And proteins would be 20 centimeters to one meter—the size of children or televisions. On this scale, an average cell would be two kilometers across—a vast, spherical, space-age metropolis. There is effectively no gravity within a cell, so this metropolis is located out in space, with its inhabitants floating around inside. The cell is bounded by a cell membrane and divided up into many compartments by internal membranes, each 0.5 meter thick on our expanded scale. The compartments include a maze of tunnels—the width of a small road on our expanded scale—connecting different parts of the cell. Attached to these tunnels and floating throughout the cell are a huge number of ribosomes, the factories that make proteins, which would be three meters across—the size of a car. And the cell is also criss-crossed by a vast number of filaments—one meter across on the enlarged scale, like steel girders or pylons—which act as the skeleton of the cell, and to which the proteins may attach. Mitochondria, the power stations of the cell, would be 100 meters across—the size of a power station—and there would be roughly 1000 of them per cell. The nucleus, a vast spherical structure about one kilometer across and a repository of eons of evolutionary wisdom, broods over the cell. Imagine then that vastly expanded cell to be a metropolis floating in space, peopled by billions of small, specialized robots, doing thousands of different tasks, making, breaking, and moving trillions of other molecules in order to feed, power, inform, and maintain the cell. All the molecules of a cell are packed in tightly, with very little free space, but movement is lubricated by water molecules that act like ball bearings. So the cell is big compared to its molecules, but note that on this outsized scale, the human body would be ten times the size of the Earth itself, so there are an awful lot of cells in the body.

Like living organisms themselves, cells come in a remarkable variety of flavors. Brown has described what might be a human cell with elaborate internal structure. However, there is no such a thing as a typical cell. A functional liver cell, a hepatocyte, is quite distinct from a nerve cell, a neuron, that, in turn, is not much like a cell of the retina of the eye. Skin cells, pancreatic cells, kidney cells, cells of the testis and ovary, red blood cells, bone cells, and on and on, are all structurally, functionally, and metabolically distinct. Indeed, there are several types of cells in the skin, pancreas, kidney, testis, ovary, and bone. Then there are the cells of bacteria and other microorganisms that have no nucleus or other membrane-limited organelles: very different. Diversity abounds.

At the same time, all cells are unified at the molecular level, as emphasized in the first chapter. There are other commonalities as well. All cells have a *plasma membrane* that surrounds and encloses them. The fundamental function of the plasma membrane is to act as a selective barrier between the cell interior and the external environment. In the Eubacteria and Archaea, single cell organisms all, the plasma membrane is the sole membrane of the cell. In contrast, in the Eukarya, there are a variety of internal membranes that define and isolate subcellular structures, as described by Brown. We have a nucleus, isolated by a nuclear membrane, that houses the genetic material of the cell. A double membrane surrounds the mitochondria, sites of cellular energy generation. Complex membrane structures define the endoplasmic reticulum, an important site of protein synthesis, and the Golgi apparatus, that regulates the intracellular trafficking of proteins.

Life is an emergent property of cells

Much scientific advance comes from a reductionist approach: take a complex system apart into its components, understand these, and then build a useful model of the more complex system. This frequently works quite well. For example, if someone gave you a clock and asked you to discover how it works, you would probably get started by taking it apart, understanding the constituent parts, and learning how they work together by a process of reassembly. Indeed, one can pretty much predict the properties of a clock based on this reductionist approach.

While reductionist approaches are surely useful, they have limitations. There is the phenomenon of *emergence*.[2] Simply stated, emergence defines the limitations of reductionism. Unlike the case of how a clock works, there are many examples of phenomena of aggregates of parts that cannot be understood on the basis of the properties of the parts themselves. Unintelligent parts can yield *intelligent* organisms. Neurons themselves do not possess intelligence. Yet the immense collection of strongly interconnected neurons in the human brain yields intelligence. There is simply no way to predict intelligence on the basis of the properties of individual neurons. Much less could we have predicted the phenomenon of *consciousness* from the properties of individual neurons. Intelligence and consciousness are emergent properties of the nervous system, a complex system of cells.

Life itself is an emergent property. The distinguishing features of living systems cannot be predicted from the properties of individual cells or organs or, in the case of single-cell organisms, from the properties of individual molecules. Cells too, as living entities, have emergent properties. There is more to cells than the sum of the capacities of individual cellular substructures. The whole is more than the sum of its parts.

What would it mean to really understand a cell?

To begin with, we should have a complete *molecular inventory*: that is, we need a list of all the molecules in the cell, together with the amount of each molecule present. We do not have a good molecular inventory for any cell, though we do have a great deal of relevant information and more is being rapidly generated.

The second thing that we need is a complete description of the *metabolic and signaling pathways* of the cell.[3] This has been a focal point of biochemistry for many years and a lot of relevant information is available. We do not have a complete picture but we do have a reasonable grip on this issue.

Beyond having the metabolic pathways, we need to understand the *control circuits* that regulate them. Of particular interest is the control of the pathways that underlie *cell division,* a critical cell function and an amazingly complex one.

To complete the list of what we need to know to really understand a cell, there are the issues of adaptive processes—those mechanisms by which cells maintain viability in the face of changing environmental circumstances—and specialized functions that may be unique to a certain cell type such as nerve conduction in neurons. Finally, when we really understand a cell, we will be able to make a definitive mathematical model for it.

Technological advances that are useful for understanding a cell arrive with regularity. Our level of understanding is increasing rapidly; the goals are ambitious and it will be some time yet before we can claim to really understand a cell, the fundamental unit of all living systems.

Living organisms are unified in terms of energy and carbon metabolism

Life is unified by the molecules of life and by the roles of cells. There are two additional, intimately connected unifying themes: energy and the metabolism of the element carbon. Here are the essentials.

There is only one overwhelmingly important source of energy to sustain life on Earth: sunlight.[4] The energy of sunlight comes to us in the form of photons, little packages of light also known as quanta. That energy warms the Earth but it does more than that. Pigments in green plants and certain microorganisms capture the radiant energy of the sun in a process known as *photosynthesis.* Here is photosynthesis in a nutshell:

$$carbon\ dioxide\ (CO_2) + water\ (H_2O) + sunlight \rightarrow carbohydrate\ (CH_2O)_n \\ + oxygen\ (O_2)$$

In words, the energy of sunlight is used to convert carbon dioxide in our atmosphere plus water into complex molecules known as carbohydrates plus oxygen, which is released into the atmosphere. The carbohydrates include sugars and starches, among other molecules. Thus, sunlight provides the driving force for the conversion of very simple molecules, carbon dioxide and water, into complex ones, while releasing oxygen at the same time.

In a certain sense, we live by reversing the process of photosynthesis. Specifically, we burn carbohydrates using oxygen and produce carbon dioxide and water in a process known as respiration:

$$Carbohydrate + O_2 \rightarrow CO_2 + H_2O + energy$$

The energy released is captured in the form of a molecule known as adenosine triphosphate (ATP) and is used to drive all our energy-requiring processes: synthesis of complex molecules, movement, nerve conduction, muscle contraction, and so on.

The net effect of all the chemistry is that light from the sun is used to drive life processes. Carbon dioxide taken up in photosynthesis is released in respiration.

Oxygen released in photosynthesis is taken up in respiration. Carbohydrates formed in photosynthesis are consumed in respiration. Sunlight drives the synthesis of ATP.

This ties everything living together: plants, microorganisms, animals. Once more, unity emerges from diversity.

Here is a useful way to think about energy

Before we get on to trying to understand exactly what life is, it will be worthwhile to get a little better understanding of energy and energy changes. This turns out to be quite useful in what follows later.

Thermodynamics is the science that deals with energy and energy changes. We will need a couple of key concepts from thermodynamics as we move forward. To begin with, we need to recognize that energy changes involve two factors: one is related to physical forces or heat exchange and the other to organization or information.

Changes in energy or enthalpy are measures of work done against physical forces and heat exchange

My wife and I take occasional hikes in the mountains of Colorado. Doing so requires energy, as anyone who hikes in the mountains will testify. Walking on a path that ascends to a mountain pass involves doing work against the force of gravity and that requires energy to overcome the physical force, gravity, involved. Pushing an automobile down a street requires that one overcome the frictional forces between tires and pavement and that also requires energy.

If you feel the need of a hot cup of coffee, you are going to need to boil some water. Even though there are no obvious physical forces to overcome in this case, you still need a source of energy in order to increase the kinetic energy of the molecules of water. Temperature is a measure of molecular kinetic energy. So you turn on your electric kettle or light your stove burner or whatever to provide the necessary energy.

We have two basic quantities here: the work done against physical forces, w, and the heat added to some system, q. The sum of these is the First Law of Thermodynamics:[5]

$$\Delta E = q - w$$

in which ΔE is the energy change of the system, q is the heat added to the system, and w is the work done by the system. Note that the Greek upper case delta, Δ, is used to denote a change in some quantity. A simple statement of the First Law is that energy is conserved. It is a fundamental statement of belief, not derivable from any more basic considerations, that sums up all our experiences about the way physical systems behave. The energy, E, is very closely related to another quantity known as enthalpy, H.[6] E is basically the total energy of a system in the sense that energy is a measure of the amount of work that the system can do. H is slightly different and is a measure of the energy of the system released as heat. It will be convenient to talk about H rather than E but remember that they are more or less the same. By convention, if values of ΔE or ΔH are negative for some process, energy is released in that process. If these values are positive, then energy is absorbed in the process. So we have half of the story.

Changes in entropy are measures of change in organization or information

The other half of the story is the organization or information part. Here is the basic idea. At craft fairs, my wife and I have seen the work of a goldsmith who generates jewelry of amazing elegance by weaving very thin threads of gold into incredibly complex patterns. It strikes me that doing that work requires a lot of energy in some sense but it is not a matter of doing work against physical forces or adding heat to some system. Rather, it is a matter of taking unorganized threads of gold and creating highly organized structures from them. It is a matter of putting information and organization into a system, in this case, a piece of jewelry. This is not a matter of E or H but of entropy, S. If values of ΔS for some process are negative, then organization or information is added to the process. Conversely, if values of ΔS are positive for some process, then organization or information is lost in the process.

Changes in free energy pull together those of enthalpy and entropy

In most real processes, including sustaining life, both factors are important: the energetic one and the entropic one. A simple example is provided by evaporation of water.

In the liquid water, the attractive forces between water molecules are stronger than they are in the gas phase. These forces are distance-dependent and the shorter the distance between molecules the greater the attractive force. Evaporating water requires that we overcome the physical forces tending to hold the water molecules together in the liquid state.

This is counterbalanced by the fact that water molecules in the liquid state are a whole lot more organized, largely by holding on to each other, than they are in the gas state. It follows that entropy will tend to drive the water into the gas state. So we have a balance between the forces of E or H, on the one hand, and S, on the other. So how do we sort this out?

We can pull everything together in terms of a quantity known as the free energy, G, a measure of useful work that can be derived from some process. Here is the definition of G:

$$G = H - TS$$

Since we are usually interested in changes, this can be rewritten as:

$$\Delta G = \Delta H - T\Delta S$$

assuming that the temperature, T, is constant. If ΔG is negative, then we can get useful energy from the process; if positive, then we need to add energy to make the process happen.

This simple equation tells us that the temperature is the quantity that influences the relative contributions of ΔH and ΔS to the overall energetics. Think once more about the evaporation of water. At low T, ΔH will be more important than $T \Delta S$ and the physical forces will outweigh the entropic ones. As T is raised, $T \Delta S$ will increase relative to ΔH and the entropic forces will dominate. As you might have guessed, when ΔH and $T \Delta S$ are exactly equal so that ΔG is zero, T is the boiling point of water.

The processes of life involve both energetic and entropic factors. When we walk, run, climb, fidget, dance, or toss and turn in bed, we are doing work against physical forces. When we ice fish on a frozen lake in Minnesota, we need to generate heat to keep our body temperature constant. That heat is lost to the environment (an ineffective way to heat Minnesota in the winter but it happens nonetheless). In contrast, when we synthesize DNA in the process of cell division with its precisely ordered sequence of bases or synthesize proteins with their precisely ordered sequences of amino acids, it is entropy that we must overcome, not physical forces or heat exchange. The bottom line is that we need a source of energy to live.

Here are some thoughts about a definition for life

Life and pornography have something in common: everyone knows exactly what they are until it comes time to define them. I will leave concocting a good definition of pornography to the courts. However, we need a good definition for life or, at the least, a good understanding of the central characteristics that distinguish life from inanimate nature.

It is easy to write down some of the characteristics that we associate with living organisms. These include growth, differentiation, reproduction, energy transduction (the interconversion of different forms of energy), and movement, all in the end directed toward procreation. But many species of bacteria do not differentiate, mules seldom reproduce, though they try and succeed rarely, and plants do not uproot themselves.[7] All of these are certainly living organisms that have survived over time. A piece of black felt placed in the sunlight will transduce radiant energy into heat energy but black felt is not living. Crystals of sugar will grow but sugar crystals are inanimate. Thus, the properties of growth, differentiation, reproduction, energy transduction, and movement, among others, are characteristics of life but they are not the characteristics that distinguish cleanly between living and inanimate matter.

There is more than one way to look at life. For example, biologists, chemists, and physicists may see the essential characteristics of life in somewhat different ways. Let us consider two examples from the thinking of Ernst Mayr and Erwin Schrödinger. I begin with Ernst Mayr, formerly a distinguished biologist at Harvard University, who offers the following as the defining characteristics of life in his elegant book: *This is Biology*.[8]

Evolved programs

Living organisms have genetic programs that result from genetic information that has evolved and accumulated over 4 billion years of the evolution of life. These evolved genetic programs are coded in the nucleic acids, generally DNA (RNA in retroviruses). We shall develop how information is stored and communicated in the structure of these molecules and how it is expressed in the form of protein and RNA molecules in living organisms. We now have in hand the detailed, evolved genetic programs, the genomes, for many organisms, from the smallest viruses to microorganisms, flies, worms, plants, mice and humans in the form of the sequences of bases along the chains of nucleic acids. We continue to accumulate such information at a prodigious rate. This information permits us to understand how the genetic programs of living organisms have evolved over time. There are no analogous evolved genetic programs in inanimate structures.

Chemical properties

Living organisms contain several classes of molecules not found in inanimate matter. These include the nucleic acids, proteins, peptides, and some classes of carbohydrates and lipids. These chemicals are characteristic of living organisms and distinguish the living from the nonliving. They are the products, direct or indirect, of the evolved genetic programs of living organisms. These are described in modest detail, together with the roles that they play in life, in later chapters.

Regulatory mechanisms

Living organisms are endowed with an amazing array of mechanisms for regulating their metabolism and physiology. These serve to maintain a steady state for the organism in response to a changing environment. We shall describe several of these regulatory mechanisms in what follows, such as the turning on and off of genes and therefore of the synthesis of key proteins and RNA molecules, the activation and inhibition of enzyme activity, and the action of hormones via signaling pathways. Inanimate objects do not possess these regulatory mechanisms. For example, the temperature of a rock reflects the temperature of its surroundings. In contrast, the body temperature of warm-blooded animals is maintained nearly constant regardless of the temperature of the surroundings, within some obvious limits.

Organization

Living systems are complex, ordered systems. This complexity and order is reflected in the molecules characteristic of life, in their interactions with each other, in the regulatory mechanisms that result from these interactions, and in the complex supramolecular structures characteristic of cells. Organization is also reflected in ordered metabolic and signaling pathways. Such complex, ordered structures and pathways are not characteristic of inanimate objects.

Teleonomic systems

Teleonomic means goal-directed or outcome-directed activities. Living systems have evolved in a way that programs them for outcome-directed activities throughout life, from embryonic development through to behavioral activities of adults. The teleonomic systems in the embryo programs the organism to differentiate in a way that will yield a functional adult, perhaps passing through larval stages, depending on species. Inanimate objects do not exhibit outcome-directed activities.

Limited order of magnitude

As developed above, the diversity of life is reflected, among other ways, in the range of sizes of organisms, from the very small unicellular species through to elephants, whales, and giant redwoods. At the same time, this range of sizes has its limitations. At the lower end, the limit is enforced by the size of individual cells, whose small size provides great evolutionary flexibility. At the least, a cell must be large enough to accommodate its genetic information in the form of a DNA molecule, together with the enzymatic machinery required for its duplication. At the upper end of the scale, size may be limited by the capacity of large organisms to supply themselves with the full range of required nutrients. Inanimate objects span an enormously greater size range, from subatomic particles, or far smaller hypothetical entities termed "strings" by physicists and cosmologists, to entities essentially without upper limit, such as galaxies or families of galaxies.

Life cycle

Sexually reproducing organisms have a defined life cycle, beginning with the fertilized egg and passing through a number of more or less well-defined stages that lead to adulthood and, finally, death. Nothing similar occurs in inanimate structures.

Open systems

Closed systems do not exchange matter or energy with their environment. Living systems, in contrast, obtain both matter (food provides an obvious example), and energy (sunlight for photosynthetic organisms, for example), as well as genes (lateral gene transfer in bacteria) from the environment. Beyond that, we contribute the end products of our metabolism to the environment. Living systems are, therefore, profoundly open systems. This fact is related to one of the ways that physical scientists view life, and is developed below.

Here are some thoughts about life from a different point of view

There are various ways of thinking about life. In contrast to the approach just described, we can come at the issue of life from one of the central laws of physics and chemistry: the Second Law of Thermodynamics.[9] Let's take this as a starting point and see where it leads us.

The Second Law is concerned with the natural direction of change. It tells us what will happen spontaneously. The First Law of Thermodynamics tells us that energy is conserved. The Second Law goes beyond the First Law and recognizes a fundamental dissymmetry in nature. Specifically, the distribution of energy changes in an irreversible manner. A rock at the top of a hill will roll down it but a rock at the bottom of the hill will not roll up it. A pizza taken out of the oven will cool to room temperature but a pizza at room temperature will not warm to oven temperature. The First Law does not forbid either process.

A simple statement of the Second Law is: *natural processes are accompanied by an increase in the entropy of the universe.* There are several other statements of the Second Law in the chapter Notes.[10] As noted above, entropy is a measure of disorder: the greater the extent of disorder, the greater the entropy. The Second Law tells us that things change spontaneously in a way that increases disorder. At equilibrium, entropy is maximized and disorder reigns.

What are the implications of the Second Law of Thermodynamics for life? Erwin Schrödinger, one of the founders of quantum mechanics and a leading figure in twentieth century physics, explored this question in a short book: *What is Life?* published in 1944.[11] Schrödinger poses the question in the following way:

> What is the characteristic feature of life? When is a piece of matter said to be alive? When it goes on "doing something," moving, exchanging material with its environment, and so forth, and that for a much longer period than we would expect an inanimate piece of matter to "keep going" under similar circumstances. When a system that is not alive is isolated. ... all motion usually comes to a standstill very soon. ... After that, the whole system fades away into a dead, inert lump of matter. A permanent state is reached, in which no observable events occur. The physicist calls this the state of "maximum entropy."

The question then is: how do living organisms "keep going" for extended periods of time; how do they maintain their entropy at reasonably low levels; i.e. keep their state of organization at a high level?

Simply stated, many living organisms, including ourselves, maintain life by converting structurally ordered food molecules into much simpler end products. As noted earlier, in the process of digestion of our food, we take the complex starch molecules in pasta or potatoes or rice and degrade them to the very simple molecules carbon dioxide and water with the release of considerable energy. Similarly, dietary fats and oils are degraded first into much smaller molecules and finally into carbon dioxide and water, also with the release of energy. That energy can then be used to take simple molecules and organize them into the complex, highly ordered ones of living systems: nucleic acids, proteins, complex carbohydrates, and lipids. This absolutely requires that living systems exchange matter and energy with the environment. The Second Law demands that the negative entropy change required to sustain life be accompanied by a larger entropy increase in the nonliving universe, so that the total entropy increases.

To make the idea more nearly concrete, here is a specific example, anticipating what follows in chapters 9 and 10. Think about creating a necklace by threading 100 beads on a chain. Let's suppose that there are 20 different beads, distinguished by color and shape. It should be clear that there are a great many different necklaces that you could produce, depending on the order in which you thread the different beads.

In fact, you can create 20^{100} (or about 10^{130}) different necklaces in this way. This is an unimaginably large number. Suppose that your task is to create a specific necklace out of all these possibilities. You would need a lot of information: which of the 20 different beads goes at each position? It would take a whole lot longer and be a whole lot more difficult to create the specific necklace than one made at random. This has something to do with proteins.

A protein molecule is a precisely ordered chain of units called amino acids. There are twenty amino acids that occur commonly. A modestly sized protein would contain a chain of 100 or more of these units. So a protein is analogous to the specific necklace cited above. The job of a living organism is to sort out one specific state from the wealth of possibilities each time it makes a protein molecule. The creation of this order comes at the expense of order in the surrounding environment.

The negative entropy, or information or organization, in living systems is expressed in several other ways. For example, nucleic acids are precisely ordered chains of nucleotides. Building molecular order from unordered building blocks involves creating order and information, which requires creating corresponding disorder, or positive entropy, in the environment.

Negative entropy is also expressed in the ordered three-dimensional structures of complex proteins, protein–nucleic acid complexes, molecular machines, biological membranes, and so on. Metabolism and chemical signaling too are highly and meticulously ordered processes. Use your brain to think about the brain: the degree of order in the creation of trillions of connections among nerve cells that is required to allow us have that most amazing facility, consciousness.

We shall see many examples of the creation of order in living systems as we move forward. Life is utterly and completely dependent on such order, both in structure and function.

Viruses are at the threshold of the living

Let's conclude this discussion of life with a short consideration of viruses. Viruses cause all sorts of problems for living organisms. The problems are the consequence of their ability to infect, and ultimately kill, many types of cells—bacterial, animal, and plant—though each virus is quite specific in terms of the type of cell that it infects. There are many types of viruses. In people, they cause measles, mumps, influenza, AIDS, polio, potentially fatal diarrhea in infants and very young children, herpes, chicken pox, shingles, the common cold, and many other diseases, that may be fatal, serious, and not so serious. In other animals, viruses also cause any number of diseases, as they do in plants. Much effort has been, and continues to be, devoted to the prevention, diagnosis, and treatment of viral diseases.

What are these things called viruses? The simplest viruses are very simple indeed. They are composed of just one molecule of a nucleic acid, either DNA or RNA, and a number of copies of a single protein molecule. The molecule of nucleic acid occupies the core of virus. The protein molecules create a coat that surrounds the core.

Virus particles are infectious. We are naturally tempted to conclude that they are living. But that is not so clear. Simple viruses can be crystallized, just as one can crystallize common table sugar, and stored in crystalline form, as one might store sugar. The structure of viruses can be determined in atomic detail, just as one determines structures for individual molecules. Viruses are not cells. The simple ones have no surrounding membrane. They do not grow or differentiate. Viruses are many times smaller than the smallest bacteria. Viruses consume no food and emit no waste products. They have no means of movement. The only thing that viruses do is infect susceptible living cells and, employing the metabolic machinery of the infected cell, replicate themselves many times over. In many cases the final outcome is lysis of the infected cell, spilling the progeny virus. This can happen quite quickly, in a matter minutes or hours. Alternatively, the virus may hide in the genome of the cell, delaying its replication for months or years. Sooner or later, replication occurs and cell death follows. Is this sufficient to consider viruses as living entities? Or are they simply inanimate particles that have the capacity to parasitize living cells and, in so doing, replicate themselves?

There is no easy answer to these questions: experts in the biological sciences will give different answers. In terms of the criteria proposed by Mayr and listed above, the viruses succeed as living organisms in terms of evolved programs, chemical properties, and organization but seem to fail in terms of regulatory mechanisms, teleonomic systems, life cycle, and open system. The issue of limited order of magnitude can be argued either way. Personally, I find that viruses fall significantly short of being alive.

In our effort to develop understanding of the defining characteristics of life in *molecular* terms, we need to continue by focusing on the central topic of chemistry: the molecule.We begin with very simple molecules. Subsequently, we will amplify our knowledge quickly and get to molecules of amazing complexity—proteins and nucleic acids. I make no effort to turn readers into chemists. I will make the effort to turn readers into knowledgeable consumers of information important to all our lives.

Key Points

1. The cell is the unit of structure and function for all life forms.

2. There is no such a thing as a typical cell; they come in numerous, distinct forms.

3. Cells have emergent properties, themselves and as collections of interconnected cells.

4. Almost all living organisms are tied together in the processes of photosynthesis and respiration.

5. Photosynthesis captures the energy of sunlight to convert very simple molecules into more complex ones.

6. Respiration captures the energy from conversion of complex molecules into simple ones in the form of ATP.

7. Free energy is the quantity that ties together energy and entropy; changes in free energy are a measure of useful work that can be done by a system.

8. It is not simple to define "life." However, there are a number of defining characteristics of living organisms.

9. The Second Law of Thermodynamics is a way of quantifying concepts around entropy, a measure of disorder.

10. Living organisms are highly ordered in a number of different ways. The order that we create in life comes at the expense of disorder in the universe. We create highly structured molecules from simple ones by degrading other highly structured molecules in nature into very simple ones.

11. Viruses are somewhere between the clearly inanimate and the clearly living.

3

Molecular structures based on carbon

The foundation for the molecules of life

Molecules are made up of two or more atoms; these atoms are held together by chemical bonds, a form of electronic glue. Chemical bonds have defined lengths and directions in space. Water and carbon-based molecules are the foundation of the molecules of living organisms.

The worst chemical disaster in history occurred on December 3, 1984, in Bhopal, India, a city of some 1.6 million people in the state of Madhya Pradesh. On that date in the middle of the night, a tank at the Union Carbide India Ltd. plant leaked between 25 and 40 tons of methyl isocyanate, a volatile colorless liquid, into the atmosphere of Bhopal. This highly toxic gas settled onto the city and its inhabitants in a silent, if odorous, cloud. The results were horrific: some 3800 people died and another 2700 experienced total or partial permanent disability. By some estimates, more than 10% of the population of Bhopal—170,000 people—suffered some adverse effect from the methyl isocyanate leak.

Methyl isocyanate is a simple molecule, composed of just seven atoms: two of carbon, three of hydrogen, and one each of oxygen and nitrogen. There is nothing particularly exotic about methyl isocyanate among molecules: but it does happen to be very toxic. Inhalation of methyl isocyanate does lasting damage to the lungs. Swelling and fluid accumulation in the lungs, pulmonary edema, caused the deaths of most of those killed at Bhopal. Many of those who survived initially later died of respiratory tract infections. The seven atoms that constitute the methyl isocyanate molecule are assembled in a way that creates an extremely toxic molecule.

30

Stories such as the Bhopal tragedy rightly color the impression of the general public about chemistry: it is frequently seen as the toxic science, just as economics is frequently referred to as the dismal science.

That impression gets of a lot of routine reinforcement from news sources. Chemistry is associated with pollution of the atmosphere by ozone and the noxious oxides of nitrogen and sulfur, oil spills, toxic pesticides and herbicides, smelly oil refineries, undesired food additives, chemical accidents, substances of abuse such as heroin, cocaine, and methamphetamine, and chemical weapons. Chemistry gets a lot of bad ink.

Chemistry has a rosy side

Like most stories, that of chemistry has another side, less often noted or remarked but a whole lot more pleasant. The world of chemistry is the world of molecules. It is a complex, critical, and fascinating world. Molecules and their constituents (atoms) make up all matter. Specific molecules affect every aspect of our lives every day, frequently for better but occasionally for worse. The simple fact is that almost everything that we use in daily life has been chemically modified in some way: consider plastics, alloys, detergents and soaps, paper, perfumes and colognes, and our drinking water. It is difficult to imagine life without the products of modern chemistry.

Think about those molecules that are pharmaceutical products: these are used for multiple purposes. They treat your infections, relieve the pain of your sore throat, control your coughing, lower your blood pressure, raise your spirits, lower your plasma cholesterol, fight your allergies, keep you awake and alert, help you sleep, alleviate your heartburn, modulate your anxieties, help diagnose your disease, and render you mercifully unconscious during surgery. In serious cases, they battle cancer, improve the chances of surviving a heart attack or stroke, alleviate symptoms of schizophrenia or bipolar disorder, and reduce the incidence of suicide in severely depressed patients. These agents that contribute to human well-being are molecules.

Small molecules provide the aromas and tastes of apples, pears, apricots, kiwis, bananas, oranges, and other fruits. They are the fragrance of lilacs, roses, and fine wines, as well as those of the perfumes, colognes, after shave lotions, and deodorants that we lavish on ourselves. The brilliant colors that we enjoy in spring flowers and fall foliage—reds, yellows, oranges—are the colors of molecules. Molecules within the complex photosynthetic machinery of green plants and certain algae capture the radiant energy of the sun, which is ultimately responsible for almost all life on Earth.

This book is about life, mostly about the molecules of life. Molecules are the focus of the science of chemistry, just as animals are the focus for zoology, plants are the focus for botany, and outer space is the focus for astronomy. A bit more broadly, chemistry is "the science of the composition, structure, properties, and reactions of matter, especially of atomic and molecular systems."[1] Let's talk a bit about chemistry, the molecular science.

Chemistry is the central science

Chemistry is the central science in the sense that it provides the tie between physics on the one hand and biology on the other. The world of physics, seen broadly, covers a wide spectrum. In general, the concerns of physics focus on entities smaller or larger than those of direct interest to chemistry. At the micro level physics unravels the mysteries of the elementary particles, known generally as fermions, which constitute all ordinary matter. Fermions include the quarks and their antiparticles, the antiquarks. There are six kinds of quarks, known as top, bottom, strange, charm, up, and down. Each of these has its antiquark: antitop, antibottom, ... The quarks combine to create protons and neutrons that, in turn, create the atomic nucleus. The fermions also include particles known as leptons. These include the electron and its antiparticle, the positron, as well the muon, antimuon, tau, and antitau particles and their associated neutrinos. Electrons, along with protons and neutrons, make up atoms.

Among the very small entities, physics is also concerned with particles that carry forces: photons, gluons, W vector bosons, and gravitons. At the extreme, the community of string theorists focuses on efforts to explain all in terms of hypothetical entities known as strings. This is a vibrant area of physics and the predictions of string theory are, in principle, testable experimentally, although that will require particle accelerators more powerful than anything currently available.

The point here is that physics pretty much leaves off where chemistry begins: with atoms and molecules. Chemistry begins with the atoms and builds molecules from them and asks how these interact. It should come as no surprise to learn that a subfield of physics is known as chemical physics and a subfield of chemistry is known as physical chemistry.

Physics is also concerned with the very large: think about cosmology and astrophysics. Issues include the beginning of the universe, known as the Big Bang, which occurred some 13.7 billion years ago, the expansion of the universe, formation and evolution of stars and galaxies, and properties of black holes. Here too there are connections between physics and chemistry: the origin of the atoms in nuclear reactions within stars and the nature of molecules found in interstellar space, for example.

At the other extreme, chemistry merges neatly into biology. Biology is, in a general way, concerned with life: animals, plants, microorganisms, viruses, ... All of life depends on the molecules of life—including the proteins and nucleic acids—and these molecules are the province of chemistry. Assemble the right mixture of the molecules of life within a structural barrier and you get a cell, the unit of function in living organisms. That is pretty much where chemistry leaves off and biology begins. Biology goes on to deal with more complex structures: tissues, organs, organisms, and collections of organisms and their interactions. Here too it will not come as a surprise to encounter chemical biologists, molecular biologists, biochemists, and so on. Just as chemistry and physics flow together at one extreme, so do chemistry and biology at the other.

Of course, chemistry makes interfaces with other areas of learning as well: these include geology and the behavioral sciences. The central point is that chemistry is the science of molecules and interactions among molecules and that these topics matter for everything.

Chemistry is part of our reality and not an abstraction

The molecules produced by living organisms, natural products, are employed in our lives as flavors, fragrances, pharmaceuticals, nontraditional medicines, dyes, and pesticides, among other uses. The products of chemistry are employed in our food as preservatives, artificial sweeteners, thickeners, dyes, taste enhancers, flavors, and texturing agents. Chemistry creates such key materials as plastics, ceramics, fabrics, alloys, semiconductors, liquid crystals, optical media, and biomaterials. Chemistry also does many kinds of analysis and these include measurements of air quality, water quality, food safety, and the search for substances that compromise the environment or workplace safety,

In sum, understanding the role of chemistry in the world provides useful and important information.

Some molecules are bigger than others

I am going to divide up the world of molecules into two parts: small molecules and big molecules. There are a great many ways to divide up the world of molecules and this is a very simple way to do so. It will meet our needs going forward.

Big molecules of life include the proteins, nucleic acids, polysaccharides, and a few other more exotic constructs of nature. Generally, it is the interactions between big molecules and small ones that underlie really interesting things: taste or smell or the beneficial actions of drugs, for example.

Most of the rest are small molecules. To provide some sense of the distinction between big and small molecules, we can refer to a metric of molecules known as the molecular mass.[2] Here, briefly, is what that means. Molecules are made up of atoms. Each type of atom has a characteristic atomic mass. If you add up the atomic masses for all the atoms in a molecule, you get the molecular mass. In general, the more atoms in the molecule, the greater is the molecular mass. In general, the greater the molecular mass, the more complex is the molecule.

Most proteins have molecular masses between 5000 and 500,000. The hormone insulin, for example, has a molecular mass of about 6000. Hemoglobin, which carries oxygen around in our bloodstream, has a molecular mass near 68,000. ApoB-100, the protein part of the low-density lipoproteins, generally known as the carrier of bad cholesterol, has a molecular mass of 513,000. Nucleic acids have molecular masses that range from a few thousand to many millions. These are clear examples of big molecules.

In contrast, the oxygen molecules, O_2,[3] in our air have a molecular mass of 32; nitrogen molecules, N_2, have a molecular mass of 28. Methyl isocyanate, C_2H_3NO, has a molecular mass of 57. Later we are going to talk about methane, CH_4, molecular mass 16, carbon dioxide, CO_2, molecular mass 44, carbon monoxide, CO, molecular mass 28, and nitric oxide, NO, molecular mass 30. Moving on to more complex small molecules, sucrose (ordinary table sugar) has a molecular mass of 342. Cholesterol weighs in at 387. Most prescription and over-the-counter drugs that you use fall in the molecular mass range 150–700. Testosterone and estradiol, important male and female sex hormones, have molecular masses of 288 and 272, respectively. These figures are typical of small molecules.

We will meet many important examples of small molecules of varying structural complexity as we move forward. One of the most important general characteristics of molecules of biological interest is their shape. The shape determines, in part, how small molecules interact with proteins or nucleic acids with amazing biological consequences. To understand all this, we need a starting place, a molecule simple enough to get a grip on and yet capable of being generalized so that we may expand our understanding in a logical way.

Methane provides an introduction to structural chemistry

The molecule *methane* is a happy place to begin. On the one hand, it is a very small molecule, composed of just five atoms. On the other hand, it is based on the element carbon. Carbon-based molecules are the foundation of all living things (and a lot of things that are not). Moreover, the chemistry of carbon is the richest of that for all the elements. Although methane has just one carbon atom, we can generalize much of what we learn about methane to molecules having many carbon atoms.

The richness of the chemistry of carbon largely reflects the facts that, first, each atom of carbon forms several chemical bonds, usually four, simultaneously and, secondly, carbon, uniquely, can form stable, endlessly long chains with itself. Much more about the wealth of molecules resulting from these simple facts follows later.

Molecules are made up of atoms. At the beginning, we need to understand two basic things: first, the atoms in molecules are held together by chemical bonds and, secondly, these bonds are oriented in space in specific directions. We are not going to be much concerned about the nature of the chemical bonds themselves. They are formed from electrons contributed by the atoms bonded together. It will suffice to know that chemical bonds exist and that they bind atoms together in molecules. Think about chemical bonds as a kind of electronic glue holding atoms together but pointed in definite directions in space.

An understanding of several aspects of the chemistry of a molecule derives from answers to questions that follow directly from the two simple statements made above:

- Between what atoms are chemical bonds formed?
- What are the sizes of the atoms involved?
- What are the distances between the atoms linked by chemical bonds?
- How are these chemical bonds oriented in space?

From our answers, we can deduce two very important properties about the molecule: its size and shape. The quantities determine, in part, the physical, chemical, and biological properties of the molecule. Consequently, we shall be concerned with sizes and shapes of molecules, some elegantly simple and some beautifully complex.

A molecule of methane contains just five atoms: one of carbon and four of hydrogen. In chemical representations of molecules, each element is identified by a symbol. Carbon is represented by the symbol C; hydrogen is represented by the symbol H. Thus, the molecular formula for methane is CH_4. This representation, or model, tells us just one simple fact: the methane molecule contains one carbon and four hydrogen atoms.[3]

A great deal of experience informs us that carbon atoms frequently form molecules in which they are simultaneously linked to four other atoms. In contrast, hydrogen atoms are usually linked to only one other atom. It follows that the methane molecule is structured with a central carbon atom to which are bonded the four hydrogen atoms. In such a structure, only carbon–hydrogen (C–H) bonds exist. We can represent methane in the following way, in which the solid lines are symbols for chemical bonds:

Methane is a tetrahedral molecule in which all the hydrogen atoms are equivalent

The four hydrogen atoms occupy equivalent positions with respect to the central carbon atom in methane. The C–H bond length in methane is 1.107 Å (1 Å [angstrom] = 1×10^{-10} meters), a very short distance indeed. The four atoms of hydrogen are arranged about the central carbon atom at the corners of a regular tetrahedron.

A tetrahedron is a three-dimensional object with four vertices and four sides, each of which is a triangle. A regular tetrahedron in which all of the sides have the same length, and this is true of methane, has equilateral triangles for each face:

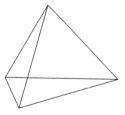

In this geometry, you need to think of the carbon atom of methane as occupying a position in the center of the tetrahedron and the four hydrogen atoms as occupying the four vertices. Here is a ball-and-stick representation of a methane molecule:

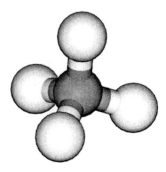

Before moving on, let's pause for a moment to recognize something that is really quite amazing. The methane molecule is tiny. Roughly, you could make up a single line of about 5 billion of them on a yardstick. That many methane molecules would weight about one-millionth of one-billionth of an ounce. Yet we know which atoms are bonded to which in methane; we know that all the hydrogen atoms are equivalent; we know how the hydrogen atoms are disposed in space; and we know the C–H bond length to within one one-thousandth of an angstrom unit. Using the most sophisticated electron microscopes, we can actually "see" individual methane molecules and their constituent atoms. It is easy to forget just how remarkable our capacities are for understanding molecules. Not only do we have this information for methane but also we have it for molecules of truly amazing complexity.

Note that I have used several models for methane, beginning with CH_4. This representation of methane is clearly incomplete but it does communicate one essential feature of the molecule: its composition. I have also introduced a ball-and-stick model for methane. There is another useful model for methane that I have not introduced—a space-filling model. Here it is:

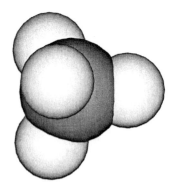

In addition to providing composition, bonding pattern, and geometry, this model tells us something about what methane molecules really look like in three dimensions. The atoms no longer appear as simple balls and the chemical bonds do not appear at all. Chemical bonding is reflected in the partial fusion of the spheres representing the atoms in the ball-and-stick model. A good sense of the relative size of atoms is provided.

Methane is a simple but really important molecule

Methane is a flammable gas, often called "marsh gas" or "natural gas." Methane, as natural gas, is widely used as an energy source for domestic and industrial purposes.

Methane has a very low boiling point $-161°C$ or $-258°F$[4] and, therefore, is typically stored in tanks under high pressure. Methane is harvested by the petrochemical industry from vast underground reservoirs of natural gas in truly enormous quantities. For example, in 2002 more than 2.5 trillion cubic meters of natural gas, almost all methane, was produced worldwide. If you are more accustomed to thinking about cubic feet instead of cubic meters, this corresponds to about 87.5 trillion cubic feet of natural gas at a pressure of one atmosphere and room temperature (1 cubic meter is about equal to 35 cubic feet). This is a lot of natural gas: 87.5 trillion cubic feet is equal to about 595 cubic miles, or a cube about 8.4 miles on a side. That is enough natural gas to fill the Houston Astrodome more than two million times.

Methane is a minor, but important, constituent of our atmosphere, contributing about 1.7 parts per million (ppm). Methane is the second most important greenhouse gas, after carbon dioxide, in our atmosphere. Atmospheric methane comes mainly from three sources:

- methanogenic (i.e. methane-producing) bacteria found in the mud of bogs and marshes;
- as a metabolic end product of ruminant animals, including cattle and sheep;
- from manmade sources—energy production, landfills, sewage treatment plants, and the like.

Methanogenic bacteria cannot tolerate oxygen and are restricted to sites that are free of it: the rumen of cows provides a good example, as does the mud in rice paddies.

Small molecules are everywhere and they matter

Moving forward, I simply want to point out some chemical names I have used here: hydrogen, oxygen, nitrogen, ammonia, methane, ethane, water, and carbon dioxide. These are all examples of small, simple molecules. Nitrogen, oxygen, and hydrogen contain only two atoms, water and carbon dioxide three, ammonia four, methane five, and ethane eight. We have described methane earlier in this chapter and will get around to the other molecules in the next one. For the present, I will simply argue that it is worth knowing something about these molecules and why they may be signals for the existence of life.

Although some molecules are exotic, the drugs that you take to prevent or cure disease, your food and food additives, vitamins, pesticides, herbicides, and so on, are not. These are molecules of daily life that are worth understanding.

An anecdote provided by James Shreeve in his fascinating book *The Genome War* is relevant:[5] "In 1999, bioethicist Arthur Caplan of the University of Pennsylvania was invited to address a meeting of state legislators who were puzzled over the issue of human cloning. Caplan asked the lawmakers if they knew where their genome was located. Roughly one third answered that it was in their brain, and another third thought it was in their gonads. The others weren't sure."

I do not know if lawmakers are more nearly ignorant of science than the general public or not. Either way, their responses are not reassuring. The fact is that your genome is in every cell in your body, with the single exception of mature red blood cells.[6] Put simply, we need to do better in terms of scientific literacy. All this great stuff is happening and too many of us either miss it entirely or see it but do not understand it.

And now, here are a few words about molecules

Here is a formal definition of a molecule. A molecule is a group of two or more atoms that are held together in a definite arrangement by strong chemical bonds. Clearly, methane fits the definition. The "definite arrangement" is composed of one carbon atom and four hydrogen atoms arranged in a tetrahedral geometry. We have described how they are bonded chemically. The building blocks are the elements carbon and hydrogen; we have met them earlier.

So far I have not made use of the word "compound," which is frequently used in a way that is almost synonymous with "molecule." Anticipating that we will talk about compounds later, here is a distinction. If we focus on one methane entity, we call it a molecule. If we talk about a whole bunch of methane molecules, we call methane a compound. Basically, a compound is any entity of fixed composition that contains two or more elements. Methane fits. The compound methane may be found in a pressurized tank. The individual entity is the methane molecule. The distinction between molecule and compound is not an important issue, but being clear is better than being muddled. Water is a compound composed of two elements: oxygen and hydrogen. We can talk about a molecule of water. Common sugar is a compound composed of three elements: carbon, hydrogen, and oxygen. The smallest unit of common sugar is the sugar molecule.

Elements and compounds constitute the world of pure substances. An element is a substance that cannot be decomposed by any chemical reaction into simpler substances. Elements are composed of only one type of atom and all atoms of a given type have the same properties. Pure substances cannot be separated into other kinds of matter by any physical process. We are familiar with many pure substances: water, iron, mercury, iodine, helium, rust, diamond, table salt, sugar, gypsum, and so forth. Among the pure substances listed above, iron, mercury, iodine, diamond (pure carbon), and helium are elements. We are also familiar with mixtures of pure substances. These include the air that we breathe, milk, molasses, beer, blood, coffee, concrete, egg whites, ice cream, dirt, steel, and so on.

Finally, let's back up to the definition of chemistry as a science having to do with matter, as stated above. So what are we talking about here? Matter is anything that has mass and occupies space. That includes everything that can be perceived by our human senses and a lot of stuff that cannot be. Matter includes, but is not limited to, the elements and compounds (pure substances) and mixtures of pure substances.

The time had come to take a little closer look at these things called elements and molecules. There are an amazing number of ways to assemble molecules. Each assembly offers the promise of useful properties. Building on this idea is a task of the next chapter.

Key Points

1. The world of chemistry is the world of molecules.

2. There are big molecules (proteins, nucleic acids) and small molecules (most of the rest). The really interesting stuff happens when big molecules and small molecules interact to produce some biological action.

3. Molecular mass is a useful measure of molecular size.

4. Molecules are made up of atoms. Methane is made up from one carbon atom (C) and four hydrogen atoms (H).

5. The structure of the methane molecule is that of a regular tetrahedron, with the carbon atom in the center and the four hydrogen atoms at the four corners.

6. Models aid the understanding of molecules.

7. The properties of molecules follow, in significant part, from their three-dimensional structures.

8. Methane is a greenhouse gas. It comes largely from methanogenic bacteria, ruminant animals, and the activities of man.

9. Methane, "natural gas," is an important energy source.

4

Building blocks and glue

Carbon-based molecules are constructed from atoms held together by the sharing of electron pairs; these are arranged in space in precise ways. Some molecules are nonsuperimposable on their mirror images and are said to be chiral; that is, they are "handed." Chirality matters.

In the previous chapter we established that the molecule methane is assembled from one carbon atom and four hydrogen atoms, arranged in a specific, precise way in space. The idea here is quite general: choose a collection of atoms of various types, connect them in a unique way and you have a new molecule. That is one of the things that chemists do, generally, in a search for molecules having useful properties, a novel drug for an important medical need for example. What follows in this chapter will establish some principles based on molecules containing only a single carbon atom that, like methane, have important meaning for us moving forward.

Atoms are the building blocks of molecules

Molecules are constructed from atoms. Atoms, in turn, are constructed from more basic particles: protons, neutrons, and electrons. Protons are very, very small objects that bear a very small positive charge.[1] Neutrons weigh just slightly more than protons but carry no charge. Protons and neutrons are clustered together in the nucleus of the atom. The diameter of a typical atomic nucleus is about 10^{-12} meters (one-trillionth of a meter), or about 0.01 Å, a very short distance indeed.

As light as they are, protons and neutrons are massive compared to the electron; one neutron weighs roughly as much as 2000 electrons. Each electron carries a negative charge exactly equal in magnitude, though opposite in sign, to the positive charge of the proton.

In an atom, the number of protons is equal to the number of electrons, so the atom carries no net charge. While the protons and neutrons are concentrated in the center of the atom and form its core, the electrons occupy the periphery. By analogy with our solar system (this analogy leaves quite a bit to be desired but will serve our purposes for the present), think of the nucleus as the sun and the electrons as the planets. Like the solar system, an atom is mostly empty space. A pro football stadium provides another mental image of an atom: a golf ball placed at the center of the 50-yard line represents the nucleus and a few mosquitoes buzzing around the upper decks represent the electrons. We can make our model of the atom somewhat more realistic by thinking about the mosquitoes as being somehow smeared out over defined areas of space. Small as atoms are, atoms can be visualized and manipulated one-by-one using very special technologies.

There are a lot of elements that we can use to build molecules

As we noted above, an element is a substance that cannot be decomposed into simpler substances by any chemical reaction. There are 92 elements that occur in nature, in widely varying amounts, and several more have been created in particle accelerators. In order to understand the molecules of life, we need to understand something about several of these elements, above all about carbon.

Elements are defined by the number of protons in the nucleus of each atom.[2] The number of nuclear protons is equal to the number of electrons orbiting the nucleus. The nucleus of carbon contains six protons. This value is known as the atomic number for carbon. In nature, carbon occurs largely in a form in which the nucleus also contains six neutrons. The atomic mass of carbon is defined as the sum of the number of protons plus neutrons. Consequently, this form of carbon is called carbon-12, or ^{12}C. About 98.9% of carbon in nature is ^{12}C. Most of the rest is carbon-13, ^{13}C, and contains seven neutrons in the nucleus. Smaller amounts of carbon occur that contain five or eight neutrons. These are known, respectively, as carbon-11, ^{11}C, and carbon-14, ^{14}C. These variations on the theme of carbon are called isotopes. Carbon-11 and carbon-14 are radioactive and decay spontaneously; carbon-12 and carbon-13 are stable.

Elemental carbon, whether it is soot, diamond, graphite, buckyballs, or graphene, contains only carbon atoms, each of which has exactly six protons in its nucleus. Lead (Pb) is a metallic element. Lead metal contains only lead atoms, each of which contains exactly 82 protons in its nucleus. Neon gas, familiar in neon lights, contains only neon atoms and each of these has just 10 protons in its nucleus. Elements are the building blocks out of which all matter is constituted.

Elements vary widely in their ability to form compounds with other elements. At one end of the spectrum are elements such as carbon, hydrogen, oxygen, and nitrogen that are constituents of an enormous, and rapidly growing, number of compounds. At the other end of the spectrum are the noble gases, named because of their reluctance to enter into intimate liaison with other elements. For example, the noble gases helium and neon exist only as the elements. No stable compounds of these two elements have been made. I know of only one unstable and exotic compound that is based on argon. A very small number of compounds have been made that contain krypton, another noble gas, and a small but more substantial number have been made from xenon, the least noble of the noble gases.

The discovery of the nature of elements had to wait for the Irish chemist and physicist Robert Boyle (1627–1691). Boyle published *The Sceptical Chymist* in 1661. This masterpiece is generally recognized as the beginning of modern chemistry (but not the end of alchemy, which continued to be practiced long after by people who should have known better). In his book, Boyle wrote the following definition of elements: "Certain primitive and simple, or perfectly unmingled bodies; which are not being made of any other bodies, or of one another, are the ingredients of which all those called perfectly mixed bodies are immediately compounded, and into which they are ultimately resolved." Given that the discovery of atoms was still hundreds of years in the future, this is a remarkably good definition. However, the conceptual basis for the subsequent search for elements was encapsulated in the ideas of Lavoisier of unique and quantifiable chemical reactivity associated with pure elements and compounds. A century later, these ideas facilitated the discovery of atoms. If you wish to learn more about early chemistry, I suggest that you have a look at John Buckingham's book: *Chasing the Molecule.*[3]

Earlier, I pointed out that things like atoms and molecules are really, really small but did not provide much insight into the matter. To provide some idea of the size range that we are talking about, let's use methane as an example.

The molecular mass of methane is 16. It has been known for a couple of centuries that the number of molecules in a sample containing the molecular mass of that sample in grams is equal to 6.022×10^{23}. This is known as Avogadro's number, after the Italian chemist who figured it out. Avogadro's number is huge, just as the diameter of an atom is correspondingly small. We can link the two. For our case, this means that 16 grams of methane, just a bit more than half an ounce, contain about 6×10^{23} methane molecules. Now let's suppose that a methane molecule were the size of an ordinary M&M chocolate candy and ask what 6×10^{23} M&M candies would look like. You can get the effective volume of an M&M by counting the number of them required to fill a container of known volume. Then a simple calculation reveals that Avogadro's number of M&M candies would cover the continental United States to a depth of about 50 miles: 6×10^{23} is a really, really big number, as this example illustrates. The fact that this number of methane molecules weighs just over half an ounce emphasizes just how really, really small a methane molecule is. The nucleus of a hydrogen atom, a bare proton for the common isotope, is perhaps a few million times smaller yet, in terms of its volume.

A single drop of water contains about 10^{20} molecules. There are a lot of drops of water in the world's oceans. These contain about 10^{46} molecules of water, a really big number!

There are several other ways to appreciate just how large Avogadro's number is. Time provides one dimension. Think about Avogadro's number of seconds: roughly 6×10^{23} seconds. Here is a simple calculation to show just how long that really is. There are 60 seconds in a minute, so that is $6 \times 10^{23}/60 = 1 \times 10^{22}$ minutes. There are 60 minutes/hour $\times$ 24 hours/day $\times$ 365 days/year $= 5.3 \times 10^5$ minutes/year. It follows that Avogadro's number of seconds is $1 \times 10^{22}/5.3 \times 10^5 = 2 \times 10^{16}$ years. In other terms, that is 20 million billion years. It has been about 2000 years since the birth of Jesus Christ. So the number of seconds in Avogadro's number is ten thousand billion (10^{13}) times as long as the time from the birth of Jesus until now. The dinosaurs became extinct about 60 million (6×10^7) years ago. Thus, the number of seconds in Avogadro's number is 300 million times longer than the time span from the death of the dinosaurs until the present. The Earth is about 4.6 billion years old. The number of seconds in Avogadro's number is five million times longer than the age of the Earth, and about two million times longer than the age of the universe, about 14 billion years! Avogadro's number is really big. There are a lot of methane molecules in half an ounce of methane.

There are a great many compounds known that, like methane, contain only carbon and hydrogen. These are collectively known as hydrocarbons. We shall encounter several more in what follows. Reserves of petroleum provide the most abundant sources of hydrocarbons. They are widely used as solvents and fuels. Octane, for example, is a hydrocarbon. Their most characteristic chemical feature is their stability. They are among the least reactive compounds known. Consequently, they may endure unchanged for very long periods of time. Hydrocarbons of apparent biological origin have been isolated from geological samples at least two billion years old, establishing a minimum age for the origin of life on Earth.

Hydrogen is the simplest possible molecule

Hydrogen is the lightest of the elements.[4] At ordinary temperatures, hydrogen exists as a colorless diatomic gas, H_2. A molecule of elemental hydrogen contains two hydrogen atoms, linked together. The H—H bond distance in H_2 is 0.74 Å. This is the simplest and lightest of all molecules, with a molecular mass 2 and a boiling point of $-252.8°C$. Molecular hydrogen is a symmetrical structure that packs easily into a crystalline lattice. Consequently, it has a very narrow liquid range of just 6°C.

The hydrogen atom is defined by having just one proton in the nucleus and one orbital electron. Hydrogen occurs in nature in three isotopic forms, containing 0, 1, and 2 neutrons. By far the most abundant form contains no neutron in the nucleus; this is hydrogen-1, ^1H, or, simply, hydrogen. The form containing one neutron in the nucleus is hydrogen-2, ^2H, or deuterium; that containing two neutrons is hydrogen-3, ^3H, or tritium. Tritium is radioactive; the other two isotopes are stable. Hydrogen is unique in having special names for its isotopes. There are no names corresponding to deuterium and tritium for isotopes of other elements.

There is very little hydrogen as H_2 in the atmosphere of the Earth. It was not always so. There is convincing evidence that the atmosphere of the very early Earth contained much hydrogen. However, the gravitational pull of the Earth for this very light gas is quite weak and atmospheric hydrogen has long since escaped into outer space.

Although there is not much hydrogen in our atmosphere, it is the most abundant element in the universe. One simple model is that the universe is a sea of hydrogen in which a few specks of nonhydrogen material bob around, mostly helium. Closer to home, the sun is composed largely of hydrogen. This hydrogen is, little by little, being converted to helium and other elements through nuclear fusion reactions, keeping us in light and heat as a result.[5] There is enough hydrogen left in the sun to sustain it, and us, for another few billion years. Eventually, however, the sun will exhaust its hydrogen fuel and turn into a very large star, a red giant. The Earth will be engulfed by the sun and burn to a crisp. Nothing lasts forever.

Hydrogen combines with many other elements, including carbon. Consequently, there is quite a bit of hydrogen on Earth, almost all in combination with other elements. It is the tenth most abundant element on Earth.

Covalent chemical bonds are generally strong

Covalent bonds are chemical bonds formed by the sharing of two electrons between the bonded atoms. Let us think about bringing two hydrogen atoms together to form a hydrogen molecule, H_2. The isolated atoms have just one electron each in a spherical home known as an orbital centered on the nucleus, which consists of just one proton in the common isotope of hydrogen. As we bring these two atoms together, the regions in space occupied by the electrons will begin to overlap. This electron density acts as electronic glue to hold the two atoms together quite tightly. The sharing of electrons between two atoms is the fundamental distinguishing characteristic of the covalent chemical bond. A covalent chemical bond is a shared electron pair.

The idea here is just the same, except for inevitable refinements and details, for the formation of all covalent bonds. So the basic ideas for chemical bonding in methane, ammonia, water, and so on, are the same.

Molecules have important symmetry properties: chirality

In relating properties of molecules to their structure, three-dimensional shape is frequently of great importance. Three-dimensional shape is a function of many variables: the nature and number of atoms composing the molecule and the nature of the chemical bonding pattern—which atoms are connected to which—are obvious factors. However, the situation can be more subtle than that. Even in cases in which the atomic composition of two molecules is the same and in which the chemical bonding pattern is the same, key differences in three-dimensional shape can arise.

Have a look at your two hands. Think of them as models of molecules. For most people, the components from which your right and left hands are constructed are the same: fingers, palm, back. Moreover, those components are linked together to form hands in just the same way. Nonetheless, your left hand and your right hand have different three-dimensional structures. You cannot superimpose them in space. Try putting a left-hand glove on your right hand. Your hands, and gloves, are nonsuperimposable mirror images of one another. This happens a lot with molecules.

If we explore substituted methanes, we will come to a very important understanding for molecules generally and the molecules of life in particular: the phenomenon of chirality. A molecule is said to be chiral when it and its mirror image are not superimposable in space.

Let's replace one of the hydrogen atoms of methane with a chlorine atom, one with a bromine atom, and one with a fluorine atom. Doing so creates a new molecule having the composition CHFClBr; that is, one atom each of hydrogen, fluorine, chlorine, and bromine attached to the central carbon atom.

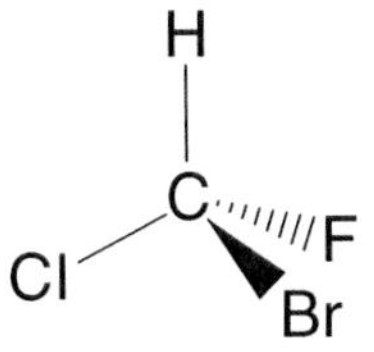

S-fluorochlorobromomethane 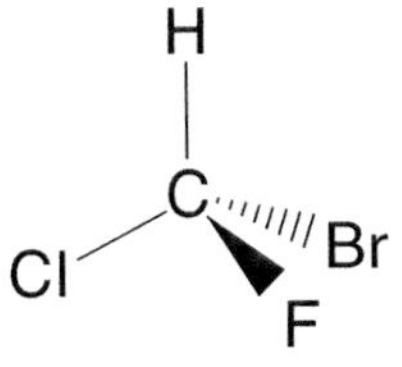   R-fluorochlorobromomethane

Let's note two important things here before moving on. First, I have used two new symbols for chemical bonds— ▼ and ′′′′′//. The former symbol is intended to tell you that the attached atom is coming at you in front of the page; the latter symbol indicates that the attached atom is receding from you behind the page. The usual simple solid lines for chemical bonds indicate attached atoms that are in the plane of the paper. Thus, in the structure on the left, the carbon, chlorine, and hydrogen atoms are to be viewed as in the plane of the paper, the bromine atom is poking out at you, and the fluorine atom is receding from you. In the structure on the right, everything is the same except that the fluorine atom is poking out at you and it is the bromine atom that is receding. These symbols are very commonly used as a way of communicating something about three-dimensional structure on a two-dimensional piece of paper.

Secondly, we have labeled one of the substituted methanes S-fluorochloro bromomethane and the other R-fluorochlorobromomethane. S (*sinister*, left) and R (*rectus*, right) are labels that are useful in designating the absolute configuration of chiral molecules in space.[6] The rules for assigning S stereochemistry in one case and R stereochemistry in the other are somewhat complex but it doesn't matter, so never mind. The point is that we have a way of talking about the two possibilities.

Careful examination of the two molecules of CHFClBr provided above reveals that these molecules are, in fact, different. They are different in a very interesting and important way: there exists no way of superimposing them in space. As a result of having four different atoms on the central carbon atom, the molecule has acquired a definite spatial orientation, or handedness. Let us imagine ourselves standing on the chlorine atom facing the central carbon atom, with the hydrogen atom above us in the two structures above. In the structure on the left, bromine will be on our right side and fluorine on the left. In the structure on the right, fluorine will be on our right side and bromine on the left. There is no way that one of these molecules can be reoriented in space so that it can be superimposed on the other. Trying to superimpose these two structures is like trying to superimpose gloves for a left hand and a right hand. In fact, the relationship between these two structures is exactly like that of a pair of gloves: each glove is composed of exactly the same structural elements but they are arranged differently in space. Like right-handed and left-handed gloves, the two fluorochlorobromomethane molecules are mirror images of each other. They are said to be enantiomers. This is the first example of stereoisomerism that we have encountered. This term is applied to those situations in which molecules having the same elemental composition (or atomic composition) and connectivity among the atoms differ in their spatial arrangement of the atoms.

As noted above, molecules that are not superimposable on their mirror images are said to be chiral. Nature makes important distinctions between "right-handed" and "left-handed" molecules.

We will encounter a number of examples of chiral molecules throughout our exploration of the molecules of life. Here are two important examples.

Thalidomide is a chiral molecule

Thalidomide is a tragic case of drug discovery gone wrong.[7] This molecule was approved in the late 1950s by several European countries for use as a sedative to help pregnant women cope with the problems of morning sickness. Thalidomide works for this purpose. Unhappily, thalidomide proved to be a potent teratogen, i.e., a molecule that causes birth defects. "Thalidomide children" were born with badly deformed arms and legs that more closely resembled flippers than human limbs. The tragedy was recognized in 1961 and the drug was withdrawn from all markets. However, a great deal of damage had been done to many children. The United States was spared this tragedy since the US Food and Drug Administration (FDA) never approved thalidomide for this use. Thalidomide is a chiral molecule.

Like fluorochlorobromomethane, thalidomide exists in right-handed and left-handed forms. Thalidomide as marketed contained equal amounts of both forms. As was learned later, the beneficial effect for pregnant women is mostly due to one of these forms. The potential to cause birth defects is largely due to the other. However, thalidomide would have been a hopeless case for morning sickness even if the marketed drug contained only the "good" molecule. The two enantiomers of thalidomide are interconverted in vivo. Any threat to the fetus is unacceptable.

In fact, thalidomide has recently been approved for two other uses in the United States and elsewhere almost 50 years after the tragedy in Europe. The first new approved use was for leprosy, also known as Hansen's disease. In addition, thalidomide has efficacy for an incurable cancer known as multiple myeloma and it has been approved as first-line therapy for this indication. This is a beautiful example of balancing risk against the benefit for the intended use in human medicine. No one will justify using a teratogen for morning sickness due to pregnancy for which other, far safer drugs are available. Using one to combat leprosy or a life-threatening cancer is another matter entirely.

The purple pill is chiral

Our second interesting case of chirality is an example of advertising trumping science. In this case, the molecule in question is known generically as esomeprazole and is marketed in the United States as Nexium, widely advertised as "the purple pill." The story begins with a closely related molecule, omeprazole, marketed as Losec or Prilosec.

Omeprazole is a potent inhibitor of gastric acid secretion and is useful for healing of gastric ulcers and for treatment of gastroesophageal reflux disease, GERD. Omeprazole has been in clinical use for a number of years and was once the largest selling drug, in dollar terms, in the world. It is now available over-the-counter, OTC. Omeprazole is a chiral compound. Omeprazole is a 50/50 mixture of the S and R forms.

Esomeprazole is, as the name indicates, just the S-isomer of omeprazole. Put another way, omeprazole is analogous to a pair of gloves; esomeprazole is simply the left-hand glove alone. There is no evidence to suggest that expensive, by prescription only, esomeprazole is any safer or more effective than much cheaper, OTC, omeprazole. Yet an intensive direct-to-the-public advertising campaign by AstraZeneca has generated a huge demand for "the purple pill" and has turned Nexium into a multibillion dollar drug. Ads for Nexium assert that it has been shown to be superior in clinical trials. However, the comparison is to another prescription pharmaceutical—Prevacid—and not OTC Prilosec. The basic differences between Prilosec and Nexium are whether or not you need to see a physician to get a prescription and the cost of the drug. Note that this is a case in which chirality does not matter. It usually does.

This example of a highly successful advertising campaign has had a very good result for AstraZeneca and its stockholders. It has had a poor result for the costs of health care in the United States without health benefit. This seems to me to be one of the better examples of the need for a chemistry-literate public.

Key Points

1. Atoms are constructed from small particles known as protons and neutrons, housed in the atomic nucleus, and orbital electrons.

2. Elements are defined by the number of protons in the nucleus of each atom of the element.

3. Compounds that contain only the elements carbon and hydrogen are known as hydrocarbons. Many of them are extraordinarily stable.

4. Molecular hydrogen, H_2, is the simplest possible molecule.

5. Strong covalent chemical bonds are formed by the sharing of an electron pair between two atoms.

6. Elements have isotopes. These vary in the number of nuclear neutrons. For example, we have hydrogen, no neutrons, deuterium, one neutron, and tritium, two neutrons.

7. Chiral molecules are not superimposable on their mirror images. They exist as "right-handed" and "left-handed" isomers, known as enantiomers.

8. Chirality is really important in determining biological properties of molecules.

5

From methane to chemical communication

Molecules are assembled from atoms of the chemical elements. Many elements form multiple chemical bonds in molecules. Among the elements, carbon is unique in its ability to form chains of atoms endlessly long. The structural chemistry of carbon is the richest of that for all the elements.

When I was a child, I had three toys that provided endless possibilities: a set of wooden blocks, a set of Tinkertoys, and an Erector Set. These three toys had a lot in common. Each had a number of distinct structural elements that could be assembled in an amazing number of ways to create something new and different. This has a direct analogy to the concept of assembling molecules from atoms: the atoms are the structural elements and they can be hooked together to make molecular constructs in a great many ways.

Consider the set of wooden blocks: some were square, others oblong or triangular or cylindrical, or something else. There was an endless number of ways to put these objects together. The only real problem with the wooden blocks was that there was no way to link them together to provide structural stability. The force of gravity was pretty much it. The inevitable consequence was that, as structures became larger, higher, and less stable, they eventually collapsed. In contrast, assembling molecules from atoms depends on formation of chemical bonds. These are generally strong, so that molecules are usually stable under ordinary conditions of temperature and pressure.

The Tinkertoy and Erector Sets were basically similar to wooden blocks in having a family of distinct parts that could be assembled in many different ways to create different outcomes. They were different in one important aspect: the pieces connected in a way that made the growing structures stable. These connections worked a good bit better than gravity. They provided stability, just as chemical bonds provide stability for molecules.

By the time that our children came along, Lego had pretty much replaced wooden blocks, Tinkertoys, and Erector Sets as the medium of construction. But the underlying idea was the same: a set of different structural units that could be assembled in a jillion different ways with a jillion different outcomes. Lego was an improvement over my toys: more diverse parts and easier to achieve structural stability.

Putting structural elements together and linking them in some way to create objects—whether with wooden blocks, Tinkertoys, an Erector set, or Lego pieces— is not so different from what happens in chemistry. Instead of making log cabins, castles, Ferris wheels, or whatever, chemists assemble molecules out of structural elements. These structural elements are of course the atoms of the chemical elements themselves: carbon, hydrogen, oxygen, nitrogen, and so on. Think of the chemical elements as nature's Lego set. The point is that we can take these elements and link them together by chemical bonds in an endless number of ways to create an endless number of molecular outcomes. I do not believe that anyone has a good count of the number of different molecules that have been made based on, for example, carbon. The number is surely in the tens of millions and can grow almost without limit.

Let's look back at chapter 3 for a moment, where we focused on methane. Methane contains only two structural elements, carbon and hydrogen. These are linked together in a way that puts the carbon atom in the exact center of a regular tetrahedron and the four hydrogen atoms at the four corners. Given that we started with only two types of structural elements and only one carbon atom, there is not a whole lot more than we can do here. To get more interesting and complex structures we are going to have to add different structural elements and more of them. We begin that task here, building on what we already know.

Before we get any further, I want to divide the chemical elements into two classes to facilitate an understanding of the structural chemistry of molecules. The first class includes those elements that form more than one chemical bond at a time. Carbon typically makes four chemical bonds and provides an example of such an element. Oxygen, nitrogen, sulfur, and phosphorus provide four additional examples of elements that typically make more than one chemical bond. Elements in this class provide for structural complexity, since, in principle at least, they can make straight chains, branched chains, cyclic structures, and so on.

The second class includes those elements that form only one chemical bond at a time. These elements terminate some aspect of the molecular structure, since, once they have bonded, there is nowhere else to go. Hydrogen provides an obvious example. We have already encountered three other elements in this class: fluorine, chlorine, and bromine. These elements, together with iodine, are related and form a useful group that is known as the halogens. The halogens typically form only one chemical bond at a time.[1]

In sum, we have one class of elements that we can employ to create straight or branched chains, cycles, and so forth, and a second class that permits us to bring the process to a conclusion by tying up all the loose ends. We shall begin to see how this works out shortly in the case of creating carbon–carbon chemical bonds.

The carbon–carbon chemical bond is really important

Among the class of elements that usually form more than one chemical bond in a molecule, carbon is unique. Carbon, among all the elements, stands alone in its ability to form chains of atoms several thousand long, perhaps endlessly long. The richness of the chemistry of carbon derives in large part from this apparently unlimited capacity of carbon atoms to form bonds with other carbon atoms. We begin small: with ethane, which contains just two carbon atoms and, hence, one carbon–carbon bond. Of all the hydrocarbons, ethane is the simplest, with the sole exception of methane.

Ethane is the most modest of beginnings. We can do much more. We can make straight chains long and short, branched chains, rings of carbon atoms large and small, rings attached to other rings, rings fused with rings, rings attached to fused rings, rings within rings, chains stuck onto rings, geometrical structures such as squares, cubes or dodecahedrons. We can link carbon atoms by single bonds (one pair of electrons), double bonds (two pairs of electrons), or triple bonds (three pairs of electrons). We can make structures that contain single bonds, double bonds, and triple bonds and structures that have double or triple bonds within rings. We can make molecules in which there is no easy way to describe whether the chemical bonds are single or double. There is very little that you can imagine in the way of hydrocarbons that chemists cannot construct. Issues of stability impose some limitations, but our possibilities are vast.

Suppose that we replace one of the hydrogen atoms of methane with a methyl group, $-CH_3$. The resultant molecule, ethane, has the composition C_2H_6, with molecular mass 30. There are several other simple ways to model the ethane molecule, as well as several rather complex and elegant ways to do so. Let's consider three simple ones. At one extreme, we can write out the bonding pattern in detail and show ethane as

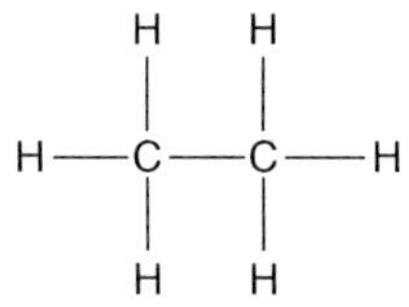

$$
\begin{array}{ccc}
& H & H \\
& | & | \\
H - & C - & C - H \\
& | & | \\
& H & H
\end{array}
$$

This model explicitly shows all of the atoms and all of the chemical bonds. The nature of each atom is made perfectly clear, as is the nature of the chemical bonds. However, there are a couple of reservations about this model: writing out everything in this detail takes quite a bit of time and quite a bit of space. These reservations are minor for ethane but major for more complex molecules. Hence, it is unusual to explicitly show the bonding pattern in this much detail. Some form of shorthand depiction of molecules is required and is generally used. Our need is to understand it, step by step.

A simpler model for ethane recognizes what we already know for methane: that each carbon atom is bonded to four other atoms. Given that knowledge, we can now write simply, $CH_3—CH_3$, showing only the carbon–carbon bond. Since each carbon atom forms four bonds and since only one is shown (the carbon–carbon bond), it follows that each carbon atom must make three bonds to hydrogen atoms. Even simpler is the model CH_3CH_3, in which none of the chemical bonds is shown directly. Once we have gained more experience, it will be clear that this simple representation contains all the information that the more detailed one does. Here are two other models for ethane:

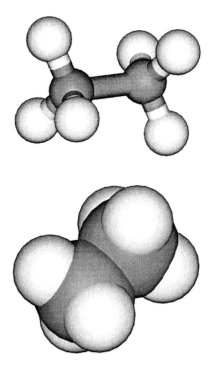

Above is a ball-and-stick model and below a space-filling model (two of the hydrogen atoms are partially obscured behind the molecule). In the ball-and-stick model, the carbon atoms are shown as dark spheres and the hydrogen atoms as light ones. The sticks represent the chemical bonds. In the space-filling model, an effort is made to depict how the ethane molecule actually looks. The carbon atoms are shown fused in part to each other and the hydrogen atoms are shown fused in part to the carbon atoms. The chemical bonds, which cannot be shown in space-filling models, pull the involved atoms close to one another, in effect creating a partial fusion. Each of these models provides insight into some aspects of ethane.

Ethane is the parent molecule for a family of inhalation anesthetics

Ethane is really quite a simple molecule: just two types of atoms, carbon and hydrogen, and eight atoms in all. Yet simple derivatives of ethane find important use in human medicine as inhalation general anesthetics.

Anesthetics fall into two broad classes: local and general. They behave as their names imply. Local anesthetics prevent pain locally. A dentist may give you a shot of procaine hydrochloride, Novocain, or other local anesthetic, to prevent the pain of filling or extracting a tooth. But if you stub your toe getting out of the dentist's chair, your toe is going to hurt. Local anesthetics may help to alleviate the pain of sunburn but they will not help with that of a concomitant headache.

General anesthetics are another matter entirely. The state of general anesthesia is drug-induced absence of perception of all sensations. The most common use of general anesthetics is to render a patient insensitive to pain, or anything else, during surgery. General anesthetics themselves fall into two classes: those that are inhaled and those that are injected. Here I focus on the former category.

The earliest inhaled general anesthetic that found significant use is chloroform, $CHCl_3$, more systematically trichloromethane, a simple derivative of methane. Chloroform was gradually replaced by diethyl ether, commonly known simply as ether, a safer and more effective molecule than chloroform.

Diethyl ether was the inhalation anesthetic of choice during my childhood. Happily, I had rather little need of it. Ether was replaced years ago by a family of superior inhaled general anesthetics, most of which are ethane derivatives. The notable exception is another very simple molecule, nitrous oxide, N_2O, frequently known as "laughing gas." N_2O has been around for a long time and was once a "party drug."

Let's have a look at some of the ethane derivatives. Perhaps the simplest of these is halothane:

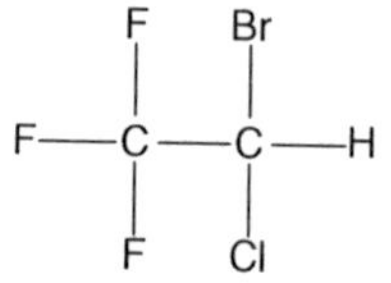

In this structure, five halogen atoms—three fluorines, one chlorine, and one bromine atom—replace five of the six hydrogen atoms of ethane.

Isoflurane and desflurane are variations on this theme:

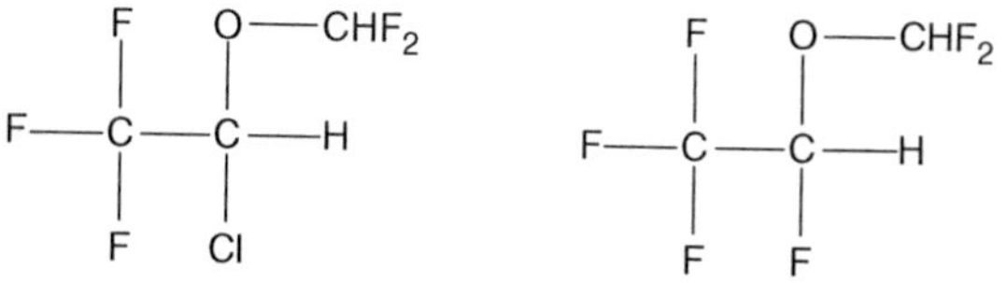

isoflurane   desflurane

as are methoxyflurane and enflurane:

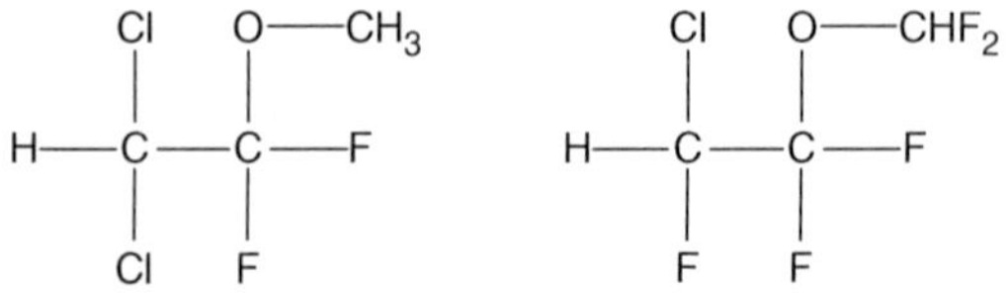

methoxyflurane   enflurane

The basic point here is clear: by taking a simple molecule such as ethane and replacing some or all of the hydrogen atoms by atoms such as chlorine or fluorine or small groups of atoms, one can create simple but novel molecules of significant use in clinical medicine. Not all molecules that are useful in life need be complex. At the same time, it is not true that simply replacing hydrogen atoms by halogen atoms—chlorine, bromine, and fluorine—randomly on the ethane framework of two carbon atoms will yield molecules that are useful as anesthetics, or anything else. Both the nature of the halogen atom and its position on the ethane framework relative to other atoms are critical for efficacy and safety. Efforts to relate chemical structure to properties are known as structure–property relationships and they play a huge role in chemistry, particularly in the field of drug design. We shall see many examples as we move forward.

Introducing more carbon–carbon bonds makes things more interesting

Let's see what new happens when we extend the chain of carbon atoms in hydrocarbons. Propane, a gas widely employed as a heating fuel, is C_3H_8 or, more explicitly, CH_3—CH_2—CH_3. There is really nothing new here. The chain of carbon atoms has simply grown by one. However, if we extend the chain again to create butane, C_4H_{10}, we do find something new. Here we can draw two molecules that correspond to this composition but that differ in the bonding pattern or connectivity (that is, in which atoms are linked to which):

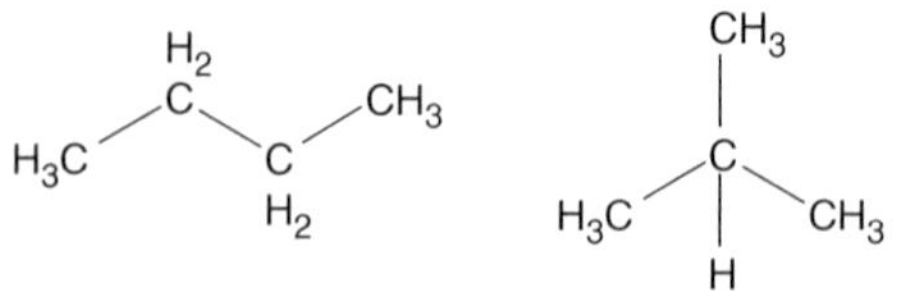

n-butane   isobutane

The first structure has the carbon atoms arranged in a linear fashion and is called normal butane or, simply, n-butane. The second has a branched structure and is termed isobutane (or, more rigorously, 2-methylpropane).

This is the second example of isomerism, compounds having the same molecular formula but different structures, that we have encountered. In our first example, the stereoisomerism of fluorochlorobromomethane, CHFClBr, the two isomers have the same bonding pattern but exhibit "handedness." The two stereoisomers of this molecule, known formally as enantiomers, and others in the same class, have the same melting and boiling points as well as many other physical and chemical properties that are the same. They are distinguished by how they interact with light and, in particular, with how they interact with other molecules that also exhibit handedness. Molecules in living systems generally fall into this category. As a consequence, two enantiomers frequently exhibit quite different biological properties: taste, smell, pharmacological actions, and toxicity, for example. Recall the thalidomide story in chapter 4.

In contrast, the *constitutional isomers* of butane have different physical properties. Thus, *n*-butane boils at –0.5°C but *iso*butane boils at –12°C. Although the composition of the two butanes is the same, their three-dimensional geometries are quite different as a result of different patterns of chemical bonding. Note, for example, that one of the carbon atoms of *iso*butane makes bonds to three other carbon atoms. No carbon atom in *n*-butane does. Thus, the chemistry of the two butanes differs in many ways, but not as a consequence of "handedness." They are simply examples of different ways to put together the same set of structural elements. Think about a Lego set.

Constitutional isomerism becomes more complex as the size of the hydrocarbon molecule is increased. For example, there are three constitutional isomers of pentane, C_5H_{12}. The number of constitutional isomers increases quite rapidly with an increasing number of carbon atoms. Thus, there are five constitutional isomers of hexane, C_6H_{14}, nine isomers of heptane, C_7H_{16}, 75 isomers of decane, $C_{10}H_{22}$, and 366,319 isomers of eicosane, $C_{20}H_{42}$. You can begin to understand why it is possible to make so many different molecules based on carbon.

Chains of carbon atoms can be closed to create cyclic structures

As the linear chain of carbon atoms grows, it can adopt a conformation in space that resembles a circle. In fact, there exist many hydrocarbons, as well as more complex molecules, which are cyclic. One of the simplest is cyclohexane, C_6H_{12}. The name makes sense: since the molecule has six carbon atoms, it is a hexane. Since it is cyclic, it is cyclohexane. In two-dimensional form, we can write the cyclohexane molecule as:

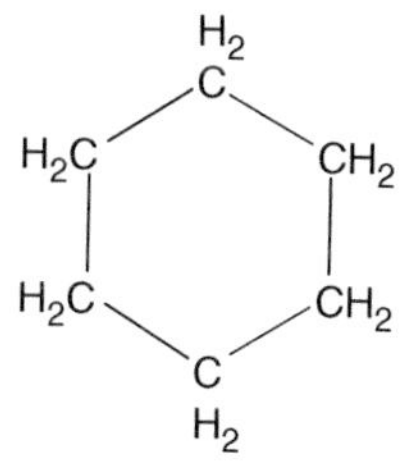

This structure reveals that the cycle is created by a series of six carbon–carbon bonds and that each carbon atom is bonded to two others as well as to two hydrogen atoms. However, this model reveals nothing about the three-dimensional structure of cyclohexane. Furthermore, it may imply that the molecule is planar, something that is simply not true. The most stable conformation for cyclohexane is the following in which the six-membered ring is flexed:

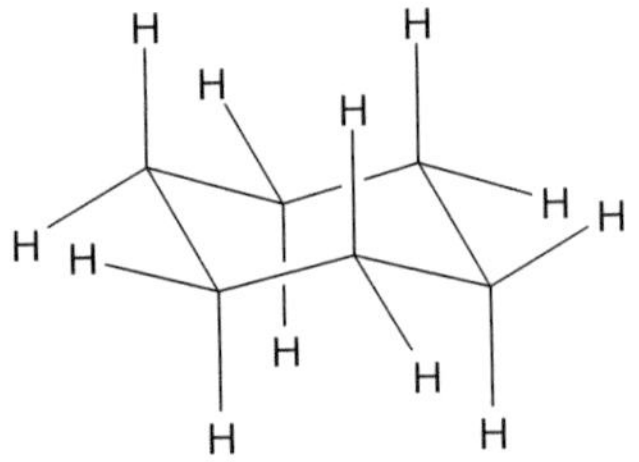

This conformation looks a little like a chair (use your imagination) and that is how it is known. This model for cyclohexane explicitly identifies the bonding pattern and provides a sense of the three-dimensional structure. Finally, we can simplify this structure even more by deleting all the bonds except those linking the carbon atoms. Now we have just:

This is the representation of cyclohexane that chemists routinely use: it is easy to write, identifies the bonding pattern, and provides an indication of the three-dimensional conformation.

At the same time, it fails to tell us that alternative conformations of cyclohexane exist and that these interconvert rapidly. All models, necessarily, leave out some information. There are always compromises to be made among the issues of clarity, information content, and ease of construction.

In writing structures in this shorthand way—and I am going to do that from now on in dealing with molecules of some complexity—we simply need to remember the following things. First, at every intersection of two bonds and at the termination of bonds, there is a carbon atom. Secondly, carbon atoms make four bonds. If four bonds to each carbon atom are not indicated in the structure, then it is understood that there are enough hydrogen atoms linked to the carbon atoms to make four bonds. Finally, atoms other than those of carbon and hydrogen are always shown explicitly. In some cases, we may choose to write out the symbols for carbon and/or hydrogen when doing so will improve clarity. To help gain some familiarity with this way of representing molecules, let's look at a number of other examples, starting with n-butane, C_4H_{10}. If we explicitly include all the atoms and show the carbon–carbon chemical bonds, we have:

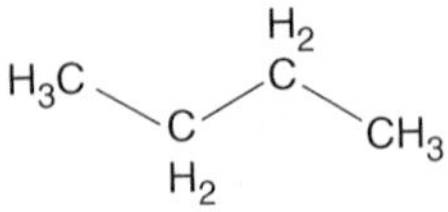

Now, if we invoke our rules about deleting carbon and hydrogen atoms and just showing the chemical bonds, we get a simpler model:

Now think about cyclopropane. Showing all the atoms and carbon–carbon bonds, we have:

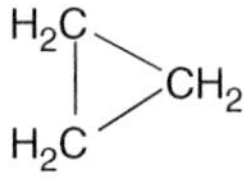

$$H_2C \overset{\textstyle \diagup}{\underset{\textstyle H_2C}{|}} CH_2$$

Confining ourselves to the carbon–carbon chemical bonds, we have:

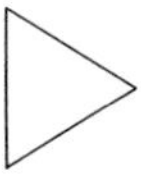

To consider a substantially more complex example, let's look at decalin, a hydrocarbon created by fusing two six-membered rings:

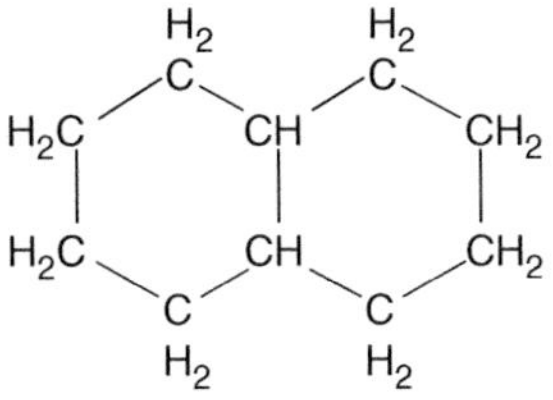

Once again, if we confine ourselves to carbon–carbon bonds, the model gets substantially simpler:

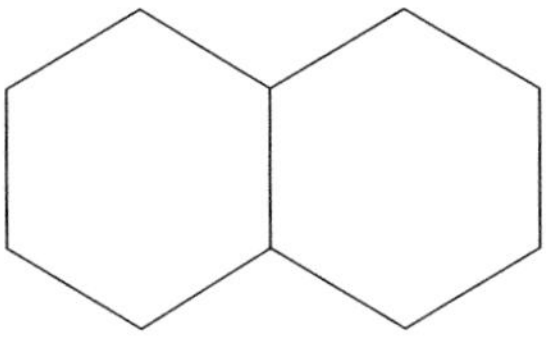

The central point here is that the two representations carry the same information. You just need to remember the rules to understand the simpler ones. To provide more examples of the meaning of these concise representations of chemical structures, I have collected a number of additional examples in the Appendix.

There are carbon–carbon double bonds and rotation around them is restricted

Now we need to face something new: carbon–carbon double bonds.[2] Here a pair of two-electron bonds link two carbon atoms. One of the simplest examples is propylene, or more rigorously propene: $CH_2{=}CH{-}CH_3$. The suffix *ene* indicates the presence of a carbon–carbon double bond while the suffix *ane* indicates that only single carbon–carbon bonds are present. These turn out to be useful in some crossword puzzles. Propylene, or propene, can be thought of as a propane molecule from which two hydrogen atoms have been eliminated and a double bond between two adjacent carbon atoms formed in their place. Note that each carbon atom of propylene still makes four bonds. The double bond counts as two.

There are also carbon–carbon triple bonds. Acetylene (or, more rigorously, ethyne—*yne* indicating the presence of a triple bond in the molecule), C_2H_2, is a common hydrocarbon that contains a triple bond: $HC\equiv CH$. Note that here too each carbon atom makes four chemical bonds: the triple bond counts as three. Acetylene finds significant commercial use in oxyacetylene torches.

Double and triple bonds linking carbon atoms differ from single carbon–carbon bonds in several ways, one of which is important for us. The rapid rotation around single carbon–carbon bonds is lost in double and triple bonds. Put another way, there is no free rotation around double and triple bonds. There is a big contrast between ethane, C_2H_6, and ethylene, C_2H_4. In ethane, the two methyl ($-CH_3$) groups spin around the carbon–carbon single bond a few million times a second. In ethylene, the two methylene ($=CH_2$) groups do not spin around the carbon–carbon double bond at all under ordinary conditions. This turns out to be critical for vision, among other things.

To see why this is important, consider a butene (2-butene)[3] in which the double bond is between the two central carbon atoms: $CH_3-CH=CH-CH_3$. If we think about it for a bit, we can recognize that there are really two of these structures; they are stereoisomers that are not enantiomers and not constitutional isomers. Such stereoisomers are termed diastereomers. Diastereomers in this class are also known by the older and largely obsolete term "geometrical isomers." They differ in the way that the two methyl groups at the ends of the molecule are disposed with respect to each other. The two possibilities are:

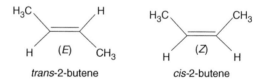

trans-2-butene   cis-2-butene

These two molecules have quite different shapes.

Since rotation around the double bond is restricted, these two isomers are stable under the usual conditions. There is no problem in isolating both *trans*-2-butene and *cis*-2-butene and in maintaining the two molecules for long periods of time. They can be interconverted at high temperature or in the presence of light. In fact, the light-induced conversion of the *cis* isomer of a pigment in the retina of the eye to the corresponding *trans* pigment is the molecular event that initiates the miracle of vision.

Some nomenclature: note that in the *trans* isomer, the two methyl groups are on opposite sides of the double bond, or across from each other. In the *cis* isomer, they are on the same side of the double bond. Like constitutional isomers, diastereomers have different physical and chemical properties.

The designations *cis* and *trans* are not completely general and there are some complex cases for which this simple distinction between *cis* and *trans* fails, but these terms will usually meet our needs. For the more complex cases, a more general nomenclature has been developed. Here the isomers are designated either E (for *entgegen*, German for opposite) or Z (*zusammen*, German for together) and these are provided for the two butene structures. For simple cases, E corresponds to *trans* and Z to *cis*, as noted in the structures above. However, life is not always simple.

Note that there are no stereoisomers around carbon–carbon triple bonds. Each carbon atom involved in triple bond formation can form only one more bond to a chemical group. For example, a butyne (2-butyne) containing a triple bond between the two central carbon atoms looks like this:

$$CH_3-C\equiv C-CH_3$$

The nature of the triple bond requires that the four carbon atoms of this molecule lie in a straight line (bond angle $= 180°$), which rules out the possibility of diastereomers.

Pheromones are really interesting

The difference between *trans* and *cis* isomers may seem arcane, a matter that could only be of interest to a professional chemist. Reality teaches otherwise. In addition to relevance to vision, let's look at just one striking example. This comes from the study of pheromones.

Pheromones are chemicals that are used for communication among organisms of the same species. There are many examples of the use of chemicals for communication in nature and I get back to this topic in much more detail in chapter 25. For the present, let's have a look at the very first pheromone ever isolated and characterized: the sex attractant of the female silkworm moth *Bombyx mori*.[4]

This signal chemical accomplishment is the work of a German chemist Adolf Butenandt and his coworkers and was completed in the late 1950s. Butenandt later won the Nobel Prize in Chemistry for his achievements, including the identification of the silkworm moth sex pheromone. This was not easy work: it required about 20 years of effort. Nearly half a million female silkworm moths had to be processed to yield a mere 6 milligrams (mg), about three ten-thousandths of an ounce, of the sex pheromone. The structure of the substance was deduced from work on this very small amount of material. Given today's powerful tools of chemistry, the work would prove far less troublesome than it did in the time of Butenandt's work.

Here are the essentials. The female silkworm moth emits a 16-carbon alcohol, bombykol, which can be detected by the male moths who are then attracted to the female in search of sex. The parallel with human behavior, specifically our use of perfumes and colognes, for example, should be obvious. Mating may ensue, leading to reproduction and maintenance of the species, for the silkworm and for humans.

Bombykol contains two double bonds along a long chain of carbon atoms. We can assign the positions of these double bonds by numbering the carbon atoms from 1 to 16, starting at the —OH end of the molecule. The two double bonds begin with carbons 10 and 12 (see figure 5.1). Each double bond may occur either as

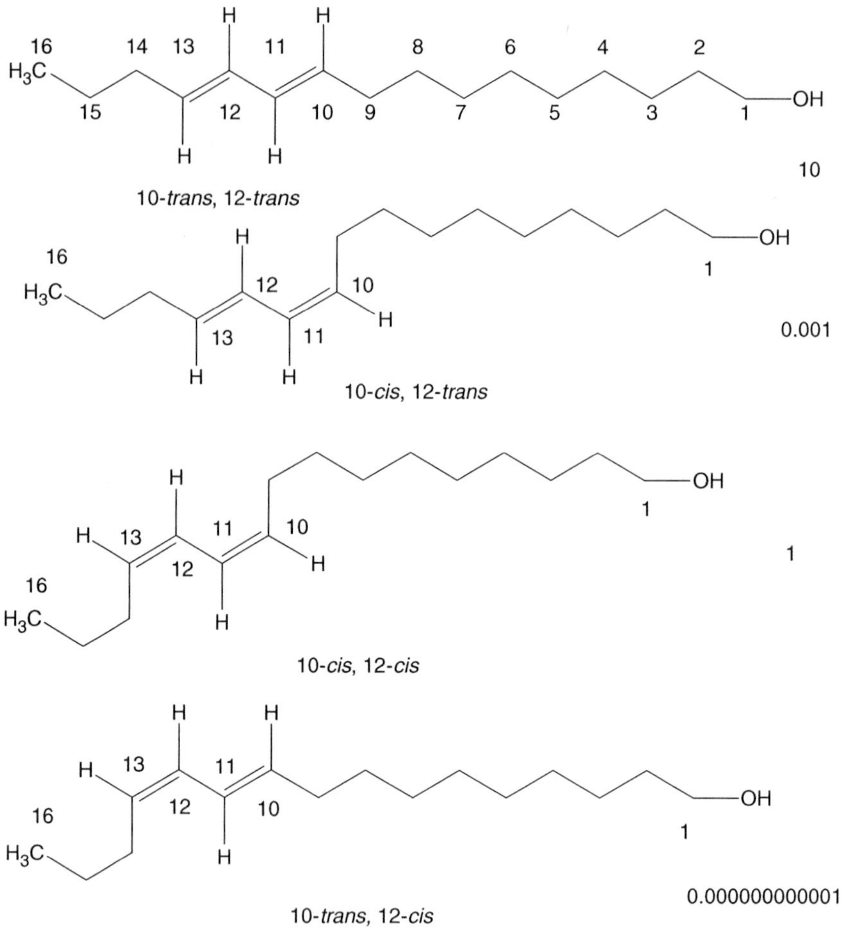

Figure 5.1 The relationship between geometry at carbon–carbon double bonds and biological activity for the female sex attractant of the silkworm moth *Bombyx mori*. The figures associated with each structure are the relative effective concentrations required to elicit a response in 50% of male silkworm moths.

the *cis* or the *trans* isomer. Consequently, there are four diastereomers possible in this alcohol, only one of which is bombykol, the molecule actually made and released by the female silkworm moth. Both double bonds may be *trans*: 10-*trans*, 12-*trans*; the first may be *trans* and the second *cis*: 10-*trans*, 12-*cis*, the first may be *cis* and the second *trans*: 10-*cis*, 12-*trans*; or both may be *cis*: 10-*cis*, 12-*cis*. Other than the geometry around the double bonds, the four molecules are identical. In figure 5.1, structures and the relative biological activity of these geometrical isomers are provided. Bombykol is the 10-*trans*, 12-*cis* isomer. The change in geometry around both of the double bonds of bombykol results in a diminution in biological activity of nine orders of magnitude, a billion fold! If the geometry about one of the double bonds is changed, the biological potency is reduced by as much as 13 orders of magnitude, 10 trillion times. Clearly, the biological activity of bombykol is amazingly sensitive to the geometry around its double bonds. These findings provide a neat example of structure–property relationships. Here the property—the utility of the molecules as sex attractants—is a very sensitive function of structure.

Benzene introduces a whole new idea into chemical structure

Here are two models for a very important molecule known as benzene:

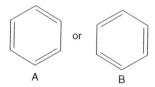

A   B

Benzene is a chemical species called an *aromatic hydrocarbon.* It turns out that rings such as the one in benzene are found in the majority of the molecules that we encounter moving forward: hormones, neurotransmitters, antibiotics, amino acids, nucleic acids, many pharmaceutical products, carcinogens, substances of abuse, and so on. So it is worthwhile to get some feeling for what is going on with benzene and its relatives.

Above, I have provided two representations of benzene in which in the three double bonds are placed in different ways, A and B. The basic point is that writing the double bonds is misleading. Surprising as it may seem, benzene does not contain any carbon–carbon double bonds; it does not contain any carbon–carbon single bonds either.

Each carbon–carbon double bond is constructed from four electrons. In benzene, the electrons that create the apparent double bonds fall into two classes. Two of the electrons are localized between two carbon atoms, just as we have come to expect. The other two electrons that contribute to the apparent double bonds are, in contrast, delocalized over the entire molecule. Since there are three apparent double bonds, we have a total of six electrons that are delocalized over the six carbon atoms. Think of these as free-range electrons. Basically, each of the carbon–carbon bonds in benzene is a 1.5 bond (technically, we say that the bond order in benzene carbon–carbon bonds is 1.5). Hence, the two models for benzene employed above, though universally used in chemistry, leave something to be desired. Benzene is better thought of as a hybrid of the two. Chemists have struggled with ways to depict the reality of benzene better than the structures A and B. The struggle has not been notably successful.

Benzene is the parent molecule of a huge family of aromatic molecules. This family includes many molecules that contain elements other than carbon in the aromatic ring. Chapter 6 and those that follow provide many examples. That is enough for aromatic hydrocarbons for the present. I want to conclude this chapter with several interesting nonaromatic hydrocarbons.

Here are some interesting hydrocarbons

Molecules containing just atoms of carbon and hydrogen (the hydrocarbons) are doubtless the least interesting class of all organic molecules. But that does not mean that they are without interest. Let's look at three examples.

First, the simplest of the alkenes, ethylene, or ethene, C_2H_4, is critically important for the ripening of fruit. This simple molecule is formed directly by fruit-forming plants. It is also used commercially to ripen fruit picked early during its storage and transport. It seems remarkable that such a simple molecule has such profound biological activity.

Secondly, consider the brown algae, a diverse group of mostly marine organisms. They span the range from microscopic plants to giant seaweeds as much as 50 meters long. The brown algae can reproduce sexually or asexually. When they do the former, the process gets started by the mature organism producing a female gamete, an egg, which falls to the ocean floor. The immediate problem for the egg is to attract a free-swimming male gamete, a brown alga sperm. It does this by producing and emitting a collection of hydrocarbons:

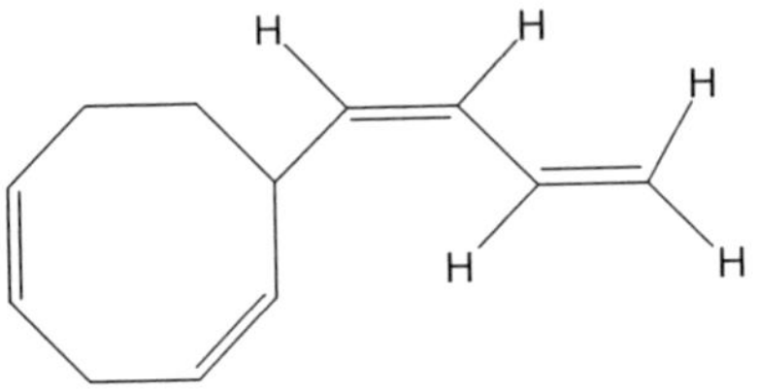

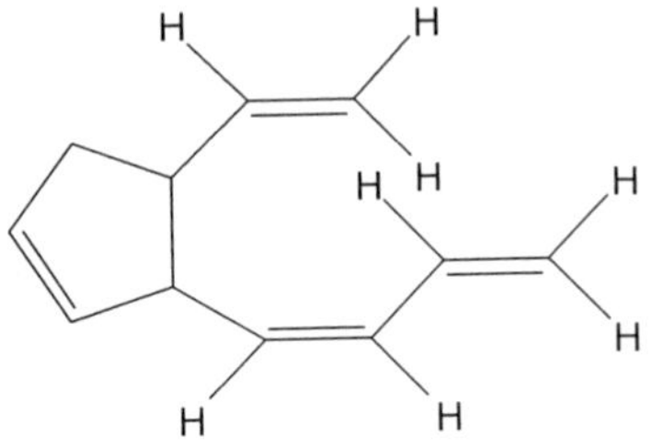

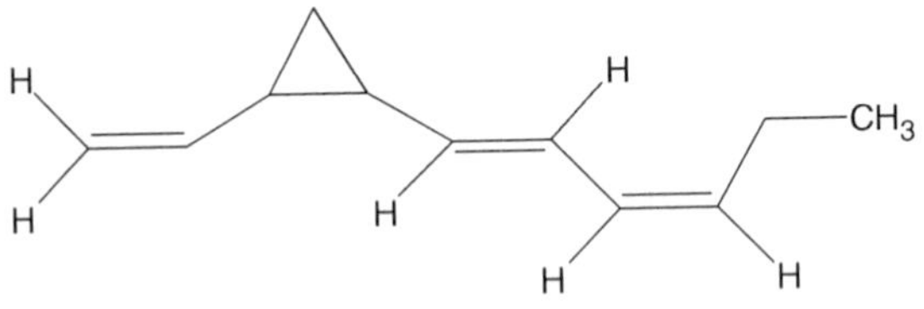

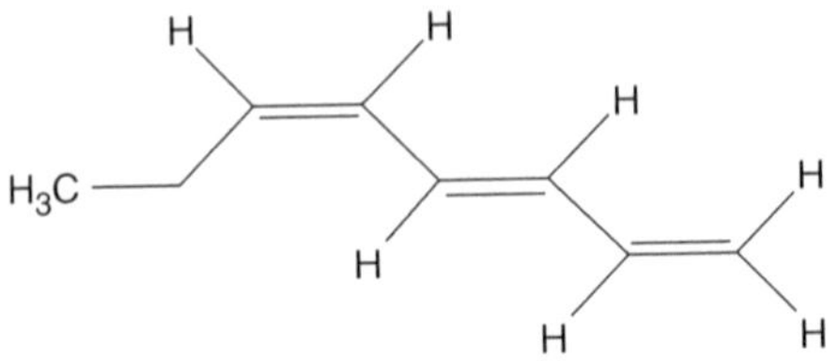

In these structures, I have elected to include, specifically, a number of hydrogen atoms for the purpose of clarity. Note that the correct stereochemistry (*cis* or *trans*) around the double bonds in these molecules is critical to the biological activity. These molecules signal the male gamete, just as bombykol signals the male silkworm moth, and it swims to the egg. Reproduction happens. This signal is very sensitive: response of the male gamete requires that about 1250 of these hydrocarbon molecules reach the male gamete each second. That may seem like a lot at first glance but consider the following. If the male gamete were to be exposed to air at room temperature and normal atmospheric pressure, it would experience about 5×10^{23} collisions with air molecules per second. The ratio of $5 \times 10^{23}/1250$ is about 4×10^{20} or about 400 billion billion. That is a big number, and it follows that these hydrocarbon molecules are really enormously potent, as is bombykol.

Thirdly let us consider ants. They are a social group and may forage across substantial distances as a group. Ants lay down a molecular trail for other ants to follow. The molecules involved are termed "trail pheromones." Now there are many species of ants and they use different trail pheromones for the obvious reason of being able to follow the right trail as opposed to that of some other species. That for the fire ant, *Solenopsis invicta,* is a hydrocarbon of moderate complexity:

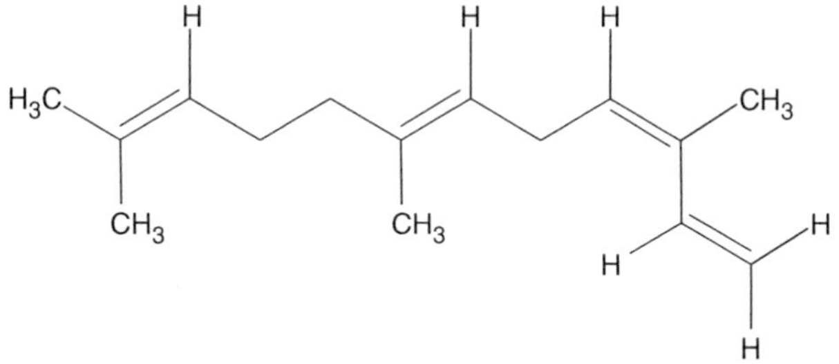

Here, too, the stereochemistry around the double bonds is critical. So, plain hydrocarbons can be pretty interesting and the geometry about carbon–carbon double bonds is frequently critical to biological activity. We will find many examples of fascinating biological activity as we move ahead to more complex molecules.

For the present, let's conclude by having a last look at a central theme of this chapter and a matter of continuing interest and importance: isomers.

Here is a summary of the key classes of isomers

Since the three-dimensional structure of molecules is so tightly linked to their chemical and biological properties, we are going to encounter multiple examples of stereoisomerism as we move forward. So it seems reasonable and useful to make a little summary of what we have said about them here.

Stereoisomers are chemical compounds having the same elemental composition but differing in structure. We have seen three subtypes of stereoisomers: constitutional isomers, enantiomers, and diastereomers.

Constitutional isomers have different bonding patterns: that is, the atoms of two structurally isomeric molecules are not identical in terms of their connectivity, the nature of the atoms to which they are bound. Constitutional isomers are clearly distinct molecules—*n*-butane is not *iso*butane. Their physical, chemical, and biological properties are different.

Enantiomers are characterized as nonsuperimposable mirror images. Enantiomers are said to be chiral (note that some diastereomers may be chiral as well). In the context of the same bonding pattern or connectivity, which atoms are bonded to which, enantiomers have handedness and are related to each other as the right hand is related to the left hand. In the specific example we saw earlier, the carbon atom is linked to four different atoms. Such molecules have non-superimposable mirror images. Stereoisomerism occurs in some molecules that do not have such a carbon atom but these cases are more exotic than we need to worry about here. Stereoisomers frequently have different, and sometimes strikingly different, biological properties, exemplified by the thalidomide case.

The third class of stereoisomers is diastereomers. One class of diastereomers arises from the disposition of equivalent groups on carbon–carbon double bonds. If the two groups are on the same side of the double bond, we refer to the isomer as *cis*; if they are on opposite sides, we refer to the isomer as *trans*. These diastereomers have different physical, chemical, and, frequently, biological properties. Examples that we have seen include the female silkworm moth sex attractant (bombykol), the sex attractants of the brown algae, and the trail pheromones of ants.

Now we are in a position to expand our command of molecules by introducing organic molecules that contain atoms of oxygen, nitrogen, sulfur, and phosphorus. That forms the substance of the next three chapters.

Key Points

1. Molecules are assembled from atoms of the chemical elements. The molecules are held together by chemical bonds between the atomic building blocks.

2. Many elements form multiple chemical bonds in molecules. Carbon, for example, typically forms four. Other elements form only one chemical bond in molecules. Hydrogen provides an example.

3. Among the elements, carbon is unique in its ability to form chains of atoms endlessly long. The structural chemistry of carbon is the richest of that for all the elements.

4. Ethane, CH_3—CH_3, is the simplest hydrocarbon that contains a carbon–carbon bond. There is free rotation around this bond.

5. Simple derivatives of ethane find important medicinal use as inhalation anesthetics.

6. Two butanes, *n*–butane and *iso*butane, provide a simple example of constitutional isomers.

7. Chemists usually write structures in an abbreviated fashion in which few atoms are explicitly shown. There are simple rules for understanding these:

    a. At every intersection of two bonds and at the termination of bonds, there is a carbon atom.

    b. Carbon atoms make four chemical bonds. If four bonds are not shown, then hydrogen atoms make up the deficit.

    c. All atoms other than carbon and hydrogen are always shown explicitly.

8. Hydrocarbons form a variety of cyclic structures.

9. There are double and triple carbon–carbon chemical bonds. These are made from two and three electron pairs, respectively.

10. There is restricted rotation around double and triple carbon–carbon bonds. The disposition of substituents on the double bond can create diastereomers.

11. The biological activity of the silkworm moth sex attractant, a pheromone, is a very sensitive function of the geometry around its two double bonds.

12. Benzene is the parent molecule of a class of molecules said to be *aromatic*. The key feature of aromatic molecules is the delocalization of some electrons over more than two carbon atoms ("free-range electrons"). Aromatic molecules are of great importance in the chemistry of life.

13. There are several hydrocarbons that have amazing biological properties.

6

Nitrogen and oxygen: atmospheric elements

Molecules based on nitrogen, the parent element of ammonia and the amines, and oxygen, the parent element of water, are key molecules of life.

Diversity adds spice to life. Think about food. Personally, I grew up with a meat-and-potatoes diet typical of midwestern food back in the 1940s and 1950s. There is nothing wrong in that but it was a bit boring in its sameness. Contrast that with the way my wife and I dine now: the wonderful diversity of regional American cuisines coupled with the foods of China, Japan, Italy, France, Thailand, and Latin America among others, specifically including fusion of regional or national cuisines. The range of aromas, flavors, and textures is amazing. Our lives are better for the diversity.

My point is that chemistry imitates life in terms of the richness added by diversity. Up to this point, our experience has been pretty much at the meat-and-potatoes level, with focus on only two elements: carbon and hydrogen. We can add interest by adding diversity in terms of additional elements. I begin doing that here.

Of the 92 natural elements, quite a few are required by living organisms, including human beings. To supplement the nutrient and caloric content of my diet, I take a multiple vitamin and mineral pill each morning. In addition to the vitamins, these pills contain the following elements in varying amounts: calcium, phosphorus, iodine, magnesium, zinc, selenium, copper, manganese, chromium, molybdenum, potassium, boron, nickel, silicon, and vanadium. Some of these elements (calcium, potassium, magnesium, and phosphorus) play multiple critical roles in living systems. Others, such as iodine, zinc, and manganese, have clear, well-defined if more limited roles. Some of the rest have more exotic, if essential, roles to play. Finally, a few of these elements may have no essential role in human nutrition at all. There they are in my vitamin and mineral pill anyway. We need to understand something about why most of these elements are there.

In this chapter, I focus on two elements that are central for most molecules characteristic of living systems: *nitrogen* and *oxygen.*

Nitrogen is a key element in the molecules of life

Elemental nitrogen is a diatomic gas, N_2, mass 28. Other elements that occur as gases under the usual conditions of temperature and pressure may also contain two atoms: H_2 and O_2, for example. Other gases occur simply as monatomic species: helium (He), neon (Ne), argon (Ar), and krypton (Kr), for example. Elemental nitrogen forms about 78% of our atmosphere; most of the remainder is oxygen.

The characteristic chemical feature of N_2 is its stability. Like acetylene, N_2 has a triple bond: $N \equiv N$. This triple bond has great strength and vigorous conditions are required to convert N_2 into ammonia, NH_3, in the chemistry laboratory. This conversion is of great importance as ammonia is employed in enormous quantities for human purposes.

In marked contrast to the conditions under which nitrogen is converted to ammonia in the laboratory, in nature atmospheric N_2 is converted to ammonia under the mildest of conditions by enzymes called nitrogenases in a process termed nitrogen fixation. This example testifies to the power of enzymes, about which much more follows in chapter 9.

It turns out that some very simple molecules have great importance for human welfare. Water constitutes 70–95% of most living organisms by weight. Methane, a key source of energy for human needs, is a member of this family in good standing. So is ammonia, a simple molecule containing just four atoms derived from two of the lightest elements. Like methane, ammonia is a gas at room temperature. Unlike methane, it is very soluble in water. Unlike water but like methane, ammonia is toxic.

Agriculture provides the principal staples of life: wheat, rice, maize (corn), oats, potatoes, vegetables, fruits .... These crops provide direct nutrition for humans as well as for livestock and our pets. While agriculture packages nitrogen and other critical elements for human consumption, it also depletes the land of nutrients, above all nitrogen. The products of agriculture provide good sources of protein and protein contains a lot of nitrogen, about 14% by weight. Crop plants sequester nitrogen from the soil, provided in the form of ammonia and other nitrogen-containing entities, including nitrate, NO_3^-, and convert it to plant protein. Natural sources of nitrogen are simply inadequate to meet the needs of intensive agriculture year after year. It follows that you cannot productively grow wheat or other crops on the same plot of land over time without replenishing required nutrients one way or another.[1] The productivity of the land diminishes as sources of nitrogen are depleted.

There have been two main ways of dealing with this problem. The first is to alternate a grain with a legume—alfalfa, clover, soybeans, chickpeas, for example. Legumes have the biological capability of fixing nitrogen: that is, converting the nitrogen in the air into ammonia. The nitrogen is actually fixed by symbiotic bacteria of the genus *Rhizobium* in the nodules of legume roots. So planting legumes one year tends to enrich the soil with ammonia that is useable by other, nonlegume crops the next year. Of course, alternating legumes with food crops cuts the productivity of the land (although the legumes, soybeans for example, are rich sources of protein and provide useful animal fodder and some human food). As the population of the Earth has burgeoned, leaving a substantial portion of arable land unproductive, production of human food becomes problematic. So, for that matter, does the use of crops as food for animals and the subsequent use of animals for human food, a remarkably inefficient process.

The second way of dealing with soil nitrogen depletion is fertilization. This has been going on for a long time. Animal waste has been employed as fertilizer for more than 2000 years and it continues to be employed for that purpose today. There is not nearly enough manure to meet the needs for soil nutrients for wide-scale, intensive agriculture. Planting legumes with other crops was a common practice in many ancient cultures. Doing so is not practical for mechanized agriculture. So we rely heavily on chemical fertilizers, including ammonia.

A central scientific accomplishment made a huge impact on the fertilizer situation. Fritz Haber developed the means to convert nitrogen to ammonia in the chemistry laboratory. The chemical reaction that leads from elemental nitrogen to ammonia involves combination with elemental hydrogen:

$$N_2 + 3H_2 \rightarrow 2NH_3$$

Although this reaction appears simple, the extreme stability of N_2 makes it tough to carry out successfully. For many years, it was regarded as impossible to carry out this reaction on a useful scale. In 1909, Haber proved that to be wrong. The reaction requires high pressures and high temperatures and a catalyst, an agent that increases the rate of a chemical reaction without being itself consumed. Haber initially employed a derivative of osmium as catalyst but later found a uranium–based molecule to be superior for this purpose.

Haber's discovery changed everything. Carl Bosch took Haber's discovery and translated it to an industrial scale. In a few short years, ammonia was produced in substantial quantities in factories, first in Germany and later elsewhere. In 1918, Fritz Haber was awarded the Nobel Prize in Chemistry for his work. Carl Bosch rose to be the head of the giant chemical cartel I. G. Farben. The industrial production of ammonia from nitrogen and hydrogen quickly came to be known as the Haber–Bosch process and it still is. Despite a great deal of work by a lot of bright chemists trying to find a better way, ammonia is still made by the Haber–Bosch process.

Ammonia is a gas that can be readily converted to liquid form and stored as such under pressure. For fertilization, it can be injected directly into the soil from capsule-shaped tanks. I grew up in the Midwest and still spend some time there for family reasons. An ammonia tank being towed through farmland behind a tractor injecting ammonia into the earth is a common sight prior to spring planting. Alternatively, ammonia can be treated with nitric acid to yield ammonium nitrate, NH_4NO_3, a solid that is useful both as a fertilizer and as an explosive. It is less toxic than ammonia. More commonly, ammonia is treated with sulfuric acid to generate ammonium sulfate, $(NH_4)_2SO_4$, also a solid but not an explosive.

The industrial production of ammonia has led directly to enormous increases in agricultural productivity. Perhaps two billion people on Earth, out of a current total of about six billion, could not be fed without the use of chemical fertilizers. Success of the high-yielding crops developed in the Green Revolution and, subsequently, by genetic engineering was made possible by chemical fertilizers. You cannot grow high-yielding crops without land that provides the required nutrients, above all nitrogen.

Fritz Haber is a major contributor to human welfare through finding the means to convert elemental nitrogen into ammonia. That is the heroic side of Fritz Haber. He is also known as the father of chemical warfare based on his development of chlorine as a lethal gas during World War I. Haber was Germany's tsar of gas warfare. He went to the front personally to oversee the placement of chlorine tanks as gas warfare weapons.

Later, Haber served for many years as head of the Kaiser Wilhelm Institute in Karlsruhe where Zyklon was developed as an insecticide. A modification led to Zyklon B, notoriously used in the death camps at Auschwitz and elsewhere to exterminate millions of Jews and other Nazi victims during the Holocaust. Among the victims of Zyklon B were several members of Haber's extended family. The life of Fritz Haber has been elegantly detailed by Daniel Charles in his book *Master Mind* which I warmly recommend to all who wish far better insights into Haber and his work and life than I can possibly provide here.[2] The life of Fritz Haber shows science in two of its manifestations: as providing the intellectual foundations that can lead to enormous good or frightening evil. There is very little new knowledge created for good purposes that someone cannot find a way to use for bad ones.

The chemistry of carbon provides insights into that for nitrogen

Let's see how much of what we know about the chemistry of carbon helps us understand that for nitrogen. The combination of one carbon atom with four hydrogen atoms yields methane, CH_4. The combination of one nitrogen atom with three hydrogen atoms yields ammonia, NH_3. The nitrogen atom in ammonia, just like the carbon atom in methane, is central. The three hydrogen atoms are bonded to the nitrogen atom. Note that the nitrogen atom makes just three bonds while the carbon atom makes four. Nonetheless, we can make a direct analogy between the geometry of methane and that for ammonia: both are in one sense tetrahedral. This is easy to see for methane as hydrogen atoms occupy all four corners of the tetrahedron. In the case of ammonia, things are a little more complicated since hydrogen atoms occupy only three corners of the tetrahedron. The fourth is occupied by something entirely new: an unshared pair of electrons. So we may write the ammonia molecule as:

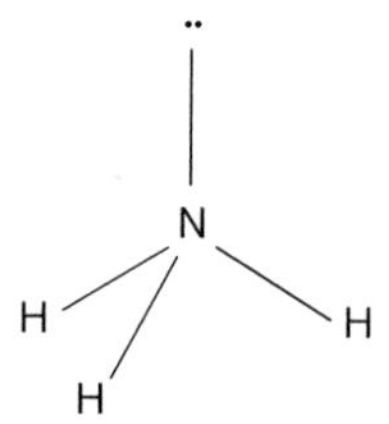

in which the two dots represent the unshared electron pair. Alternatively, if we just consider the atoms and ignore the unshared electron pair, we can describe the ammonia molecule geometry as a trigonal pyramid with the three hydrogen atoms forming the base of the pyramid and the nitrogen atom the apex.

From methane we derive the methyl group: $-CH_3$. In an analogous manner, from ammonia we derive the amino group: $-NH_2$. Just as we replaced one of the hydrogen atoms of methane with a methyl group to yield ethane, we can replace one of the hydrogen atoms of methane with an amino group to yield aminomethane, or methylamine, CH_3-NH_2, an analog of ethane. This is a simple example of a molecule formed by making a chemical bond between a carbon atom and a nitrogen atom. Such molecules are generically known as *amines*. There are a great many molecules of life that contain carbon–nitrogen bonds

In creating models for nitrogen-containing compounds of life, we shall proceed as we did for molecules containing just carbon and hydrogen. We will create a skeleton of the carbon and nitrogen atoms, explicitly indicating only the nitrogen atoms, and delete the hydrogen atoms connected to carbon. We shall show those connected to nitrogen. Let's look at a few examples of molecules typical of living systems that contain only carbon, nitrogen, and hydrogen.

Spermidine is a simple but important molecule; it is essential for the packing of DNA into compact structures within bacterial cells, for example:

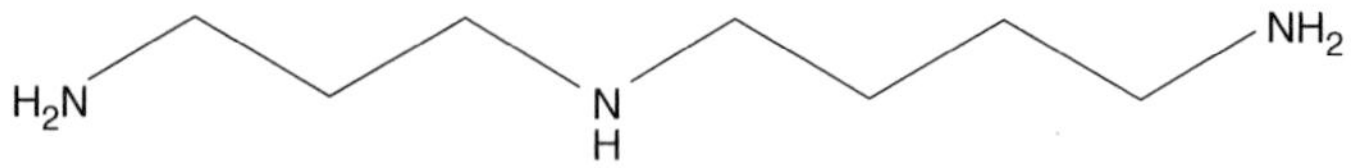

Spermidine also occurs in many of our tissues and was first isolated from semen, hence the name. Note that spermidine contains two simple amino groups, one at each end of the molecule, and a substituted amino group near the molecular center. There are three $-CH_2-$ groups between the amino group at the left end of the molecule and the central substituted amino group and four $-CH_2-$ groups between the central amino group and the one on the right. There are two more molecules in this general class that may prove of interest: they have the utterly unlovely names of putrescine and cadaverine:

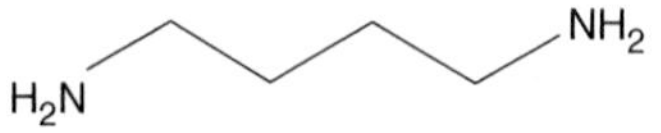

putrescine

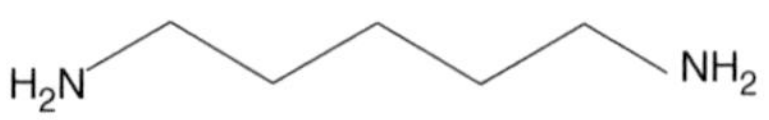

cadaverine

As you might guess from their names, these molecules have notably unpleasant odors. They were initially identified as degradation products of the action of bacteria on animal tissue. Cadaverine is also notably toxic. Putrescine is found in most cell types and is a precursor to spermidine. Note that both molecules have two amino groups.

Amino groups have the capacity to add a proton, a hydrogen atom that has lost its single electron, H^+. We can write this in the following way:

$$R-NH_2 + H^+ \rightarrow R-NH_3{}^+$$

in which "R" indicates that the amino group is attached to a carbon atom of some group, say a methyl group. The proton binds to the unshared electron pair on the amino nitrogen atom. Functional groups that can add a proton are known as bases and those that donate a proton are known as acids.[3] For example, HCl, hydrochloric acid, is a proton donor. It can donate a proton to ammonia, a base:

$$HCl + NH_3 \rightarrow NH_4{}^+ + Cl^-$$

As noted earlier, positively charged species such as the ammonium ion, $NH_4{}^+$, are known as cations and negatively charged species such as the chloride ion, Cl^-, are known as anions. The point here is that spermidine (and putrescine and cadaverine) is a base: that is, it has the capacity to add a proton. In fact, it has three sites to which a proton might be added, the three amino nitrogen atoms. In principle, it could add three protons and end up with three positive charges. In fact, under the conditions of most living cells, it adds two protons and exists as a dication (dye-cation):

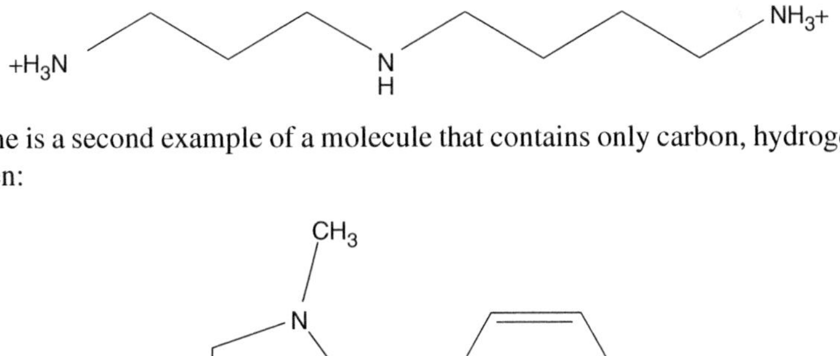

Nicotine is a second example of a molecule that contains only carbon, hydrogen, and nitrogen:

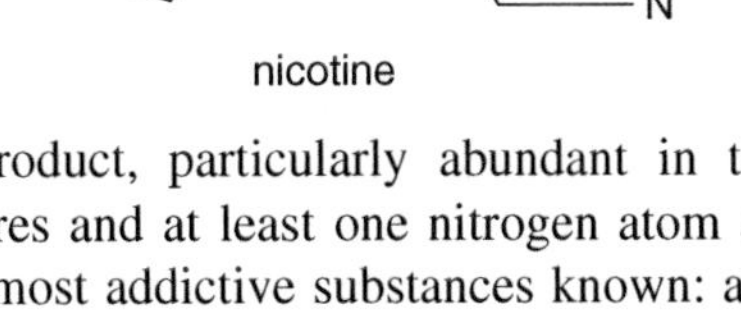

nicotine

Nicotine is a plant product, particularly abundant in tobacco. Plant products containing ring structures and at least one nitrogen atom are known as *alkaloids*. Nicotine is one of the most addictive substances known: ask any cigarette smoker who has tried to quit.

In passing, note that nicotine contains two rings that contain a nitrogen atom: these are known as heterocycles. This term denotes a cyclic molecule or group in which at least one of the ring atoms is something other than carbon. Other common examples include oxygen and sulfur. The noncarbon atoms are known as the heteroatoms. We are going to encounter a number of additional examples in what follows. For example, the nitrogenous bases of the nucleic acids are all heterocycles as are two of the amino acids of proteins. And now, back to our molecules derived from carbon, hydrogen, and nitrogen.

Phenethylamine is an interesting example:

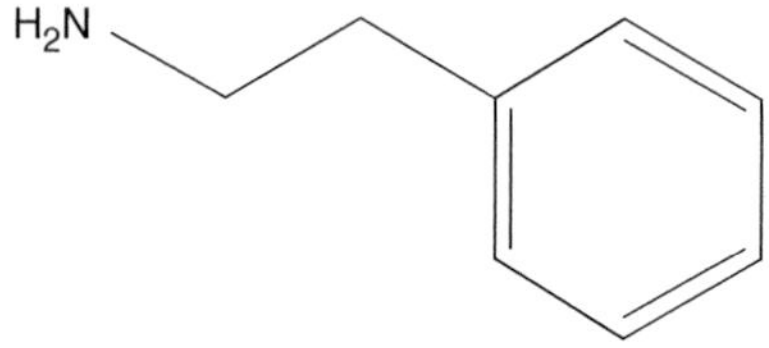

Note that the benzene ring when it occurs as a substituent in a molecule is known as a phenyl group. So here we have a phenyl group, the amino group, connected by the $-CH_2-CH_2-$ group, related to ethane. These groups suggest the origin of the name: phen-ethyl-amine.

Phenethylamine is structurally related to a number of psychedelics, including mescaline. It also occurs in chocolate and is thought to be one of the substances responsible for the "feel good" effect of chocolate (other molecules contribute as well, including theobromine). Chocoholics will understand.

Elemental oxygen is vital for many forms of life

The usual form of elemental oxygen is a diatomic gas, O_2, mass 32, that is indispensable to life for many organisms, including human beings. Elemental oxygen as O_2 forms about 21% of our atmosphere. You can read all about oxygen in Nick Lane's book: *Oxygen: the Molecule that Made the World.*[4]

When we inhale, we take in mostly nitrogen and oxygen from our atmosphere. The nitrogen we simply exhale; the oxygen we use. In the lungs, oxygen attaches to molecules of hemoglobin in the red cells of the blood and is transported in the blood to the tissues. There, some oxygen detaches from the hemoglobin and diffuses into cells. Once in the cells, it is employed as an electron acceptor. More about that follows a bit later.

Pure elemental oxygen is administered to tired athletes to help them recover their energy more quickly and to patients unable to meet their needs for oxygen from atmospheric oxygen alone, such as those with emphysema, lung infections, or other limitations to getting adequate oxygen intake from normal breathing. Note, however, that normal individuals whose breathing is not impaired cannot breathe pure oxygen indefinitely without organ damage or, eventually, death. Even this life-sustaining substance, in excess, is toxic.

There is a second, rarer, form of elemental oxygen, ozone, O_3, consisting of three linked oxygen atoms. Ozone is formed in the atmosphere in electrical discharges, such as lightning, and accounts for the smell of the air following a thunderstorm. Though both forms of oxygen contain only atoms of oxygen, their chemical properties are very different.

Ozone is important to us in two general ways, one positive and one not. The ozone layer in the upper atmosphere absorbs a great deal of the ultraviolet light from the sun, protecting us from damage by this high-energy light. O_2 does not have this property. The absorption of light is a sensitive function of molecular structure. O_3 is well-suited to absorbing ultraviolet light while O_2 is transparent to this radiation.

Lower in the atmosphere, ozone is a component of photochemical smog and a cause of oxidative damage to lungs. So ozone protects us in the upper atmosphere, where we are not in direct contact with it, but threatens us at ground level, where we are. We need to keep it but keep it where it belongs.

In cells, oxygen is a terminal electron acceptor

Human beings derive energy from oxidizing the food in their diet. This includes oxidation of molecules of carbohydrates, lipids, and proteins. In the process of oxidation, electrons are extracted from these molecules, a process that generates useful chemical energy, and the electrons need some place to go: that is, they need an electron acceptor. In human beings and many other species, though not all by any means, that electron acceptor is molecular oxygen, O_2. That is why we need to inhale oxygen, and is the role of oxygen released from our hemoglobin into cells.

At the end of the process of losing electrons, the carbon atoms in our carbon-based foods end up as carbon dioxide, CO_2. This diffuses into the blood, is carried back to the lungs, and then exhaled. At the end of the process of accepting electrons, molecular oxygen is converted into water. This can be exhaled or eliminated through the kidneys or as sweat.

To make the concept of oxidation a bit more concrete, here are several molecules containing just one carbon atom at various states of oxidation. At the lowest state, we have methane, CH_4. There are eight shared electrons around the carbon atom, two for each C—H bond. They are assigned equally to carbon and hydrogen. It follows that carbon is assigned four bonding electrons in methane. Now if we replace one of the hydrogen atoms of methane by an OH group, we get a molecule known as methanol, CH_3OH (see chapter 7). In the case of a carbon atom bonded to an oxygen atom, both electrons are assigned to the oxygen atom. Hence, carbon is assigned only three bonding electrons, one for each C—H bond. Since the carbon atom has lost one electron in going from methane to methanol, it has been oxidized.

We can push this to completion. In formaldehyde, $H_2C{=}O$, only two bonding electrons are assigned to the carbon atom so it has been oxidized again. In formic acid, HCOOH, only one electron is assigned to the carbon atom and in carbon dioxide, CO_2, none are. So the states of increasing oxidation are: methane, methanol, formaldehyde, formic acid, and carbon dioxide.

For a couple of centuries, scientists have been making an analogy between human metabolism and the burning of a candle. The analogy is really pretty good. A candle is basically lipid in nature and most of its carbon atoms are attached to hydrogen atoms. This is a low oxidation state, similar to that of methane. Much of our food is as well; we call it fat. When a candle burns, it converts the candle wax into carbon dioxide using molecular oxygen as electron acceptor. Thus we are basically going from a low oxidation state of carbon to the highest, carbon dioxide. The oxygen is reduced (i.e. it has accepted electrons) and ends up as water in the form of water vapor since burning candles are hot. When we metabolize (burn) fat molecules, the carbon atoms of the fat are converted into carbon dioxide using molecular oxygen as electron acceptor. Here too the oxygen ends up as water, in the form of liquid water in this case since we are not so hot. An essential difference is that the energy released when a candle burns takes the form of heat and light. The energy released when we metabolize fat takes the form of useful chemical energy in the form of the molecule adenosine triphosphate, ATP. Unlike the burning candle, we do not get hot or emit light. Fireflies and other bioluminescent organisms do not get hot either but they do convert a portion of the energy released by the metabolism of food into light. Finally, note that the details of the molecular processes involved in candle burning and human food metabolism are entirely distinct.

The chemistry of oxygen can be related to that of carbon and nitrogen

We have found that carbon generally makes four bonds and nitrogen three. Oxygen makes just two. Hence, the combination of oxygen with hydrogen yields H_2O, water. The central oxygen atom of water bears two unshared electron pairs in addition to the two hydrogen atoms. These four substituents are directed, again, toward the four corners of a tetrahedron:

If we consider the geometry created just by the atoms of water, ignoring the unshared electron pairs, then the molecule is described as bent.

Water covers about 70% of the surface of the Earth. It has some amazing properties. Here they are.

Water is a truly amazing substance

Water lies at the very heart of life. Oceans, seas, rivers, lakes, ponds, streams, intertidal pools, and puddles provide the environment for an abundance of living organisms. Those of us who reside on land are composed largely of water and are utterly dependent on water for life. Some species can do without water longer than others but, sooner or later, we all need it.

This simple substance has an amazing collection of unexpected properties; these prove to be ideally suited for the support of life. Let's consider a few of them.

The first surprise is that water is a liquid at room temperature. Water is the hydride of oxygen: H_2O. Let's consider molecules that are hydrides (molecules formed from some element and hydrogen alone). We have methane, CH_4, ammonia, NH_3, and hydrogen sulfide, H_2S. All of these substances are gases at room temperature: methane boils at $-161°C$, ammonia at $-33°C$, and hydrogen sulfide at $-100°C$. Yet water boils at $100°C$. In fact water freezes at a temperature significantly higher than that at which its hydride neighbors boil. Two things are clear: this property of water is (a) unexpected and (b) absolutely essential for life. We need to understand this and I will get to it shortly.

A second unexpected property of water is that it expands when it freezes. Water has its maximal density, mass per unit volume, at $4°C$. As it is cooled further, it begins to expand. Ice at $0°C$ occupies about 11% more volume than does liquid water at the same temperature. In this respect water is nearly unique. Almost all other liquids contract when they freeze, as we would expect since the solid phase is generally more compact and more ordered than the liquid phase and, hence, is denser. This behavior is not just a laboratory curiosity: the fact is that our life on this planet is dependent on this remarkable property. This point has been elegantly stated by L. J. Henderson, a leading biochemist in the early twentieth century, in his thoughtful book: *The Fitness of the Environment* which he wrote in 1913[5] Here are his words.

> And so it would be with lakes, streams, and oceans were it not for the anomaly and the bouyance of ice. The coldest water would continually sink to the bottom and freeze there. The ice, once formed, could not be melted, because the warmer water would stay at the surface. Year after year the ice would increase in winter and persist through the summer, until eventually all or much of the body of water, according to the locality, would be turned to ice. As it is, the temperature of the bottom of a body of fresh water cannot be below the point of maximum density; on cooling further, the water rises; and ice forms only on the surface. In this way the liquid water below is protected from further cooling, and the body of water persists. In the Spring, the first warm weather melts the ice, and at the earliest possible moment all ice vanishes.

So it is critical that ice floats on water, a consequence of its expansion below $4°C$ and, especially, when it freezes.

Two additional interesting and important properties of water are its high heats of vaporization and fusion. The heat of vaporization is a measure of the amount of energy required to convert one gram of liquid to vapor.[6] The heat of vaporization of water is among the highest known for any liquid. It takes a lot of energy to convert one gram of liquid water to water vapor. Similarly, the heat of fusion is a measure of the amount of energy required to convert one gram of a solid to the liquid state. The heat of fusion of water is very high; it takes a lot of energy to melt ice. The energy required to melt ice or boil water is far greater than that required to melt or boil other hydrides. This is another fact that we need to account for.

The key to understanding the properties of water lies in the nature of something called the hydrogen bond. I introduced covalent bonds, the sharing of a pair of electrons between two atoms, back in chapter 3. Hydrogen bonds are another matter.

Hydrogen bonds are weak but really important: water

The unusual properties of water are largely accounted for by the formation of hydrogen bonds. These interactions also prove to be critical for the structure and function of the proteins and nucleic acids.

A simple solid line linking, for example, a carbon atom to a hydrogen atom has symbolized chemical bonds that we have introduced thus far: covalent bonds. In general, these bonds are strong so that the molecules that they hold together are stable under ordinary conditions. Hydrogen bonds are different: they are substantially weaker than the bonds we have introduced thus far and are, hence, far easier to make and break. On an arbitrary scale where we might give a carbon–hydrogen bond a strength of 100, a hydrogen bond would have a strength of perhaps 5. Dotted lines, rather than solid lines, conventionally symbolize hydrogen bonds. The dotted lines serve to suggest their relative weakness and ease of formation and disruption.

A hydrogen bond can be formed any time one of the following conditions is met: when an oxygen atom, nitrogen atom or fluorine atom bearing a hydrogen atom meets an unshared electron pair on an oxygen atom, nitrogen atom, or fluorine atom. Symbolically, we can write:

$$X-H\bullet\bullet\bullet : Y$$

in which $\bullet\bullet\bullet$ indicates the hydrogen bond and where X and Y are either O, N, or F. The hydrogen atom is shown strongly associated with the X atom and weakly associated with the Y atom in this model.

Water molecules meet the necessary condition for formation of hydrogen bonds. In fact, they do so in an ideal way. A hydrogen atom on one molecule of water can form a hydrogen bond to an unshared electron pair on a second water molecule: $H-O-H\bullet\bullet\bullet:OH_2$. Since each water molecule contains two hydrogen atoms bound to oxygen and two unshared electron pairs on oxygen, it is clear that large, highly ramified networks of hydrogen bonds can be formed in water.

This network is perfected in ice: the maximum number of hydrogen bonds, 4, is formed to each water molecule. Although the formation of each hydrogen bond is energetically favorable by a modest amount, the formation of a great number of them is energetically compelling. It is a little like collecting tolls on a heavily traveled highway: each toll is small but the total amount of money collected is substantial since a great many tolls are collected.

In ice—including ice crystals, hail, snow—each water molecule is surrounded tetrahedrally by four other water molecules through the formation of four hydrogen bonds. This structure is shown in figure 6.1. The oxygen atoms form a structure that is akin to that formed by the carbon atoms of diamond. The key difference is that hydrogen atoms separate the oxygen atoms in ice while nothing separates the carbon atoms in diamond. The structure of ice is therefore "open." The favorable energy change brought about by forming the maximum number of hydrogen bonds imposes a structure in which the molecules of water are not as closely packed together as possible.

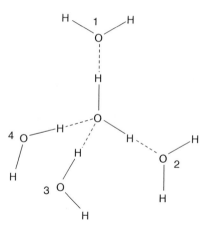

Figure 6.1 A view of the tetrahedral coordination of water molecules in ice is shown here. Molecules (1) and (2), as well as the central molecule, lie entirely in the plane of the paper. Molecule (3) lies in front of the plane of the paper and molecule (4) lies behind it, so that oxygens (1), (2), (3), and (4) lie at the corners of a regular tetrahedron.

Now we can begin to understand the properties of water. Let's start with ice and ask what happens when it melts. Liquid water is not as well-ordered as is ice. Although a majority of the hydrogen bonds of ice are retained in liquid water, many are broken and the precise, well-ordered crystalline structure of ice is lost. This accounts for three things. First, the destruction of the open structure of ice causes the water molecules to adopt a more closed structure. Thus, the melting of ice is accompanied by contraction in volume and an increase in density. Conversely, expansion happens when ice forms, as we noted above. Secondly and thirdly, it takes a lot of energy to break a number of hydrogen bonds in converting ice to water. This accounts for the high melting point of ice and the high heat of fusion of ice. The ice structure is highly stabilized by the three-dimensional lattice of hydrogen bonds. Disrupting that lattice takes energy, energy in the form of heat.

As we increase the temperature of the water, we tend toward an increasingly disordered system. This involves loss of additional hydrogen bonds as the structure of water gradually loses order. It takes energy to disrupt hydrogen bonds. Finally, if we convert liquid water to steam, we lose all, or very nearly all, of the remaining hydrogen bonds. Water molecules in the gas phase at ordinary pressures are quite far apart and the opportunity for hydrogen bond formation is largely lost. Consequently, vaporizing water involves breaking of many hydrogen bonds and, hence, requires a lot of energy. Thus, we account for the high heat of vaporization of water.

So far, we have just considered formation of hydrogen bonds among identical molecules. Hydrogen bond formation can occur anytime the basic requirements are met (X—H•••:Y, where X and Y are O, N, or F) and this includes hydrogen bonds between unlike molecules. For example, water forms hydrogen bonds with ammonia: H—O—$H•••:NH_3$ or H_2N—$H•••:OH_2$.

Water is essential to life on Earth as we know it. Water is the only substance on Earth that is present simultaneously and in large quantities as gas, liquid, and solid.

Interesting very small molecules come from just carbon and oxygen

There are two interesting molecules that are derived from carbon and oxygen alone. Earlier, we encountered carbon dioxide, CO_2, as a constituent of our atmosphere, a product of the combustion of fossil fuels, and a major contributor to global warming. An even simpler molecule is carbon monoxide, CO, a toxic gas and a minor product of incomplete combustion of fossil fuels.

The toxicity of CO derives from its ability to compete with oxygen. For example, CO has substantially greater affinity than does oxygen for hemoglobin. To the extent that CO sits on hemoglobin where O_2 should be, the availability of O_2 to the tissues is diminished. Clearly, this could be fatal. In fact, fatalities due to CO inhalation derive from its ability to compete with O_2 at the site of terminal electron acceptance. The idea is just like that for hemoglobin but at a different molecule; this one known as cytochrome oxidase. Since CO is toxic and has the potential to be produced in our homes through faulty furnaces or heaters, there are commercially available home CO detectors. If CO levels rise above a threshold level, an alarm rings. We have one in our home, just in case.

It may come as a surprise that CO is synthesized by the human body and has roles in human metabolism. Specifically, an enzyme that degrades heme, a constituent of hemoglobin, our oxygen-transporting protein, makes CO, which is a neurotransmitter. Much more about neurotransmitters follows in chapter 21 when we talk about the central nervous system. For the present, just understand that specialized cells known as neurons are the conduits for communication in the nervous system. Neurotransmitters are small molecules that relay information from one neuron to another. Neurotransmitter CO is made, functions, and is quickly destroyed. Personally, I find it surprising that CO has such a critical role in the nervous system. Surprised or not, there it is and there is no doubt about it.

Finally, what don't you understand about NO?

Oxygen and nitrogen are two of the commonest elements on Earth and in living systems. As is developed in this chapter, both are colorless, nontoxic gases. Oxygen and nitrogen combine in a number of ways to make more complex molecules. The simplest of these is to link one nitrogen atom to one oxygen atom to yield a molecule having the composition NO, nitric oxide. Nitric oxide is a gas that is toxic in high concentrations but, nevertheless, is a key molecule in living systems. NO has a lot of important and interesting chemistry.

Until the last couple of decades, the main interest in NO was in the context of air pollution. The basic energy-generating reaction of internal combustion engines is the explosive combination of oxygen with the hydrocarbon molecules of gasoline to produce, largely, carbon dioxide and water. The source of the oxygen is, of course, the air drawn into the engine. At the high temperatures inside the cylinders of an internal combustion engine, there is some combustion of the nitrogen: that is, some of the oxygen combines with the nitrogen in addition to the combination with the gasoline hydrocarbons. One product of that reaction is NO.

Once expelled into the atmosphere as a component of exhaust gases, NO reacts with oxygen in the air to form nitrogen dioxide, NO_2, a toxic brown gas. Nitrogen dioxide, in turn, reacts with water in the atmosphere to form nitric acid, HNO_3, one of the constituents of acid rain. NO, NO_2, and HNO_3 all contribute to air pollution. These molecules are frequently lumped together under the designation NO_x. Measures taken in an effort to reduce atmospheric levels of NO_x have been partially successful but more work remains to be done.

More than a decade ago, it became clear that the human body makes NO. It is made in the brain, in the muscle cells which exist in the interior of the blood vessels, by macrophages (white cells that form an important part of the immune system), by the corpus cavernosum of the penis, and perhaps elsewhere. NO plays an important role in each of these tissues. The source of the atoms for the synthesis of NO is the common amino acid arginine (chapter 9). Under the influence of an enzyme termed NO synthase, arginine is converted to NO (and other products). The lifetime of NO in the tissues is quite short, a few seconds, but it lasts long enough to be effective.

In the brain, NO appears to be involved in the reinforcement of learning, a key function. In the vasculature, it is clear that NO released by smooth muscle cells of the veins acts to relax these muscle cells, expanding the vasculature, facilitating circulation of the blood and lowering blood pressure. Nitroglycerin and amyl nitrite have been used for many years to relieve the pain of angina pectoris, pain due to constriction of the coronary arteries that provide blood flow to the heart. It is now clear that these drugs work by releasing NO. Thus, they act to augment the amount of NO available to relax smooth muscle cells of the veins and arteries.

Macrophages are cells that fight invasion by foreign organisms. One way that they do this is to zap them with NO in concentrations that are toxic to the invaders. So NO is important as a component of the defense system of human beings. As it turns out, sodium nitrite, $NaNO_2$, a commonly used preservative, works by generating NO, which is responsible for its ability to preserve foods.

Finally, erotic thoughts trigger the release of NO in the corpus cavernosum of the penis. This causes the corpus cavernosum to relax, permitting the inflow of blood, resulting in a penile erection and the capacity for turning the erotic thoughts into erotic action. It is now clear that the widely prescribed drugs Viagra, Levitra, and Cialis work to increase the release of NO by the corpus cavernosum. That has given rise to a number of popular phrases: NO sex! NO wonder! NO way! NO news is good news!

Key Points

1. A substantial number of elements are required in the human diet for the maintenance of good health.

2. In addition to carbon and hydrogen, the key elements in the molecules of life include nitrogen, oxygen, phosphorus, and sulfur.

3. Elemental nitrogen exists as N_2. Its hydride, ammonia, NH_3, is employed in enormous quantities as fertilizer in agriculture.

4. Many molecules, including ammonia and water, possess one or more pairs of unshared electrons.

5. Molecules known as amines possess at least one amino group, $-NH_2$, or a simple derivative of this group.

6. Plant natural products containing ring structures and at least one nitrogen atom are known as alkaloids.

7. Molecules containing rings that have at least one atom in the ring other than carbon are known as heterocycles.

8. Elemental oxygen exists as O_2 and as O_3, ozone.

9. Removing electrons from an element, molecule, or ion is known as oxidation. Addition of electrons to an element, molecule, or ion is known as reduction.

10. Water is an amazing substance, with a set of properties required to sustain life on Earth as we know it.

11. Hydrogen bonds are formed by the sharing of a hydrogen atom between the following atoms: N, O, F. These bonds are weak and break and reform rapidly. They are important for determining properties and structures of substances in which they occur.

12. Carbon monoxide, CO, a toxic gas, is also a neurotransmitter.

13. Nitric oxide, NO, plays a number of roles in human physiology, including acting as a smooth muscle relaxant. Several drugs employed in clinical medicine act by supplying or increasing the levels of NO.

7

More about oxygen-containing molecules

Alcohols, carboxylic acids, and esters are oxygen-containing molecules that play multiple roles in living organisms. Oxygen-centered radicals are thought to contribute to the aging process.

The Bible is a bit racier than it is generally given credit for. Alcohols are among the classes of molecules that are the focus of this chapter, specifically including ethanol, the basis of alcoholic drinks. Let's begin with the earliest recorded use of ethanol as an intoxicating agent.

In my copy of the King James version of the Bible, verses 20 and 21 of the 9th chapter of Genesis report the following: "And Noah began to be an husbandman, and he planted a vineyard: And he drank of the wine, and was drunken; and he was uncovered within his tent." Perhaps Noah felt that a small celebration was warranted after his experiences on the ark at the time of the great flood. The Bible is silent on his motive so I am reduced to speculation.

People have been drinking wine ever since, perhaps encouraged by the longevity of Noah, who lived to be 930. The 15th verse of Psalm 104 notes: "And wine that maketh glad the heart of man," At the same, in several places the Bible cautions us to avoid overindulgence. That seems like very good advice.

Oxygen is a defining element in several classes of the molecules of life. These include alcohols, carboxylic acids, and esters.

Alcohols are characterized by possession of one or more hydroxyl groups

Let's go back to methane and ask what happens if we replace one of the hydrogen atoms of this molecule by a hydroxyl group, —OH. We get methanol: CH_3—OH.

We derive the methyl group from methane, the amino group from ammonia, and the hydroxyl group from water. Molecules containing this functional group are called alcohols. Methanol is the simplest of the alcohols. It is a toxic liquid in substantial industrial use. It is also an occasional contaminant of "moonshine." Methanol is particularly toxic to the optic nerve and ingestion can lead to blindness. "Blind drunk" is more than just a phrase. One can think of methanol as derived from methane by replacement of one hydrogen atom by the hydroxyl group or as derived from water by the replacement of one hydrogen atom by the methyl group.

Ethanol is a rock-star among alcohols

More common than methanol is ethanol: CH_3—CH_2–OH.

Ethanol is the basis of all alcoholic drinks. Ethanol is a derivative of ethane in which a hydroxyl group replaces one of the hydrogen atoms. Ethanol is a colorless, flammable liquid that mixes with water in all proportions.

Ethanol has a lot of uses: fuel, solvent, antifreeze agent, and mood-modifying substance for people. Automobiles, if suitably modified, can run on pure ethanol or, more commonly, on ethanol mixed with gasoline. In the United States "gasohol" is typically 10% ethanol/90% gasoline. Less common is E-85, which, as the name implies, is 85% ethanol.

In the United States, ethanol for fuel is derived from the fermentation of the starches and sugars of corn. Fermentation is a process in which organic molecules such as sugars are converted to simpler substances, such as ethanol, by living organisms, frequently yeasts. In 2005, about 4 billion gallons of ethanol were employed as fuel for automobiles in the United States. In 2006 about 2.15 billion bushels of corn were converted to ethanol for use as fuel; this amounts to about 20% of the entire United States corn crop. Current estimates suggest that the United States may eventually produce as much as 15 billions gallons of ethanol from corn per year. The attractiveness of ethanol as fuel derives from a modest reduction of the dependence of the United States on foreign sources of petroleum as fuel and from the creation of a new market for corn grown by the nations' farmers.

Ethanol for fuel or human consumption is made by fermentation of sugars or starches found in plants. Starches are large complex molecules consisting of many sugar molecules linked together chemically. These starch molecules must be converted to simple sugars prior to fermentation. There are enzymes, protein catalysts, which carry out this transformation rapidly in many microorganisms. Once we get sugars, yeasts rapidly convert the sugars to ethanol and carbon dioxide.

Ethanol has been known to humans and consumed by humans in the form of alcoholic beverages since prehistory. As noted at the beginning of this chapter, the Bible makes several references to wine and its consumption.

Enormous quantities of ethanol are consumed each year in the form of alcoholic beverages such as beers and ales, wines, and hard liquors. The feedstock for beers and ales is generally corn, wheat, or barley; other flavoring substances, such as hops, may be added. Beers generally contain 4–7% alcohol. The feedstock for wines is usually grapes, though fruits and other sources of sugars (e.g., dandelion leaves) are also employed. Unfortified wines usually contain 12–16% alcohol, though a few fall outside these limits. Fortified wines such as sherry and port contain 20% ethanol. An amazing 2000 varieties of grapes are employed around the world to make wine. Among the more common types are Chardonnay, Sauvignon blanc, Pinot gris, Pinot blanc, and Riesling, all used to make white wines. The prominent grapes for red wines include Cabernet sauvignon, Merlot, Syrah (known as Shiraz in Australia), Pinot noir, Malbec, Zinfandel, Tempranillo, Sangiovese, and Brunello (a clone of Sangiovese). There are many more.

For hard liquors, those containing a high concentration of alcohol, usually 40–50%, bourbon, Scotch whiskey, Canadian whiskey, Irish whiskey, gin, vodka, tequila, cognac, armagnac, . . ., almost anything containing sugars has been employed. These include multiple grains, many varieties of grapes, most fruits, potatoes, sugar cane, sugar beets, and some cacti. The result is an amazing variety of products.

Although I have not done a thorough survey, and cannot think of a good reason to do one, there is clearly an association between certain varieties of hard liquor and nations. Scotch whiskey, Irish whiskey, and Canadian whiskey provide three embarrassingly obvious examples. Beyond these we have bourbon and the United States, tequila and Mexico, vodka and Russia, gin and the United Kingdom, ginever and the Netherlands, slivovice and Poland, pisco and Peru, sake and Japan, rum and several Caribbean countries including Jamaica and Venezuela, and cachaça (aka pinga) and Brazil. This list can doubtless by expanded effortlessly but enough is enough.

There is a good bit of scientific evidence to support the idea that moderate, one or two drinks a day, intake of alcoholic beverages is healthful. Moderate drinking seems to strengthen the cardiovascular system and to prolong life. Maybe it is worth saying here what a "drink" is. An alcohol drink is one 12 oz beer, 4 ounces of wine, or one ounce of 100-proof (50% ethanol) hard liquor or about 1 1/2 ounces of 80-proof liquor. Each of these contains 0.5–0.6 ounces of ethanol. A "drink" is not a tumbler full of gin or half a bottle of wine.

Excess consumption of alcohol is not healthful, as many people will testify. Ethanol is a depressant and can be a mild tranquilizer or a general anesthetic, depending on how much is consumed over what period of time. At low doses, ethanol depresses some of the brain's inhibitory systems and acts as a social lubricant. It can also exacerbate seizure disorders such as epilepsy by depressing the inhibitory systems in the brain that suppress seizures and convulsions. At higher doses, alcohol leads to the classical symptoms of intoxication: unsteady walk, slurred speech, altered sensory perception, slow reaction times, bizarre behavior, and finally, loss of consciousness. Consumption of a fifth of a gallon of hard liquor over a short time period can be fatal.

Before leaving these simple alcohols for more exotic structures, let's return again to the issue of hydrogen bonds. Alcohols, like water, can form hydrogen bonds: the hydroxyl hydrogen atom of one molecule links to an unshared electron pair on the oxygen atom of another molecule:

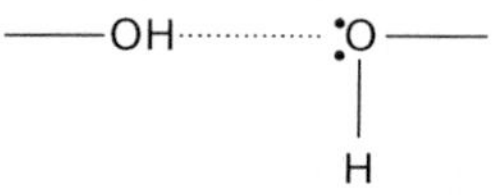

These interactions have profound effects on the physical properties of alcohols. Let's compare the boiling points of ethane and methanol: a hydroxyl group in methanol replaces a methyl group in ethane. One of the important determinants of boiling point is molecular mass. Heavier molecules tend to have higher boiling points, all other things being equal. In this case, there is little difference: ethane has molecular mass 30 and methanol has a corresponding value of 32. However, ethane is a gas at room temperature and boils at –88°C while methanol is a liquid at room temperature and boils at 65°C, a difference of 153°C, substantially larger than the temperature range over which water is a liquid. Methanol can form hydrogen bonds; ethane cannot. You need to break the hydrogen bonds in methanol when it boils; no such energy cost is required to boil ethane.

Phenol is a novel alcohol

A novel type of alcohol is derived from benzene. If we replace one of the hydrogen atoms on the benzene ring with a hydroxyl group, we get phenol:

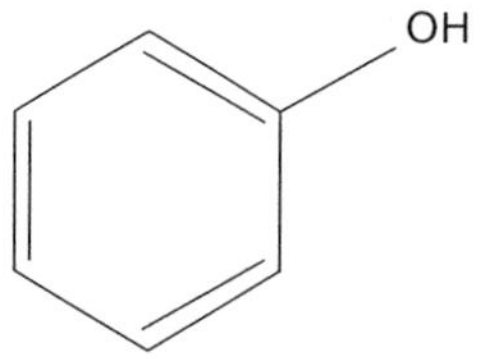

Phenol is a low-melting solid with a somewhat peculiar but not really unpleasant odor. Though not a potent toxin, it must be handled with care as it does damage in contact with the skin. Derivatives of phenol in which a methyl group is added to the benzene ring either adjacent to the hydroxyl group, 2-methylphenol, or at the position directly opposite to the hydroxyl group, 4-methylphenol, are components of human sweat:

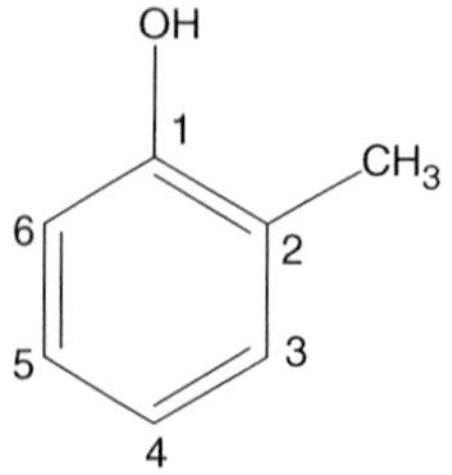

2-methylphenol

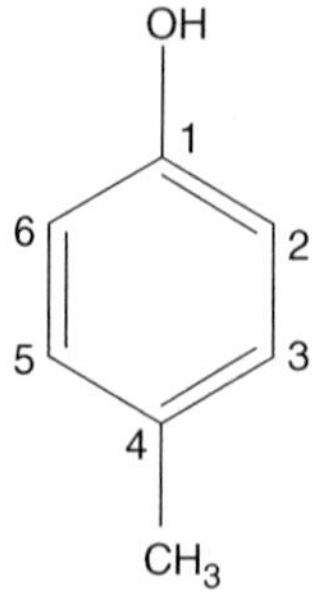

4-methylphenol

It turns out female *Anopheles* mosquitoes, the most important vectors for *Plasmodium falciparum* malaria, possess odorant receptors for these molecules. This permits the mosquito to home in on humans, at least those that sweat. Malaria is responsible for more than a million deaths per year. This finding suggests novel ways of creating mosquito repellants and/or traps.[1]

Carboxylic acids are important oxygen-containing molecules

Now let's consider acetic acid, the acidic substance in vinegar and an important molecule of life: CH_3COOH. Molecules such as acetic acid that contain the —COOH group are termed carboxylic acids.

The carboxyl group, —COOH, has one of the oxygen atoms linked to the carbon atom by a double bond. The other oxygen atom is linked to the carbon atom by a single bond and holds the hydrogen atom:

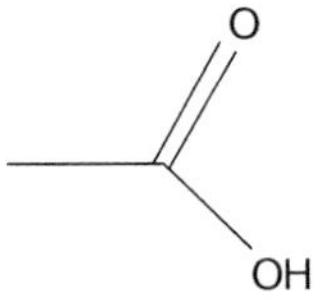

The carboxyl group can easily lose a proton (as implied by the designation acid; remember that compounds that donate protons are acids) and form a carboxylate anion:

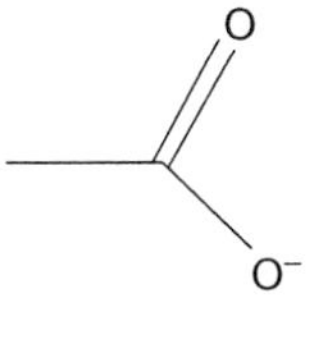

In living systems, carboxylic acids exist largely in the form of their corresponding carboxylates. We shall encounter many more carboxylic acids, among them the amino acids, the building blocks of proteins, as we move forward.

The simplest carboxylic acid is formic acid (more systematically methanoic acid): HCOOH. Formic acid contains a single carbon atom and can be thought of as a derivative of methane or methanol obtained by oxidation:

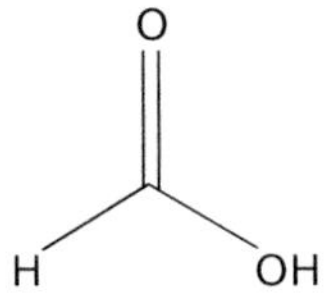

It was first isolated from ants and takes its name from the Latin word for ant, *formica*. An ant bite injects a small amount of formic acid into the victim, accounting for the sting. The venoms of both bees and wasps contain formic acid as well, among other toxic chemicals.

A particularly important carboxylic acid is benzoic acid, a derivative of benzene, with the formula:

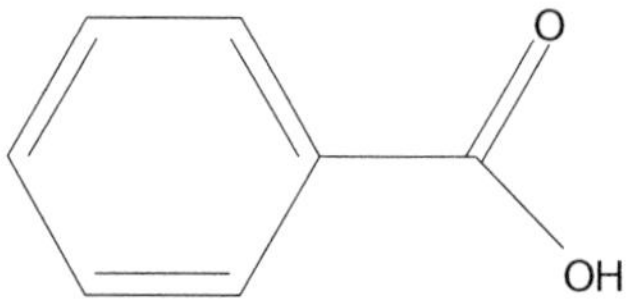

The sodium salt of benzoic acid, sodium benzoate, is a very commonly employed preservative. Let's pause here for a moment to re-emphasize an important point: biological activity is a sensitive function of chemical structure. Benzene, the parent molecule of benzoic acid, is a serious toxin; in contrast, the sodium salt of benzoic acid is sufficiently safe to be added, in modest amounts, to a great many foodstuffs. The addition of the carboxyl group to benzene has created a far safer molecule.

Some molecules contain more than one carboxylic acid group. The simplest is oxalic acid, which contains two such groups back-to-back:

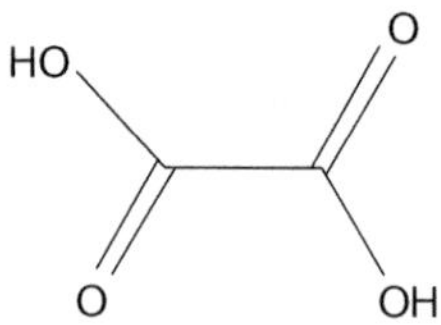

Oxalic acid occurs in high concentrations in pineapple and rhubarb, among other plants, and is responsible for the sharpness of the fresh fruits. Ingestion of too much oxalic acid can cause gastroenteritis, commonly recognized as a stomach ache. A salt of oxalic acid, calcium oxalate, is the stuff of kidney stones.

A final, and more complex, example is citric acid, which contains three carboxylic acid groups, as well as a hydroxyl group:

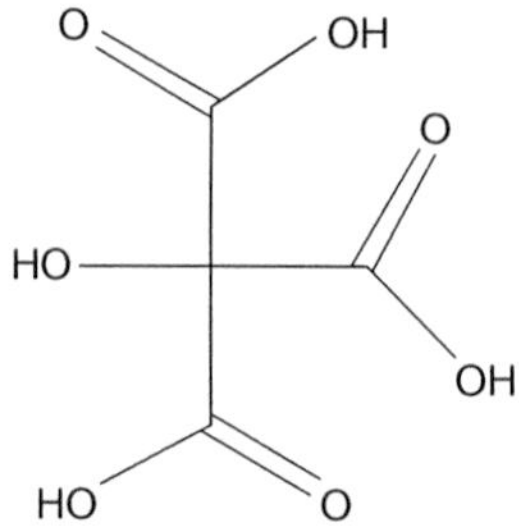

Citric acid gives its name to citrus fruits (lemons, limes, oranges), all of which contain high concentrations of this acid. Citric acid is also a principal player in the metabolism of carbohydrates and the generation of ATP, the energy currency of the cell.

There is much more to be said about carboxylic acids and life and it is worth pausing here for a bit to do some exploring. To be specific, I want to provide a couple of examples of the use of carboxylic acids for purposes of defense or offense among the insects.

Two people working together can frequently do things neither can do alone: making babies is perhaps the most obvious example. A somewhat more exotic example comes from a long-term collaboration between a biologist and world-class authority on insects, Thomas Eisner, and an outstanding organic chemist, Jerome Meinwald, both of Cornell University. Many of the scientific fruits of their collaboration have been collected by Eisner in his lovely book: *For the Love of Insects.*[2] The two examples that follow are taken from this book.

The whip scorpion is a 300 million-year-old insect that you may encounter in the Arizona desert. Full-grown, the whip scorpion will cover the palm of the hand of an adult. The whip scorpion defends itself from predators by repeated, as necessary, discharges of a repellent spray from two glands on the bottom rear of its body through a revolvable (i.e. aimable) "gun" at the base of its whip. This spray is remarkably effective in deterring predators, even more remarkable considering its simplicity: no complex molecules here.

The composition of the defensive spray is 84% acetic acid, which we have already met, 5% octanoic acid, an eight-carbon acid, and 11% water. An 84% solution of acetic acid in water is pretty strong stuff; vinegar is about a 5–10% solution of acetic acid in water, plus some flavoring agents. Unlike acetic acid, octanoic acid is quite hydrophobic (that is, water-hating and poorly water-soluble), thanks to its long hydrocarbon tail:

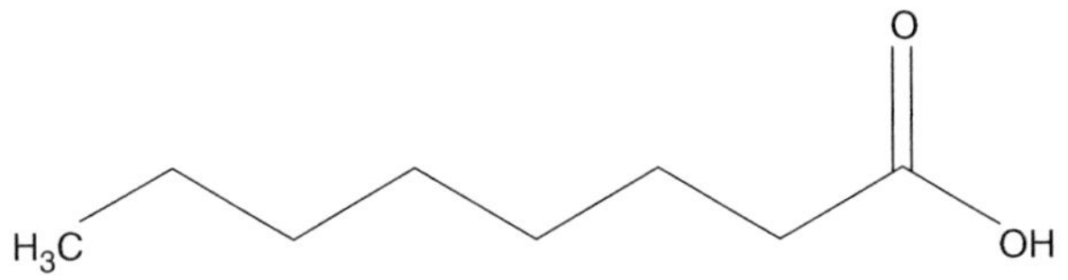

This property lets it adhere to and penetrate the waxy surfaces of many whip scorpion predators, providing a route of entry for acetic acid as well. The combination is, therefore, synergistic and serves the whip scorpion very well indeed as a defensive weapon.

The formacine ants possess a formic acid gland from which they may eject a spray of concentrated formic acid. Eisner refers to a formacine ant as "a spray gun on legs." The formic acid spray may be employed either defensively to ward off predators or offensively to incapacitate prey.

Esters are among the pleasantest of molecules

We shall encounter a multitude of other acids as we move forward through our exploration of the molecules of life. For the present, let's move on to another group of oxygen-containing molecules, the carboxylic acid esters. These are formed from the combination of a carboxylic acid and an alcohol, with loss of a molecule of water. Schematically we have:

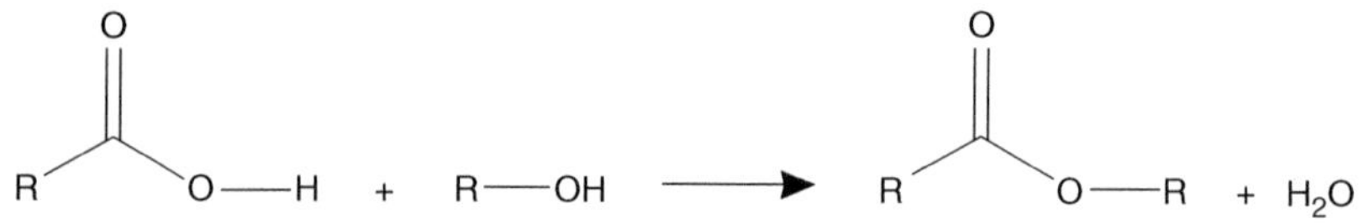

Esters can be recognized by the designation –*ate* in their names. Here are a few simple examples:

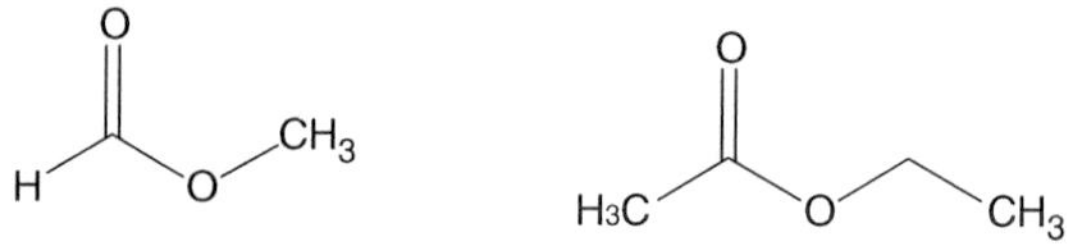

methyl formate

ethyl acetate

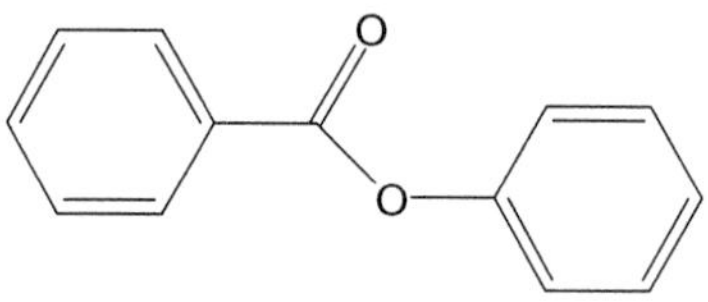

phenyl benzoate

Note that the group derived from the alcohol (methyl, ethyl, phenyl) comes first in the name and the group derived from the acid comes second with the –*ate* designation.

Esters are among the pleasantest of molecules. They frequently have pleasing odors and tastes and are widely used as flavorings or fragrances. Let's have a look at a few examples.

Although butyric acid is a foul-smelling substance, its methyl ester, methyl butyrate, is largely responsible for the attractive smell of apples:[3]

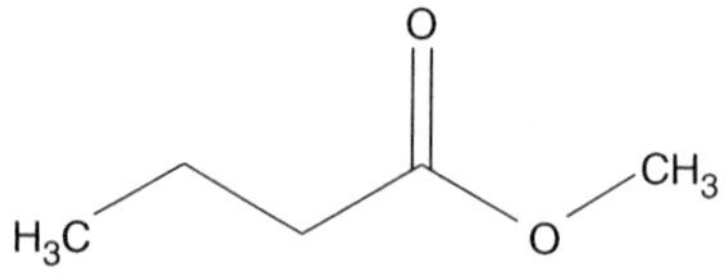

Here we have another example of the sensitivity of biological activity, in this case aroma, to chemical structure: replacing a hydrogen atom in butyric acid by a methyl group to yield methyl butyrate moves us from foul to pleasing.

We can do more. We can make a simple structural change to this ester and obtain a new fragrance. If we replace the methyl group by an ethyl group, to get ethyl butyrate and the smell of pineapple:

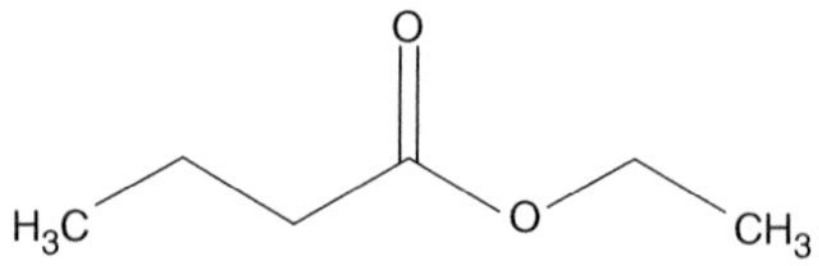

Continuing with this theme, if we extend the alcohol part of the ester to five carbon atoms, to get pentyl butyrate and the aroma of apricots:

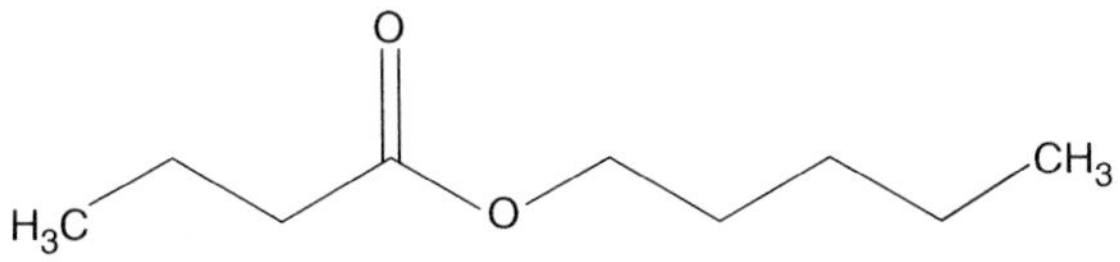

In these three cases, we have kept the structure of the acid component constant and varied the alcohol part with surprising results. This is another example of the sensitive relationship between chemical structure and biological properties. The relationship between structure and aroma can be understood pretty well at the molecular level. We return to this issue in chapter 25.

Here are some additional examples of esters as aromas or flavors: propyl acetate is the aroma of pears; pentyl acetate is the aroma of bananas; octyl acetate is that for oranges.

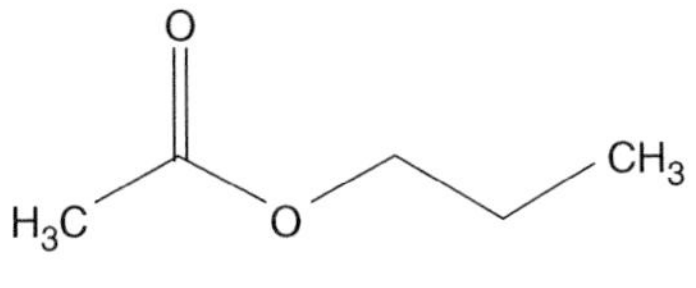

propyl acetate

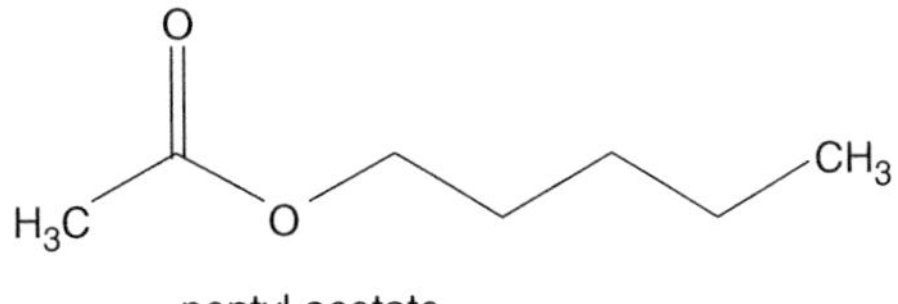

pentyl acetate

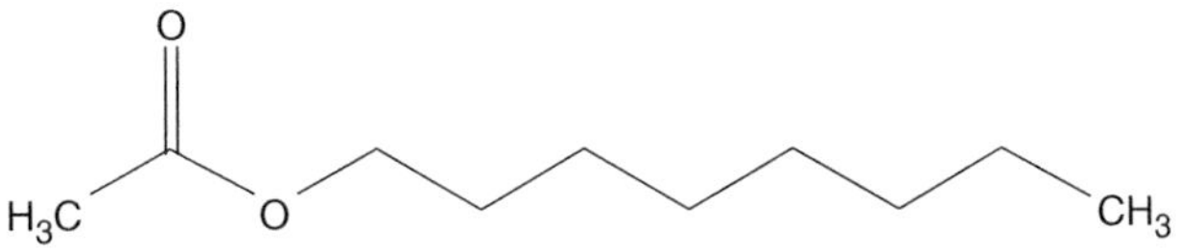

octyl acetate

Here too we have a common contributor to the acid component of an ester, acetic acid in this case, coupled to various alcohol components to yield a variety of distinct biological responses, the aromas of different fruits.

Ethyl formate provides much of the smell of rum and methyl benzoate that for kiwis. I could cite a bunch of additional examples but the point should be clear.

Oxygen-centered free radicals: elements in aging?

We need to pause for a moment to introduce something new and important: oxygen-centered free radicals. Free radicals have nothing at all to do with politics. The term refers to chemical species, frequently quite reactive, that possess an unshared electron.

Suppose that we begin with a simple molecule of oxygen, O_2, and ask what happens if we add an electron to that structure. Since the electron is negatively charged and we are adding it to the oxygen molecule, what is produced must have a single negative charge. We denote the product by $O_2^{\bullet-}$. The superscript "dot" is taken to mean that we have added an electron to the parent oxygen molecules and the superscript "−" indicates that this entity possesses a negative charge. This structure is known as the superoxide anion or the oxygen radical anion or, most simply, as superoxide. Superoxide is produced as a normal metabolic product, largely by the mitochondria.

Superoxide is interesting because it is a highly reactive oxidizing agent. It is capable of pulling an electron out of a number of other molecules, creating a dianion, which, after picking up a couple of protons, makes hydrogen peroxide, and creating a new molecular species in the process. That process can be either protective or injurious to the human body. On the protective side, superoxide is toxic to microorganisms and may help to protect us from infection. On the negative side, superoxide is also toxic to normal human cells and can create a number of problems such as mutations.

There is a related species, the hydroxyl radical, which we write as OH•, which is even more reactive than superoxide and also capable of being protective or injurious. You can think of the hydroxyl radical as being generated from a water molecule by the removal of a hydrogen atom or as being generated from the hydroxide ion, OH^-, by removal of an electron.

Generally included in this list is hydrogen peroxide, H_2O_2, which is also an effective oxidizing agent. Strictly speaking, hydrogen peroxide is not an oxygen-centered free radical since it does not possess an unshared electron. However, it is also produced by a number of cells in the human body, perhaps most notably by neutrophils. These cells of the immune system guard against infection and use oxygen-centered free radicals as killing agents.

On balance, these oxygen-centered radicals are generally considered to be worrisome things. Many scientists believe that molecular damage by oxygen-centered radicals underlies, in part, the process of aging. In fact, the human body has generated a number of defenses against them, including enzymes such as superoxide dismutase and catalase, which are effective in destroying superoxide and hydrogen peroxide, respectively. In addition, there are a number of small molecules that trap and neutralize oxygen-centered radicals. These include vitamin C, vitamin E, and uric acid. The central rationale for taking vitamins C and E in the form of nutritional supplements is precisely due to their antioxidant properties.

Although it seems highly likely that oxidative damage contributes to the aging process, there is ongoing debate about the mechanism. The simplest, and earlier, suggestion is that oxidative damage itself is the issue. Nick Lane has developed a far more complex picture in which reactive oxygen species are actually required as feedback messengers to the mitochondrion and the nucleus: see *Power, Sex, Suicide: Mitochondria and the Meaning of Life*.[4] More work needs to be done here. At the very least, these reactive oxygen species have provocative roles in disease, aging, and death.

Key Points

1. Alcohols are molecules containing the hydroxyl group, —OH. Ethanol, CH_3CH_2OH, provides a common example.

2. Carboxylic acids are molecules containing a carboxyl group, —COOH. Acetic acid, CH_3COOH, provides a common example.

3. Esters are molecules formed by adding an alcohol to a carboxylic acid, with elimination of a molecule of water. Esters are among the pleasantest of molecules. Many of the odors and tastes of fruits are from esters.

4. Oxygen-centered free radicals may be important contributors to aging. Antioxidants destroy these species.

8

Now for the rest of the elements in vitamin pills

The major elements phosphorus and sulfur and the trace elements sodium, potassium, magnesium, calcium, chlorine, iodine, manganese, iron, cobalt, nickel, copper, and zinc, and a few others, play specific, critical roles in life. Several others occur in living systems but may not be essential for life.

Of the 92 elements that occur in nature, 28 are found in living organisms. Most of these are trace elements: that is, they occur in small or very small amounts. The major elements in living organisms are carbon, hydrogen, oxygen, nitrogen, phosphorus, and sulfur. One criterion for the existence of life as we know it is a source of these elements. These derive ultimately from nuclear reactions in stars. Together, these six elements account for about 92% of the dry weight of living organisms: of these six, we have so far discussed carbon, hydrogen, nitrogen, and oxygen. We begin this chapter by examining some of the roles of phosphorus (P) and sulfur (S) in living organisms. Having disposed of six elements, that leaves 22 to go: of these, four seem not to be essential for human life—aluminum, arsenic, boron, and bromine—and I will have nothing more to say about them except to note that boron is essential for some plants and that arsenic is essential for some animals and possibly for humans. Bromine occurs in a number of marine natural products. That leaves us with 18 more trace elements to worry about: sodium, potassium, magnesium, calcium, fluorine, chlorine, iodine, silicon, selenium, molybdenum, manganese, iron, chromium, cobalt, nickel, copper, vanadium, and zinc. I will have something to say about some of these.

Elemental phosphorus is unpleasant stuff

In a sense, each element seems to have its own personality. The noble gases—helium, neon, argon, and the rest—seem aloof, independent, uninterested. Precious metals, including silver, gold, and platinum, impress me as serene, quiet, confident. Chromium, a shiny metal that forms many highly colored compounds appears to be a rock star among the elements, as does neodymium, the key element in early lasers.

Primo Levi, an Italian chemist and Auschwitz survivor, has pulled together additional examples in his inventive and amusing book: *The Periodic Table.*[1] Here are his thoughts about zinc:

> They make tubs out if it for laundry, it is not an element that says much to the imagination, it is gray and its salts are colorless, it is not toxic, nor does it produce striking chromatic reactions; in short, it is a boring metal. It has been known to humanity for two or three centuries, so it is not a veteran covered with glory, like copper, or even one of those newly minted elements which are still surrounded with the glamour of their discovery.

Zinc appears to lack any sort of elemental charisma. It is, nonetheless, essential for life.

Sodium, which is also essential, gets treated only slightly better:

> Sodium is a degenerated metal: it is indeed only a metal in the chemical sense of the word, certainly not in that of everyday language. It is neither rigid nor elastic, rather it is soft like wax; it is not shiny or, better, it is shiny only if preserved with maniacal care, since otherwise it reacts in a few instants with air, covering itself with an ugly rough rind: with even greater rapidity it reacts with water, in which it floats (a metal that floats!), dancing frenetically and evolving hydrogen.

Among the less attractive elements is bromine—a reddish-brown liquid that evolves a toxic brown vapor that one must absolutely avoid inhaling. The liquid itself is nasty stuff, causing burns that are slow to heal. Elemental bromine is foul stuff. It has the personality of an ill-treated pit bull.

However, the Lord Voldemort of elements is phosphorus. The whole story has been captured by John Emsley in: *The Thirteenth Element: The Sordid Tale of Murder, Fire, and Phosphorus.* Note that the identification of phosphorus as the 13th element refers to the fact that it was the 13th one to be isolated in pure form. In the periodic table of the elements, phosphorus is element number 15.

Here is the story of the discovery of phosphorus in Hamburg, Germany in 1669, in the words of John Emsley:[2]

> Like many before him (an alchemist) he had been investigating the golden stream, urine, and he was heating the residues from this which he had boiled down to a dry solid. He stoked his small furnace with more charcoal and pumped the bellows until his retort glowed red hot. Suddenly something strange began to happen. Glowing fumes filled the vessel and from the end of the retort dripped a shining liquid that burst into flames. Its pungent, garlic-like smell filled his chamber. When he caught the liquid in a glass vessel and stoppered it he saw that it solidified but continued to gleam with an eerie pale-green light and waves of flame seemed to lick its surface. Fascinated, he watched it more

closely, expecting this curious cold fire to go out, but it continued to shine undiminished hour after hour. Here was magic indeed. Here was phosphorus.

What the alchemist had isolated from his urine[3] was white phosphorus, one of two common forms, allotropes, of the element. White phosphorus is a waxy solid that is very poisonous and very reactive. As our alchemist observed, when it is exposed to air, a source of oxygen, it bursts into flames. White phosphorus is usually stored under water, protecting it from contact with atmospheric oxygen.

The composition of white phosphorus is P_4. Each of the phosphorus atoms lies at the corner of a regular tetrahedron, just as the four atoms of hydrogen in methane do. The phosphorus atoms feel strain; reacting to form more stable structures relieves that strain and releases energy. People behave in basically the same way: finding ways to relieve strain.

If white phosphorus is heated at about 400°C for a period, it is converted to a different allotrope, red phosphorus. In red phosphorus, the P_4 tetrahedra are linked together in long chain-like structures. This evidently satisfies the needs of the individual phosphorus atoms better, as red phosphorus is far less reactive and may be stored in air. It is also substantially less toxic than is white phosphorus.

Despite the high toxicity of phosphorus, it was formerly a widely used pharmaceutical, happily in quite small doses (though happier would have been no dose at all). It was recommended for "nervous breakdown, depression, migraine, epilepsy, stroke, pneumonia, alcoholism, tuberculosis, cholera, and cataracts."[4] *Free Phosphorus in Medicine*, published in 1874, extolled its benefits. By 1930, elemental phosphorus was eliminated from the practice of medicine. That is entirely appropriate since it has absolutely no medical benefits.

Phosphorus in living organisms is mainly important as phosphate

In spite of its ugly personality, phosphorus is an important element for life. The biochemical interest in phosphorus derives from its compounds with oxygen. Of these, the most important are phosphates. Indeed, in nature, phosphorus occurs mainly in the form of phosphates.

Phosphates derive from phosphoric acid: H_3PO_4. There is a lot of phosphoric acid in soft drinks, particularly colas. This accounts for their high acidity. In this molecule, the hydrogen atoms are bound to three of the oxygen atoms, so the molecule has a tetrahedral shape:

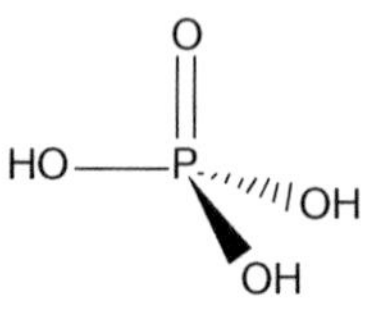

As noted briefly back in chapters 6 and 7, acids are molecules or ions that have the capacity to donate one or more protons to some base, the proton acceptor. As the name implies, phosphoric acid falls into this category. It is a somewhat complex example, donating up to three protons.

If you add phosphoric acid to water, what you get in solution is mostly phosphate anions that bear one or two negative charges, derived from phosphoric acid by loss of protons. These species are found in high concentrations in our cells and body fluids. Blood has about 350 parts per million (ppm) of phosphorus, as phosphate. The human body contains almost a kilogram of phosphorus, mostly in bones as a salt with calcium (known as hydroxyapatite.) In general, the phosphates that we encounter among the molecules in living systems also bear one or two negative charges.

Perhaps the most important signaling molecule in human biochemistry is the phosphate diester cyclic adenosine monophosphate, commonly known as cAMP. cAMP is part of many signaling pathways, a sequence of steps in which some physiological signal—say the arrival of a hormone molecule at a cell surface—is translated into a physiological outcome. Here is its structure:

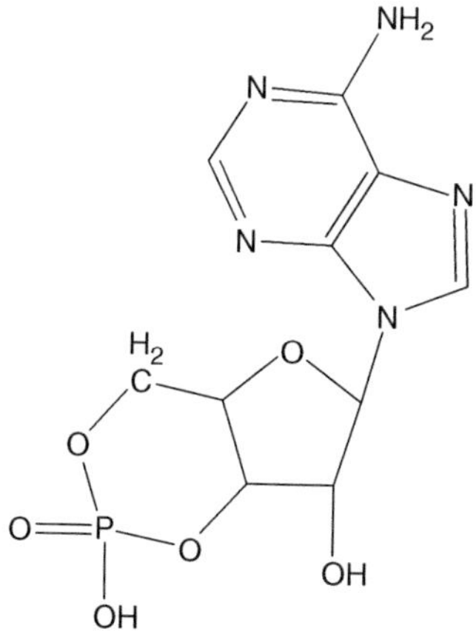

cAMP

Note that this molecule is a phosphate diester. Phosphate diesters also hold the subunits of nucleic acids together.

Compounds containing carbon–phosphorus bonds occur in living organisms but are rare. Such compounds are known as phosphonates. Phosphonates are important in human medicine as the most widely used drugs to treat osteoporosis, a loss of bone calcium that can lead to fractures, particularly those of the spine and hip. More about osteoporosis follows a bit later. They are also employed in the treatment of Paget's disease and bone cancers. These phosphonate drugs contain two carbon–phosphorus bonds and are, in consequence, known as bisphosphonates. Prominent among them are alendronic acid (Fosamax) and risedronic acid (Actonel), although there are several more on the United States market.

Phosphate is of enormous importance for the chemistry of bone. The basic structure of bone is provided by hydroxyapatite, $Ca_5(PO_4)_3OH$, and a protein, collagen. As indicated by the formula, each molecule of hydroxyapatite contains five atoms of calcium, Ca, three phosphates, PO_4, and one hydroxyl group, OH. The mineralization of bone occurs by the crystallization of hydroxyapatite within a three-dimensional matrix provided by collagen, an insoluble protein. Thus, bone contains an organic component, collagen, and an inorganic one, hydroxyapatite.

The development and maintenance of healthy bone depends in part on an adequate supply of dietary calcium. Although phosphate is absolutely required for the mineralization of bone, it is abundant in the diet. If one gets enough to eat, one gets enough phosphate. The same is not necessarily true for calcium and many people supplement their diet with a preparation of calcium, frequently calcium gluconate or calcium citrate. More follows below about the chemistry of bone when we get to a consideration of calcium.

Sulfur is more important than pleasant

Elemental sulfur is not nearly so exotic as elemental phosphorus. In air, sulfur just sits there while white phosphorus bursts into flame. Sulfur seems sort of dull, like an aging uncle.

Elemental sulfur is a yellow solid, in which eight sulfur atoms form a crown-shaped ring: S_8. Unlike elemental white phosphorus, elemental sulfur has very low toxicity. Years ago, people in the United States took a preparation of sulfur and molasses as a "spring tonic." Taken orally, sulfur has a mild laxative effect. Beyond that, it is not clear to me what favorable effects this tonic may have had, if any. There is a lot of sulfur around. When taken in all its forms, it accounts for nearly 2% of the weight of the crust of the Earth.

The hydride of sulfur is H_2S, hydrogen sulfide, related in composition to water, H_2O. While water is an extraordinary molecule that is absolutely required for life as we know it, as developed in chapter 6, hydrogen sulfide is an exceedingly foul-smelling and toxic gas. Surprisingly, this pariah among molecules has significant promise in human medicine.[5] Potential uses are from minimizing damage caused by heart attacks to inducing a hibernation-like state to help victims survive serious trauma. This is a story that will take time to play out. If the promise is realized, hydrogen sulfide will join two other toxic gases, carbon monoxide and nitric oxide, as a player in human health. Sometimes you can find value in strange places.

Sulfur is a constituent of two amino acids that occur in proteins, cysteine and methionine (chapter 10), and occurs in a large number of additional natural products, several of which we will encounter in what follows.

If one replaces one of the hydrogen atoms of H_2S with a methyl group, one obtains methylmercaptan, CH_3SH,[6] analogous to methanol, CH_3OH, and methylamine, CH_3NH_2. Methylmercaptan is distinguished by being, quite possibly, the worst-smelling substance known.

Among the truly obnoxious odors in nature are those associated with the defensive secretions of the skunk. As you might expect, these are sulfur compounds. The striped skunk accumulates a store of these and, when threatened, can eject them rather accurately in the direction of a potential predator within a range of about 15 feet. The resulting stench is enough to discourage all but the most determined predators.

Striped skunk defensive secretion contains a number of compounds but two are largely responsible for the odor: *trans*-2-buten-1-thiol and 3-methylbutane-1-thiol:

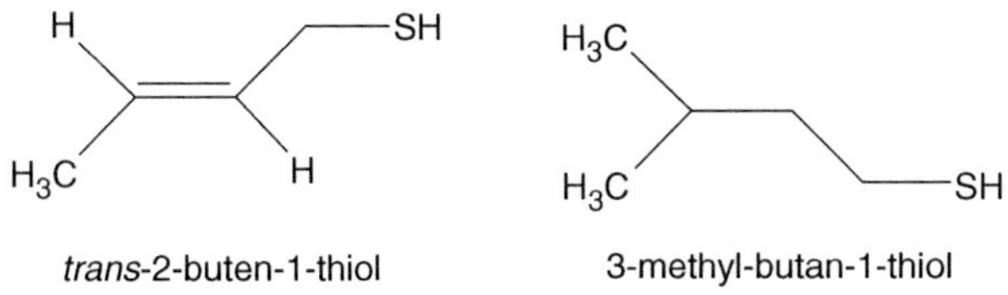

trans-2-buten-1-thiol   3-methyl-butan-1-thiol

These are volatile compounds and do have rather dreadful smells. They also occur in skunk spray in the form of their acetate esters.[7] These are much less volatile and so contribute less to the immediate odor. However, on contact with water, they are slowly converted to the free thiols, a reaction that accounts for the persistence of skunk odor for days or weeks. The chemical components of skunk defensive secretion vary somewhat from species to species but thiols play a prominent role in all cases.

Preparations that are used to eliminate skunk odor are oxidizing agents that convert the sulfhydryl groups to sulfonic acids:

$$R{-}SH \xrightarrow{\text{Oxidizing agent}} R{-}SO_3H$$

These lack the offensive odor and are water-soluble and can be washed away. The best move is to avoid threatening skunks in the first place but it is easier for you to do that than for your pet dog to get the idea in a timely way.

Garlic has a long history of putative medical use

Garlic has a long history of warding off fictional vampires. It also has a long history that espouses a number of medical uses.[8] Clinical trials of garlic as a cholesterol-lowering agent or antitumor agent have been inconclusive, although the effects of a garlic-derived compound, ajoene, on platelet aggregation and as an antimycotic agent are well-established. There are sensible reasons for believing that garlic itself or one or more of its constituents may have a role to play in human health. So let's have a look at some of chemistry of garlic before you thunder down to your local health food store and buy a preparation of it.[9]

Here are three molecules from garlic. First, we have a very unusual amino acid, allin:

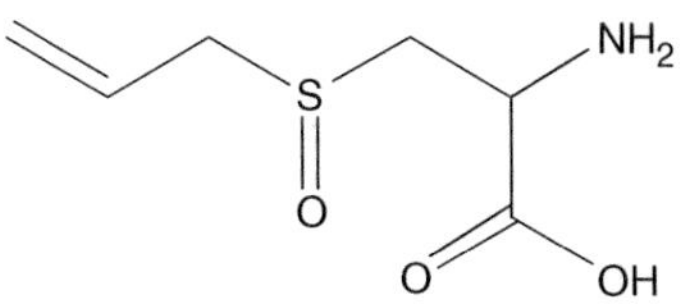

Allin is converted to allicin by an enzyme known as allinase when garlic is mashed (as recommended by many Italian cooks) rather than just cut:

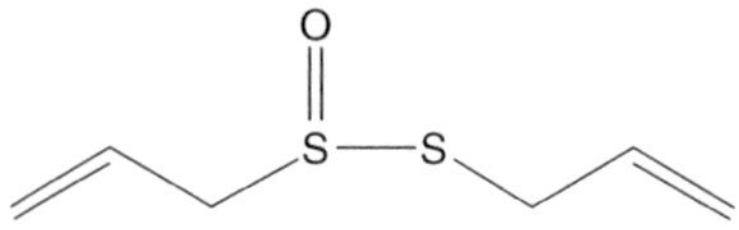

Note that we have something new here: a sulfur–sulfur bond. Allicin is a disulfide. Finally, allicin is converted to a molecule largely responsible for "garlic breath"— dipropenyl disulfide:

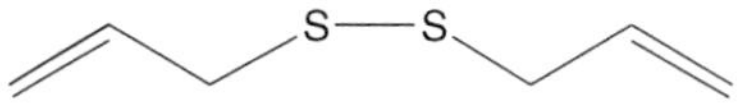

However, the most interesting molecule is ajoene, which is derived from garlic when it is cooked, particularly under slightly acidic conditions such as in tomato sauce:

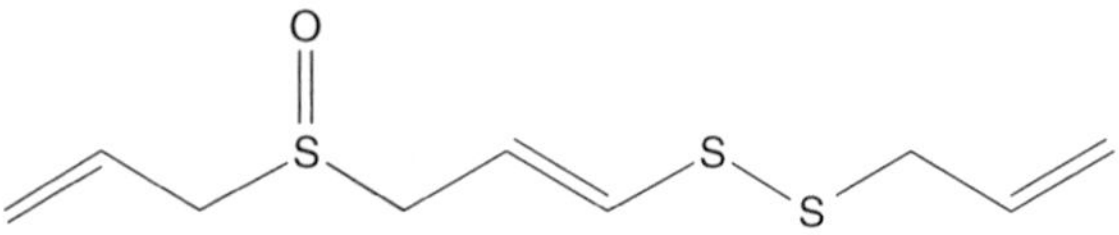

Ajoene is formed from three molecules of allicin and is responsible in part for the odor and flavor of garlic.

However, the real interest in ajoene lies in its pharmacological properties. It is a potent inhibitor of platelet aggregation by interfering with the fibrinogen receptor on the platelet surface. This inhibits the formation of blood clots. Such compounds are generically known as blood thinners. It is also an antioxidant and has substantial antifungal activity. In addition, ajoene is an inhibitor of a key enzyme on the metabolic pathway to cholesterol and, therefore, of cholesterol synthesis. It also exhibits antiproliferative activity. More work is needed to define the role of ajoene in human physiology but the story thus far is surely interesting and provocative.

During the early part of my scientific work, I spent 17 years as a professor in the Department of Chemistry at Indiana University. During that period, I had, among others, two talented coworkers: Mahendra Jain as a post-doctoral fellow and Rafael Apitz-Castro as a visiting professor from the Instituto Venezolano de Investigaciones Cientificas (IVIC) in Caracas, Venezuela. These two former coworkers did the most important work on ajoene. It is always great to see former colleagues go on to do important work.

Sodium and potassium are major players in all cells and body fluids

Sodium and potassium are among the alkali metals: lithium, Li; sodium, Na; potassium, K; rubidium, Rb; and cesium, Cs. All these elements are metals and all react with water, explosively, with the exception of lithium.

The alkali metals have in common the ready formation of ions bearing a single positive charge, cations. For example, we have sodium ion, Na^+, and potassium ion, K^+. We have an interesting contrast between the alkali metals and the halogens (fluorine, chlorine, bromine, and iodine). The former readily form cations bearing a single positive charge and the latter readily form anions bearing a single negative charge. This is important as it provides insights into the formation of simple salts. Salts are compounds formed from a cation and an anion. They are said to be ionic and are basically held together by the electrostatic attraction of positive for negative.

Sodium and potassium ions are everywhere in living systems: there are substantial concentrations in both cells and body fluids. Both have multiple roles to play in life. For example, sodium and potassium (along with calcium, magnesium, chloride, and bicarbonate, see below) are major ions in blood, where they have several roles: maintenance of ionic equilibrium with interstitial fluids, regulation of the pH (a measure of acidity) of blood, and regulation of membrane permeability. Levels of sodium and potassium in blood are regulated within rather narrow limits. This turns out to be important. For example, too high a concentration of potassium in blood stops the heart (as in lethal injections in executions).

Of the greatest importance is the role of these ions in the conduction of signals along neurons. The flow of sodium and potassium ions across the neuronal membrane is the basis for transmission of neural impulses (chapter 21).

Calcium ion influences numerous physiological processes

Calcium is a member of the alkaline earth family of elements, which also includes beryllium, magnesium, and strontium. These elements are all metals and all tend to form cations bearing two positive charges. Thus, we have the magnesium ion, Mg^{2+}, and the calcium ion, Ca^{2+}. Both of these ions are important in living systems. These elements frequently form salts with the halogens: so we have calcium chloride, $CaCl_2$, and magnesium iodide, MgI_2, and so on. When we look at the label on a bottle of vitamin and mineral pills and see the content of sodium, potassium, calcium, magnesium, and so on, these elements are present as salts, not as the elements themselves.

It is the divalent metal cation Ca^{2+} that is absolutely critical in human physiology. It is important both structurally and functionally. I noted above that hydroxyapatite, a phosphate salt of calcium, is an integral component of bone. As a component of bone, calcium is quantitatively one of the most abundant elements in higher organisms, such as humans.

In addition to its structural role in bone, and in teeth, the calcium ion is a carrier of chemical messages. It influences secretion, contraction of muscle, cell division, growth, transcription, as well as other key physiological processes.

The importance of the calcium ion in human physiology is reflected by the emphasis placed on this element in human nutrition, which is important throughout life. When we are young and growing rapidly, adequate dietary calcium is required for proper bone growth and development. Children are encouraged to drink plenty of milk, an excellent source of calcium. In my childhood, I drank a quart of whole milk every day, sometimes more, with the benign approval of my parents. My bones seem fine but I sometimes wonder a bit about the effects of all the fat in whole milk. I still drink a good deal of milk but have switched to 1% milk, a choice not available when I was a youngster (skim milk was available but it was not to my taste).

When we are older and begin to lose bone mineral density, we are encouraged to continue to get adequate calcium to retard bone loss. A lot of people supplement dietary calcium with calcium preparations available OTC. Have a look at the variety of these available in your local pharmacy. Both men and women begin to lose bone mineral density during middle age. This is particularly marked for postmenopausal women when the loss of estrogen is followed by loss of bone calcium. This can lead to osteoporosis, weakening of bone, and its sequelae—bone fractures including those of the vertebrae and hip. An adequate intake of calcium and vitamin D, coupled with bone-strengthening exercise, will minimize the probability of developing osteoporosis.

Osteoporosis is a serious matter. Cells known as osteoblasts make bone; cells known as osteoclasts degrade bone. It may not be obvious but the action of osteoclasts is important. We do small amounts of damage to our bones in the course of our daily activities. These damaged areas need to be repaired to ensure continued bone strength. The osteoclasts remove the small areas of damaged bone and the osteoblasts fill in the gap with new bone. The damage is repaired and bone strength maintained.

As we age, the activity of osteoclasts tends to outrun that of osteoblasts, leading to gradual loss of bone and increasing susceptibility to bone fracture. In the elderly, a hip fracture has about the same mortality rate as a heart attack. Crush fractures of vertebrae lead to an abnormal curvature of the spine and an inability to stand up straight. So, exercise and get enough calcium and vitamin D in your diet.

When diet and exercise have not been adequate to maintain good bone mineral density, the bisphosphonates are frequently prescribed. These molecules are actually incorporated into bone where they inhibit the action of osteoclasts. The biochemical consequence is an improvement in bone mineral density over time and the clinical consequence is a lessened frequency of bone fractures.

The importance of calcium is also revealed in the elegance of physiological controls on its absorption and metabolism. Two hormones regulate calcium concentration: calcitonin, a peptide hormone (secreted by the thyroid gland) and parathyroid hormone (PTH), another peptide hormone (this one secreted by the four parathyroid glands embedded in the thyroid gland). These two hormones have opposite effects. PTH stimulates calcium ion reabsorption in the kidneys and stimulates osteoclasts, cells that decompose bone and, thus, release calcium stored in bone. Both mechanisms tend to increase calcium ion concentration in the blood. vitamin D is essential to the action of PTH. Calcitonin, in direct contrast, tends to drive blood calcium ion levels lower. The net effect on blood calcium ion levels is determined by the balance between the action of these two hormones. A mechanism of this type, with opposing actions of effectors of physiology, is quite common. It provides for better control than mechanisms that rely solely on activation or inhibition.

Chlorine and iodine are required human nutrients

In living systems, chlorine is present as chloride, Cl^-. In part, chloride is there to balance out the charges contributed by cations such as sodium and potassium ions. There are mechanisms to get chloride across biological membranes as required to maintain charge neutrality. When such a mechanism goes wrong, it is a problem.

A pertinent case is provided by the life-shortening genetic disease cystic fibrosis. Cystic fibrosis is among the commonest of genetic diseases, affecting about one person in 3900 in the United States. The disease affects the entire body but the effects are most dramatic on the pancreas and lung. There is inadequate production of pancreatic enzymes involved in digestion, among other issues. In the lung, there is excessive production of thick mucus that is difficult to expel. Patients frequently have breathing difficulties. In addition, frequent lung infections, treated with various degrees of success with antibiotics, are a cardinal symptom of cystic fibrosis.

The genetic defect in cystic fibrosis is well-known: there is a mutation in a protein known as the cystic fibrosis transmembrane conductance regulator or, far more simply, as CFTR.[10] People generally have two copies of the gene that codes for this membrane protein but even one normal gene is adequate to prevent cystic fibrosis. Cystic fibrosis patients have no normally functioning CFTR proteins. The CFTR protein is a chloride channel: that is, a protein that promotes the transport of chloride across biological membranes. This transport is required for the normal production of digestive juices, sweat, and mucus. The pathology of cystic fibrosis testifies to the importance of the chloride ion in human physiology.

Elemental iodine, I_2, is a purplish solid. Like solid carbon dioxide (dry ice), iodine sublimes. That is, it goes directly from the solid to the gas phase without ever becoming liquid. So a bottle of iodine has a nice purple haze above the solid element.

Iodine is a required human nutrient. We obtain it in the form of the iodide ion, I^-, largely from iodized salt or vegetables grown in soils containing high levels of iodide. A nutritional deficiency of iodine results in the enlargement of the thyroid gland, known as a goiter. It is a rare condition in the United States.

Iodine is required for the formation of amino acids that act as thyroid hormones. The amino acids thyroxine and triiodothyronine, both derived from tyrosine, a protein amino acid, provide examples.

Thyroid hormones generally stimulate metabolism in most tissues, brain being an exception. People exhibiting unusually high levels of activity are sometimes referred to as hyperthyroid (whether they are or not). Too little thyroid hormone, hypothyroidism, is associated with slow metabolism, fatigue, weakness, intolerance to cold, weight gain, and other symptoms. Taking thyroxine in pill form is usually adequate to treat hypothyroidism.

Iron is specifically important for storage and transport of oxygen

In the human body, iron is present as Fe^{2+} and Fe^{3+}. About 10 mg of iron per day are required for optimal human nutrition. As some iron is lost from the body each day, a regular supply of dietary iron is important. However, the body can store about 1 gram of iron complexed with specific proteins.

Most of us get an adequate supply of iron in our daily diet. In fact, my vitamin pill does not contain any iron among its family of supplemental metals. A deficiency of iron causes anemia.

Iron plays a special role as a key constituent of hemoglobin and myoglobin. These proteins carry, hemoglobin, or store, myoglobin, oxygen in the human body and in those of other mammals. It is the iron atom, as Fe^{2+}, to which molecular oxygen binds in these proteins.

Iron is also a key constituent of many enzymes involved in electron transfer reactions, including those involved in the mitochondrial electron transport chain coupled to the synthesis of ATP.

Here is a grab bag of additional essential elements required for human health

At the beginning of the chapter, I made a note of several other metals that are important for living organisms: among them are zinc, magnesium, manganese, copper, and cobalt. Here are a few words about the role of these elements in life.

Zinc, magnesium, manganese, and copper occur, as the corresponding ions, linked, more or less tightly, to many of the key catalysts of life: the enzymes. The ionic forms that are most important are Zn^{2+}, Mg^{2+}, Mn^{2+}, Cu^+ and Cu^{2+}. They play two roles for the most part: first, these metals sometimes participate directly in the chemical reactions of life, secondly, they may play a structural role in the proteins in which they occur.

Cobalt occurs in vitamin B_{12}. This vitamin is very complex and is required for some exotic reactions of metabolism. A deficiency of vitamin B_{12} results in pernicious anemia. More about this vitamin follows in chapter 15.

I will get back to issues in human nutrition in chapter 15 when I explore the role of vitamins in metabolism. This brings us logically to our next topic, proteins. This is an amazing group of large molecules. Interactions of small molecules with proteins elicit many fundamental manifestations of life, including the senses of taste and smell and much, much more, including the actions of drugs. Developing the chemistry of proteins and all that follows from it will occupy the next three chapters.

Key Points

1. A substantial number of elements are required in the human diet for the maintenance of good health.

2. In addition to carbon and hydrogen, the key elements in the molecules of life include nitrogen, oxygen, phosphorus, and sulfur. Also, a family of trace elements is required: sodium, potassium, magnesium, manganese, calcium, chlorine, fluorine, iodine, iron, copper, nickel, cobalt, zinc, molybdenum, silicon and vanadium.

3. Salts are ionic compounds formed from anions and cations and are held together largely by electrostatic forces. Sodium chloride, NaCl, provides a common example.

4. Phosphorus is important in human biochemistry, mainly as phosphate. Phosphate is an important component of bone.

5. ATP, a phosphate derivative, is the energy currency of the cell.

6. Cyclic AMP, cAMP, is perhaps the most important signaling molecule in human biochemistry.

7. In human biochemistry, sulfur is important as a constituent of many key molecules.

8. The calcium ion is a key factor in human nutrition. It has an important structural role in bone and teeth and is a regulatory factor in many aspects of metabolism.

9. Sodium and potassium ions play multiple roles in human physiology and are required for nerve conduction.

10. A mutation in a chloride channel protein is the underlying defect in cystic fibrosis.

11. Iodine is a required human nutrient and is a component of thyroid hormones.

12. A number of metal ions are required as functional or structural components of proteins.

13. Cobalt is a component of vitamin B_{12}.

9

Proteins

*An amazing collection of
multifunctional properties*

---

*Proteins are a diverse family of large molecules composed of 20 amino
acid building blocks. They are an integral part of our diet. Our genetic
blueprint codes for proteins produced inside our cells that promote the
chemical reactions of life, protect us from infection, act as
communication devices, facilitate transport of molecules and ions,
mediate movement, and provide structure for our cells and bodies.*

I grew up in a solidly middle-class family, largely of German descent, in a city of modest size in central Nebraska. Like a lot of such families, our diet was based on meat and potatoes. It was an unwritten but religiously observed law in our home that two meals each day would include both meat and potatoes. The meat was turkey twice a year, ham on occasion, chicken or pork from time to time, but mostly beef. The potatoes were usually boiled or boiled potatoes subsequently sliced and fried. My brother and I also drank a lot of whole milk, at least a quart a day each and frequently more (skim milk was available but no one gave much thought to "reduced fat" or "low fat" milk back in those days). On farms, a lot of people just drank what the cows had on tap. Between the meat, potatoes, and the whole milk, we got a lot of protein and starch in our diet, which is good; we also got a lot of saturated fat in our diet and that is not so good.

Adequate protein in our diet is essential for good health. Proteins provide *essential amino acids* that our bodies cannot make or cannot make in adequate quantity for optimal health. For proteins, two things matter: amount and quality. The amount of protein is a simple quantitative matter: it is measured in grams per day. The amount that you need depends on a number of factors: your gender, age, size, level of exercise and other physical activity, and whether you are pregnant or lactating, for example. The quality of protein is not so easy to evaluate. Proteins are made from units termed amino acids and the essential amino acids are more important than others. The highest-quality proteins are those that contain a relative abundance of all of the essential amino acids. The central lesson is clear: proteins are important for human health. Indeed, they are essential for all forms of life.

To understand how small molecules work their biological magic, we need to understand some basic things about the big molecules of life, particularly the proteins and nucleic acids. The point is that *all the wonderful things that small molecules do derives from their interactions with large molecules.* So we need to understand both the small and the large.

Talking about big molecules does not require that we lose focus of small ones. The fact is that the big molecules of life are constructed from small ones. If you link enough small molecules together you eventually get big ones. Let's see how this works for proteins.

The building blocks of proteins are the L-α-amino acids (L-alpha-amino acids). The amino acids that occur commonly in proteins have molecular masses that vary from about 75 to a bit more than 200. Proteins have molecular masses that vary from a few thousand to a few hundred thousand. If we wanted to make models of proteins, we would need a really big set of Tinkertoys, Lego, or whatever. We need to get acquainted with the L-α-amino acids, understand how they are linked up to create protein molecules, and how this relates to the many functions of proteins in living organisms. Then we will be in a position to understand the consequences of small molecules interacting with proteins: metabolism; chemical communication; mechanisms of drug action; how vitamins work; taste and smell; and on and on. It makes sense to set the stage by focusing first on the proteins themselves.

Proteins are critical components of all forms of life. Proteins form the coat of the simplest viruses, though one may or may not choose to include viruses among living forms (chapter 2). Proteins are the only detectable component of mysterious infectious particles termed prions. Proteins serve multiple functions in all organisms, simple and complex, that clearly fall into the category of the living. These functions include catalysis, defense, signal transduction, metabolic regulation, movement, and architecture.

Enzymes are truly remarkable catalysts

Enzymes are a subset of proteins whose role in life is catalysis. Catalysts are agents that increase the rates of chemical reactions without undergoing chemical change themselves. Let's consider a model reaction in which some substrate, S, is converted into some product, P. We can write this simply as S → P. The point is the difference in the rate of this reaction in the absence and presence of a catalyst. If we make the necessary measurements, we will find that the rate in the presence of the catalyst is greater, perhaps very much greater, than in its absence. At the same time, the catalyst,

E, for enzyme, emerges from the reaction unaltered and ready to facilitate the reaction again and again. To illustrate, we can write the reaction in the presence of a catalyst as $S + E \rightarrow P + E$, showing that the catalyst, E, emerges unchanged as the substrate, S, is converted to product, P. As is emphasized below, the enzyme catalyst specifically recognizes its substrate, S, and is selective in terms of promoting the conversion of S to its product, P.

Enzymes are proteins

The phenomenon of enzyme catalysis in living organisms was known a long time before anyone really knew what an enzyme was in chemical terms. Enzymes are exceptionally efficient catalysts. So efficient in fact that chemists working in the early twentieth century could observe the phenomenon of catalysis under conditions in which the analytical techniques available at the time could not detect any protein. Proteins were there but just not detectable. Many chemists of that era considered enzymes to be carbohydrate in nature.

The protein nature of enzymes was established through the seminal work of James Sumner. In 1926, Sumner succeeded in isolating the enzyme urease in crystalline form from jack bean meal. This was the first time in history that an enzyme had been obtained in crystalline, though not completely pure, form. Subsequently, Sumner established that the crystalline enzyme was a protein. Urease is an enzyme that degrades one of the human end products of nitrogen metabolism, urea, to ammonia and carbon dioxide:[1]

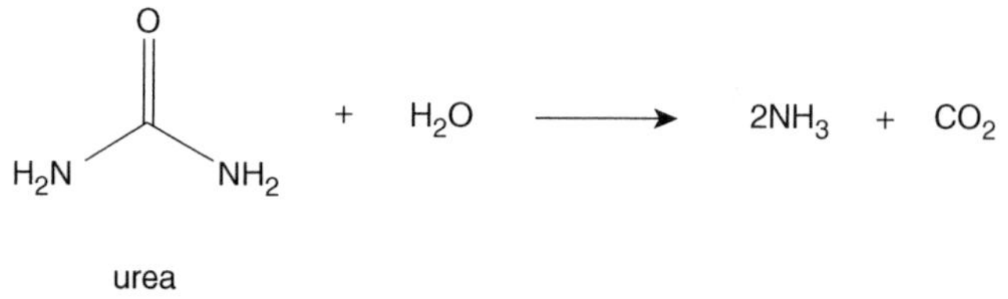

urea

Like many classic contributions to science, this spectacular advance was initially met with criticism and even derision. However, Sumner was not deterred: he took his show on the road and, through demonstration, convinced many scientists of the protein nature of enzymes. Subsequently, a number of enzymes were crystallized in the 1930s by John Northrup and Moses Kunitz and were also shown to be proteins. These later preparations were essentially pure. Northrup and Kunitz demonstrated that enzyme activity paralleled the amount of protein present, laying the issue to rest. Sumner and Northrup shared the 1946 Nobel Prize in Chemistry.[2]

Many years later, Tom Cech and Sidney Altman established that there is a second class of catalytic entities in living systems that are nucleic acid in nature: the ribozymes.[3] I get around to these in a later chapter.

Enzymes are potent, specific, and subject to regulation

There are a large number of chemical reactions that are essential to life. The sum of these reactions is termed metabolism. The metabolism of human beings, for example, includes the digestion of foodstuffs, the conversion of the products of digestion into other molecules, both very simple and very complex, as required to generate energy, create the structures of cells and tissues, provide for regulation of metabolism, support movement, permit us to sense our environment and react appropriately, and provide for reproduction.

The problem to be solved with respect to the chemical reactions that constitute metabolism and sustain life is that, without the action of catalysts, they are far too slow. Let's consider the digestion of the proteins themselves, an important constituent of our diet. In an environment similar to that of our digestive system, several tens of thousand years would be required to digest half of the protein content of a typical meal in the absence of a catalyst. Clearly, this will not do. In reality, the stomach secretes one protein catalyst, the enzyme pepsin, and the pancreas secretes several enzymes that catalyze the digestion of proteins. In the presence of these enzymes, dietary proteins are fully digested and reduced to their basic constituents, the amino acids, in a matter of hours. Obviously, these enzymes are enormously potent catalysts.[4]

In fact, enzymes are remarkable catalysts for three reasons: potency, specificity, and susceptibility to regulation.

There are other ways to appreciate the catalytic potency of enzymes in addition to that provided above. A second way to understand the same point is to accurately measure ratios between rates of enzyme-catalyzed reactions and the corresponding uncatalyzed reactions under the same conditions, a refinement of the qualitative argument just made.[5] These ratios are frequently not easy to obtain since the rates of the uncatalyzed reactions may be so slow as to make them exceedingly difficult to measure. Nonetheless, a number of these ratios are known and they typically vary between about 10^3 (one thousand) and 10^{15} (one quadrillion), truly enormous values. To help understand just how large 10^{15} is, consider a chemical reaction begun at the time of the creation of our solar system whose progress would be barely detectable at the present day. That same reaction would be nearly complete in 1 minute if catalyzed 10^{15}-fold.

Enzymes (E) work by forming a complex (E•S) with their substrate (S):

$$E + S \rightarrow E \cdot S \rightarrow E + P$$

in which P denotes the reaction product. The point is that E•S is very much more reactive than S itself. That is another way of saying that enzymes are highly effective catalysts. In figure 9.1, I provide a molecular model example of a complex between an enzyme and a small molecule, in this case an inhibitor, I. The E•I complex is a model for an E•S complex. This molecular model shows the enzyme human imunodeficiency virus-1 (HIV-1) protease complexed with a specific inhibitor, amprenavir. Note that amprenavir occupies a specific site partially exposed on the enzyme surface. Amprenavir is employed in the therapy of acquired immunodeficiency syndrome (AIDS).

In addition, we have the issue of enzyme specificity. The formation of E•S may be thought of by analogy to a key fitting a lock: here the lock is the enzyme and the substrate is the key (this analogy leaves something to be desired but will do for the present). Just as a lock is specific for a key, so are enzymes specific for their

Figure 9.1 A molecular model showing HIV-1 protease complexed with an inhibitor, amprenavir. (This figure kindly provided by Vertex Pharmaceuticals Inc.)

substrates. They are specific for what they bind and they are specific for the chemistry that follows.[6]

The formation of an E•S complex is one example of an enormously important phenomenon that we shall encounter throughout this book: *molecular recognition.*[7] The enzyme is said to recognize its substrate just as a lock may be said to recognize its key. The phenomenon of molecular recognition underlies the structure of proteins and nucleic acids, catalysis, the immune response, the senses of sight, taste, and smell, hormone action, the mechanism of action of drugs, and so on. It is one of the key underlying phenomena of life.

The catalytic power of enzymes is no more surprising than their specificity. As a general rule, each enzyme catalyzes only one type of reaction and, even then, with a very limited number of substrates: that is, enzymes are specific for both the type of reaction and the structure of the substrate. There are not many keys that fit these locks and of those that fit, few open them. Only those that are able to form E•S complexes have the potential to go on to form products. Enzymes are very particular about what they do and with whom they do it.

Typically, a specific enzyme catalyzes each chemical reaction in the metabolism of an organism. This specificity is required to properly regulate metabolism. One needs to be able to independently control the rates of all, or almost all, metabolic reactions, enzyme by enzyme.

Enzymes are susceptible to a variety of control mechanisms. Some of these control the rates of their synthesis or degradation, and so control the amount of enzyme present. Others control, through activation or inhibition, the activity of the enzyme that is present. The first mechanism works on the time scale of hours to days. The second works on the time scale of seconds.

Finally, there is a class of molecules—some large and many small—termed enzyme inhibitors. These molecules bind to enzymes, generally quite specifically, and prevent them from carrying out their catalytic function. These are keys that fit the lock but do not open it. This is another example of molecular recognition. In the simplest cases, the inhibitor of an enzyme is structurally related to the normal physiological substrate for the enzyme. The inhibitor looks enough like the normal substrate to bind to the enzyme at the site where the substrate normally binds but is sufficiently different so that no reaction subsequently occurs. The key fits in the lock but cannot open it. It follows that the enzyme is captured in the form of an enzyme–inhibitor complex, $E \cdot I$, where I denotes the inhibitor. The point is that $E \cdot I$ cannot make products. The enzyme has been rendered nonfunctional as long as I is bound to it.

Many of these enzyme inhibitors are important as therapeutic agents. Enzyme inhibitors serve as antibiotics, lower our plasma cholesterol and blood pressure, relieve pain and inflammation, help heal our ulcers, reduce our fevers, and treat cancer, among many other uses.[8] Many examples of the therapeutic uses of enzyme inhibitors follow in later chapters.

In summary, enzymes are potent, specific, controllable catalysts for almost all the chemical reactions of life. The ability to design and produce potent, specific inhibitors for enzymes provides a family of important drugs for human health.

Antibodies form part of the effective human immune system

Human beings, and other higher organisms, have developed highly effective and very complex mechanisms of defense against infection. In people, the first line of defense is our skin and the exposed mucous tissues of the mouth, eyes, and nose. This line of defense takes care of the great majority of organisms that we encounter daily.

The second line of defense against invasion by microorganisms includes a cellular component and a component that circulates in the blood, known as the humoral component. The latter is mediated by a second group of critical proteins, the antibodies or immunoglobulins.[9]

The generation of antibodies to neutralize, for example, a typical bacterial infection is amazingly complicated, enormously important, and generally works very well. Individuals having a compromised immune function are susceptible to numerous, repetitive infections. A compromised immune function can result from diseases such as AIDS, from certain types of inborn (genetic) errors, or from some drugs, including many commonly employed in cancer chemotherapy. In organ transplantation, drugs that suppress the immune system are deliberately administered to prevent rejection of the new organ, a foreign tissue.

The principal events that follow invasion of a human being by a bacterium (or a virus or a fungus ...) and lead to antibody formation are well known. Briefly, here is what happens. Molecules on the bacterial surface, the antigens, are recognized as foreign by receptors on the surface of B cells, a class of white blood cells. These receptors are proteins themselves. Once binding of an antigen to the antigen-specific B cell receptor occurs, the receptor and bound antigen are engulfed by the B cell. Once inside the B cell, the antigen is partially digested by enzymes into processed antigen. This is then transported to and bound by proteins known as Class II major histocompatibility complex (MHC) molecules and displayed on the cell surface. The Class II MHC•processed antigen complex is recognized by a second class of white blood cells, the helper T cells. The helper T cells are then induced to secrete yet other protein molecules (called interleukins) that cause the B cells to proliferate and to differentiate into plasma cells. These plasma cells make and secrete antibody to the bacterial antigen.

The antibody binds to the antigen exposed on the surface of the invading bacterial cell. This forms an antigen–antibody complex: Ag•Ab. This is another very specific molecular recognition process. Like enzymes, antibodies are very particular with whom they associate. Once an antigen–antibody complex is formed on the bacterial surface, this is a signal for phagocytic cells, such as macrophages, to ingest and kill the bacterial cell. That is the end of infection.

One more point about the immune system is very important. The cellular machinery responsible for antibody production is capable of distinguishing "self" from "non-self." This means that those molecules which occur normally in the human body do not act (usually) as antigens. Specifically and happily, they do not elicit antibody formation. Those B cells that have receptors specific for normal human macromolecules (proteins, nucleic acids, polysaccharides) are deleted from the repertoire of B cells in a cell suicide process known as apoptosis. Thus, the antibody response is directed almost exclusively against foreign molecules presented by invading organisms. However, in some instances the ability to recognize "self" breaks down and the body creates antibodies against a normal cellular component. When this occurs, it gives rise to "autoimmune diseases" such as myasthenia gravis, type 1 (juvenile-onset) diabetes, systemic lupus erythematosus, and rheumatoid arthritis, among others.

Insulin, growth hormone, and glucagon are key protein hormones

Hormones control, in part, metabolism. In general, hormones are molecules synthesized at one site in the body, travel to another site, and influence metabolism at the second site. Some of these hormones, though by no means all, are proteins.

Perhaps the simplest example is insulin, a small protein made by β cells (beta cells) of the islets of Langerhans of the pancreas. Insulin action has many consequences. A critical one is control of blood glucose levels. Following ingestion of carbohydrates (sugars and starches) in a meal, these substances are converted to glucose, which enters the blood. In response, the pancreas secretes insulin. Insulin then causes glucose in the blood to be taken up by muscle and fat cells, reducing the blood glucose levels to their normal value, about 100 mg per deciliter (100 milliliters; one-tenth of a liter).

Liver cells also take up glucose from the blood but this process is not dependent on insulin. In muscle and the liver, insulin also promotes the conversion of glucose to glycogen, a starch-like storage substance for carbohydrates.

Type 1 diabetes is an autoimmune disease in which antibodies are made to proteins of the β cells of the pancreas. These antibodies gradually destroy these cells and, with them, the body's source of insulin. Untreated, insulin-dependent diabetes is a fatal disease.

The discovery of insulin is one of the great stories in human medical history. I think it is well worthwhile to provide a brief account of the discovery of this hormone and its impact on human health before we proceed with our story about proteins in general.

In 1982, Michael Bliss published an intimately detailed history of the discovery of insulin, updated in 2007: *The Discovery of Insulin: 25th Anniversary Edition.*[10] The opening words of this book follow:

> The discovery of insulin at the University of Toronto in 1921–22 was one of the most dramatic events in the history of the treatment of disease. Insulin's impact was so sensational because of the incredible effect it had on diabetic patients. Those who watched the first starved, sometimes comatose, diabetics receive insulin and return to life saw one of the genuine miracles of modern medicine. They were present at the closest approach to the resurrection of the body that our secular society can achieve, and at the discovery of what has become the elixir of life for millions of human beings around the world.

Patients with type 1 diabetes mellitus make no insulin. The classic symptoms of Type 1 diabetes are excessive hunger, constant thirst, and frequent urination. Prior to the availability of exogenous insulin, a diagnosis of type 1 diabetes was a death sentence. The optimal therapy was to restrict food intake, usually to a few hundred calories a day. This extended life. However, toward the end, the only question was whether death would come as a consequence of the disease or through starvation.

To provide a sense of the impact of insulin on these diabetics, here are the words of Frederick Banting, one of its discoverers, cited in Michael Bliss's book:

> Another striking example ... was the case that came to Toronto early in June 1922. I was in my office on this afternoon and a man carried his wife into the office under his arm. He was a very handsome man of 30 & he deposited in the easy chair 76 lbs of the worst looking specimen of a wife that I have ever seen. She snarled and growled and ordered him around. I felt sorry for him. I placed her in the hospital more in pity for him than in regard for her.
>
> She was one of the most uncooperative patients with which I have ever dealt. It was in those early days when insulin was very scarce & precious and we endeavoured to get as much experimental knowledge as possible from each dose. She would steal candy or any kind of food she could lay hands on. She demanded that her poor husband come to the hospital early in the morning and every night he must not leave until she was asleep, yet she scolded him, cursed him, and treated him like nothing all day long. I could never understand why he took it all patiently, unruffled and even cheerfully. She was a terrible looking specimen of humanity with eyes almost closed with edema, a pale and pasty skin, red hair that was so thin that it showed her scalp, and what there was was straight and straggling. Her ankles were thicker than the calves of her legs and

her body had sores where the skin was stretched thin over the bones. Above all, she had the foulest disposition that I have ever known. I could not understand and I marveled at & sympathized with her poor husband.

She was in hospital some weeks and improved considerably and then he took her home. I was frankly glad to see the last of her. For his sake I had been kind. As a case to follow she seemed hopeless. I did not write to them nor did I hear from them.

A year later I was at my desk early one morning when the phone rang. A cheerful chuckling voice asked if I would be there ten minutes. I said I would. The receiver was hung up. I went on with my correspondence.

In a few minutes I heard that the outer door open and a moment later my office door was thrown wide open as in rushed one of the most beautiful women that I have ever seen. She was a stranger. I had never seen her before but she threw her arms around my neck and kissed me before I could move from where I stood. Over her beautiful head I saw the laughing face of the patient husband. I stood back. The three of us stood hand in hand. I looked at them. The husband said: "Doctor I wanted you to see her now. This is the girl I married—before she had diabetes." We laughed and talked. She was a devoted wife. He was no longer the slave but did most of the talking. I asked them many questions. As they went out he whispered "I'll have to take some insulin myself doctor."

Months later I received a tiny envelope with the name and a pink ribbon. A daughter. And I have wondered if the little one had red hair, and I prayed that she would never have diabetes.

Insulin is not a cure for diabetes. Patients with type 1 diabetes must take insulin throughout life. Proper treatment includes adherence to a diet and a program of physical exercise. There are serious long-term sequelae to diabetes if blood sugar levels are not adequately controlled: blindness (retinopathy), kidney failure (nephropathy), and microvascular disease that can lead to heart attacks or amputations.

Historically, insulin to treat type 1 diabetics was isolated from the pancreas of animals, usually pigs. This saved many lives but, in some cases, proved less than optimal. One problem is that pig insulin is not exactly the same as human insulin and the immune system may recognize it as foreign and make antibodies to it, neutralizing its effect. The first protein to be made available for medical practice through recombinant gene technology, the manipulation of genetic material to create designed proteins, was human insulin. Thus, human insulin was the first medical product to emerge from the biotechnology industry. It has proved to be a boon to many insulin-dependent diabetics. There are now more than 100 useful products of the biotechnology industry employed in clinical practice.

There are many more examples of the control of metabolism by protein hormones. Growth hormone is secreted by the pituitary gland and is responsible for providing normal growth in the young; it also has other activities.[11] A deficit or absence of growth hormone leads to retardation of growth and may result in dwarfism. For many years, growth hormone to treat growth-hormone-deficient children was isolated from the pituitary gland of cattle. Since these glands contain very, very little growth hormone, isolation proved both expensive and inadequate to meet the medical need. The majority of children in need of growth hormone supplementation simply went without. This need was met through determining the structure of human growth hormone, followed by cloning of the gene for this hormone and expression of the growth hormone protein in large amounts. This was the second medical success for recombinant gene technology.

Glucagon is a second pancreatic hormone that, like insulin, influences carbohydrate metabolism. However, most of its actions oppose those of insulin. The enkephalins and endorphins are the body's natural painkillers: they are thought to be responsible for "runner's high." These, among many others, are proteins or are closely related to them.

Receptors are essential for chemical signaling in cells

Let's think a little about the action of the protein hormone insulin. As noted above, insulin is secreted into the bloodstream by specialized cells of the pancreas in response to a dietary carbohydrate load. The insulin then travels to fat and muscle cells that respond to the insulin by taking up sugar. Blood glucose taken up by these tissues can either be used to generate energy or stored in the form of glycogen, depending on the needs of the organism.

Now this sequence of events raises two questions. How does the pancreas know that carbohydrate has been ingested? How do the fat and muscle cells know that insulin has arrived?

Although these questions have been posed for two specific examples, they reflect a general truth: cells do not live in isolation. Human beings are multicellular organisms: that is, our bodies contain an enormous number of cells (perhaps 100 trillion) and about 200 different cell types.[12] Brain cells are different structurally and functionally from liver cells, which are different from muscle cells, and so on. In fact, brain, liver, and muscle each contain a variety of clearly distinguishable cell types. Each cell type has one or more specific roles to play in the life of the organism. Sustaining that life and providing for reproduction absolutely requires that these cells work together in a highly cooperative and coordinated way. It is a little like an orchestra playing a symphony: there are many different instruments and each has a specific role in the symphony. When all the instruments play together in a highly cooperative and coordinated way we get music. Otherwise, the outcome is noise.

The requirement that cells act in a coordinated way implies that cells be linked through a complex communication network. There are a number of ways that cells communicate with each other. These include exchange of small molecules by direct cell-to-cell contact, cell-to-cell interaction via cell-surface receptors, as in the case of B cell/helper T-cell interactions leading to interleukin production, and communication locally or at a distance by means of extracellular signaling molecules. Insulin action provides one such example. It is a signaling molecule and acts at a distance via the bloodstream, termed endocrine signaling. Other signaling molecules may act directly on the same cell that releases them (autocrine signaling) and still others act on adjacent cells (paracrine signaling).

Signaling molecules initiate a process of signal transduction. "Transduction" implies that a signal received in one form is translated into a signal of another form. Typically, here is what happens. A signal from a molecule, insulin for example, arrives at the surface of a cell, the plasma membrane. This signal is transferred across the plasma membrane, through the cell cytoplasm and into the nucleus. There it may influence the synthesis of proteins that, in turn, alter the metabolism of the cell or organism. The net effect is that a message from the external environment, a carbohydrate-rich meal for example, is communicated to specific cells that respond with an appropriate adjustment to the metabolism of the organism. Communication happens.

Communication requires a signal and a signal receiver. Signal receivers are termed *receptors*. We have seen two examples: the antigen-specific receptors on B cells and the receptors on T cells which recognize Class II MHC•processed antigen complexes on the surface of B cells, two more examples of molecular recognition.

Receptors have at least two things in common with enzymes. In fact, many receptors, though not all, are enzymes-in-waiting and become catalytically active through binding to a signaling molecule. Both enzymes and receptors are protein in nature and both are highly specific: enzymes for their substrates and receptors for their signaling molecules, known as ligands. Thus, each receptor will generally interact with one or very few physiological signaling molecules. The insulin receptor is quite specific for insulin, the growth hormone receptor for growth hormone, and so forth. Thus, each signaling molecule carries some specific message to cells bearing receptors for it.

Receptors fall into two general classes. Many are proteins embedded in biological membranes. These proteins usually span the membrane. The extracellular domain of the receptor interacts with the signaling molecule, changing the structure of the receptor in a way that transmits the information that a signal has arrived to an intracellular domain, which then acts to alter some aspect of cellular metabolism.

Other receptors are soluble proteins that exist in the cell cytoplasm. When complexed with their signaling molecules or ligands, they migrate to the nucleus where the receptor•ligand complex binds specifically to sites on the nuclear DNA and controls protein synthesis.

In summary, receptors are proteins, membrane-bound or intracellular, that act as specific receivers for chemical messages and initiate metabolic changes in cells and organs. Specific and potent agonists and antagonists at receptors comprise several classes of drugs useful in human medicine.

Proteins form ion channels

Biological membranes, whether they enclose cells or organelles within cells, are basically barriers. Those that enclose cells, plasma membranes, isolate the cell interior from the external environment. Those that enclose subcellular organelles—the nucleus or mitochondria for example—provide for compartmentalization of function within cells. The barrier function of membranes is critical to life.

At the same time, there is clear need for some forms of communication and material exchange between the cell interior and the external environment and between different membrane-enclosed structures within cells. To meet these needs, living organisms have developed a number of specialized, membrane-centered mechanisms. As noted above, membrane-localized receptors provide one means for message exchange between cells and the external world. Ion channels provide another.

As detailed in chapter 17, biological membranes are basically lipid—think fat or oil—in nature with some attached proteins. As such, these thin sheets of phospholipids and proteins are nearly impermeable to charged particles such as sodium, potassium, or chloride ions. While the isolation of the cell interior from the exterior ionic environment is critical in many ways, it is also true that controlled permeability to ions may be critical. In fact, it is the near-impermeability of biological membranes to ions that permits control of ion transport across them by certain, specific proteins.

Ion channels fall into three distinct categories. These are all proteins that open and close gates in response to suitable signals. The categories reflect the nature of the agent responsible for causing the ion channels to open and close, known as gating. Think about a gate in a picket fence. The gate needs to be open for people to pass through. Ion channels work the same way: in the resting state they are closed and nothing gets through. Some impetus is required to open these gates and permit ions to flow through them.

A change in voltage across the cell membrane housing an ion channel may open that ion channel. Restoration of the original voltage may close it again. Channels that behave in this way are known as voltage-gated ion channels. The sodium and potassium ion channels of nerve cells fall into this category.

The second category includes ion channels that open in response to the binding of some molecule, known generically as ligands. The resting state of the ion channel is closed. When an appropriate molecule binds to the channel proteins, it opens. When that molecule leaves the channel proteins, the ion channel will close again. Such channels are known as ligand-gated ion channels. The ligands communicate a signal from the external environment and have the effect of opening and closing ion channels. This will alter the internal workings of the cell. For example, nerve cells contain ligand-gated ion channels: these respond to the action of neurotransmitters at junctions between nerve cells or between nerve and muscle cells.

Finally, there are ion channels that respond to intracellular (as opposed to extracellular or environmental) chemical stimuli. These form part of cellular signaling pathways and may be opened by cellular messengers such as calcium ion, Ca^{2+}. Such channels are known as signal-gated ion channels.

Proteins underlie movement

Movement is an inherent attribute of human life, and of most other forms of life as well. Most obviously, we walk, run, crawl, dance, stagger, swagger, climb, fidget, stretch, flex: examples of movement. Slightly less obviously, our heart beats, lungs expand and contract, the stomach churns, intestines propel, the walls of our veins and arteries expand and contract, the uterus contracts during childbirth, and so forth. Even less obviously, cilia in intestines and lungs beat, sperm swim, molecules move from their place of synthesis to their place of action or storage, nutrients are distributed to their site of utilization, and electrical impulses travel along nerve cells. There is a lot of movement in human life and proteins mediate it.

What we commonly call movement depends on muscle cells. There are several types of muscle in the human body: heart muscle; striated skeletal muscle, which includes the large muscles of the body; and smooth muscle of the stomach, intestines, arteries and veins, and uterus. There is a fundamental difference between striated and smooth muscle. You can control the former but not the latter. You can pretty much will what you do with your arms and legs, for example. The activities of your stomach, veins and arteries, and intestines are pretty much beyond your ability to control by force of will. Heart is an exception: heart muscle is basically a striated muscle but we are at a loss to willfully control the rate or force of its contractions.

In animal cells, the beating of cilia is used largely to move fluids. Protozoa employ ciliary beating for propulsion. The structural element of cilia that underlies their ability to beat is the microtubule. Microtubules, in turn, are built up of tubulins, key proteins of movement.

Finally, many cells, including spermatozoa, use flagella for propulsion. Flagella too are energized by microtubules.

Proteins provide the internal architecture of human cells

Let's begin on a small scale: the architecture of cells. The cells of plants and bacteria have strong cell walls that provide for maintenance of shape and protection against outside forces. The cells of animals, including those of the human body, in contrast, lack cell walls. Animal cells make do with a fragile cell membrane. Our cells, consequently, have need of an internal architecture to meet the needs supplied by cell walls in other life forms. Cells have three types of internal architectural elements: microtubules, intermediate filaments, and actin filaments. Each of these structures is composed of protein.

On a larger scale, we have ligaments and cartilage, structures that hold joints together and tie muscle to bone. These too are composed of proteins: tough, water-insoluble proteins. Principal among them are the collagens.

Finally, we have other water-insoluble proteins that constitute such structures as skin, hair, the wool of sheep, fur in other animals, silk, kangaroo tail, and the like.

In summary, proteins play an amazingly diverse set of roles in life. We need to understand the basis of their ability to exhibit such diversity of function. We begin that task in the next chapter.

Key Points

1. Adequate dietary protein is essential for good health.

2. The highest-quality proteins are those containing an abundance of essential amino acids.

3. Proteins are critical components of all forms of life.

4. In living systems, proteins function as catalysts (enzymes), for defense (antibodies, immunoglobulins), signal transduction (hormones, receptors), metabolic regulation (hormones, enzymes, receptors, ion channels), movement (microtubules), and architecture (structural proteins such as collagen).

5. Enzymes are remarkable catalysts: potent, specific, subject to metabolic control.

6. Molecular recognition is critical for life: enzymes recognize substrates, antibodies recognize antigens, receptors recognize signaling molecules, and so on.

7. Some hormones are proteins: insulin, growth hormone, and glucagon provide three examples.

8. Hormones act through protein receptors: hormones are signals; the receptors are the receivers of the signals. This is known as signal transduction.

10

Amino acids

The building blocks of proteins

Proteins are polymers of amino acids linked together by peptide bonds;
the same family of 20 amino acids is employed by all living organisms
to construct their proteins. The sequence of amino acids along the chain
defines the primary structure of the protein, a species-specific quality.

In the previous chapter, I introduced proteins as an amazing family of large molecules
serving multiple roles in life, from catalysis to defense, to structure, and much more.
Proteins are also a great example of one facet of our definition of life: chemical
properties. Proteins are found in all living organisms and in many products of living
organisms. They are also key constituents of viruses, at the edge between the living
and nonliving. However, proteins do not occur in inanimate nature.

   The task now is to develop an understanding of the molecular basis of the ability
of the proteins to function beautifully in a variety of roles. That is the principal goal
of this chapter.

Proteins are polymers of amino acids

Proteins are polymers. Polymers are molecules, frequently very large molecules,
which are made up by the repetition of one or more structural elements, termed
monomers. Think about making long chains of Lego pieces. The simple process
of hooking up unit after unit, the monomers, permits you to create a large, if
uninteresting, structure, the polymer. Other trivial examples include making a string

of beads or a paperclip chain. In each case, the underlying idea is the same: take some element or elements and hook them up in a linear fashion to create a chain.

Amino acids, as the name suggests, contain two fundamental units: an amino group, —NH_2, and a carboxylic acid group, —COOH. For those amino acids that occur in proteins, the amino and carboxylic acid groups are attached to the same carbon atom. This carbon atom is termed the α (alpha) carbon atom. These amino acids are, reasonably, called α-amino acids.

The parent molecule of the group is glycine (*gly-seen*): H_2N—CH_2—COOH. In addition to the amino group and the carboxylic acid group, the α carbon atom is bonded to two hydrogen atoms. These are the simplest possible substituents. Thus, glycine is the simplest possible α-amino acid and is a very common constituent of proteins:

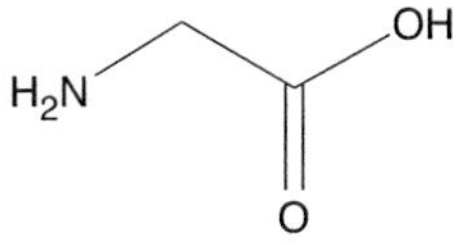

The other 19 amino acids that commonly occur in proteins can be thought of as derivatives of glycine. These are formed, with one exception, by replacing one of the hydrogen atoms on the α carbon atom (that atom that is adjacent to the carboxylic acid carbon) by some group of atoms. For example, if we replace one of the hydrogen atoms of glycine by a methyl group, we get a new amino acid, alanine (*ala-neen*):

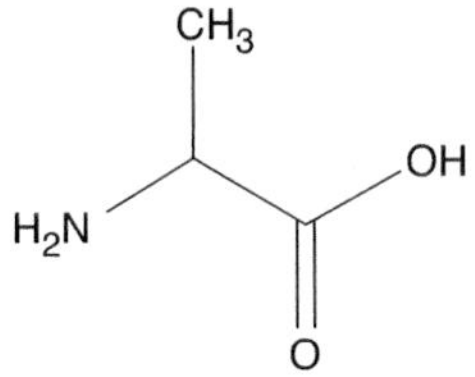

Before we elaborate on this theme, we need to take a closer look at alanine. Notice that the central carbon atom, the α carbon atom, has four different groups linked to it: the carboxyl group, the amino group, the methyl group, and a hydrogen atom. It follows that alanine is chiral. I introduced the idea of chirality back in chapter 4. There we noted that molecules in which a carbon atom contains four different substituents linked to it exists in two different forms, corresponding to left- and right-handed gloves. These are enantiomers, nonsuperimposable mirror images, and are typically designated D and L or, alternatively, R and S. Some of the properties of the two enantiomers are distinct.

So there are two alanine enantiomers. These are generally called L-alanine and D-alanine:[1]

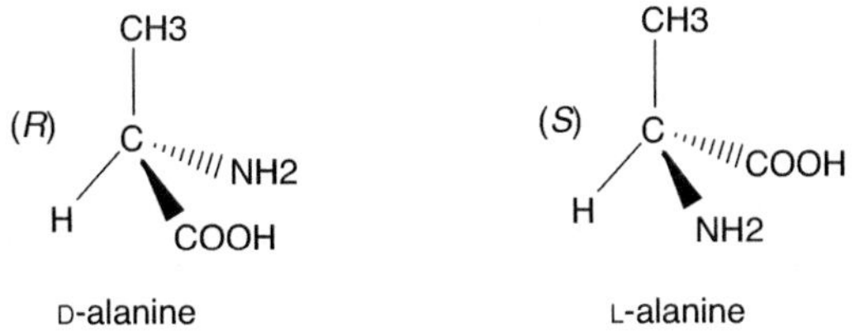

D-alanine

L-alanine

Now look back at glycine. Glycine has only three different groups attached to its α carbon atom: the carboxyl group, the amino group, and two hydrogen atoms. So glycine is not chiral. It is the only amino acid that commonly occurs in proteins that is not chiral. As noted above, all the others are derived from glycine by replacing one of the hydrogen atoms with another group: this will generate chiral molecules.

Here is an important point: almost all chiral amino acids that occur naturally in proteins throughout all of nature have L-stereochemistry. Why has nature uniquely selected L-amino acids for the construction of proteins? No one knows for sure. In passing, we note that D-amino acids do occur in some living systems. The cell walls of bacteria possess both D- and L-amino acids, for example. However, these are introduced in a manner distinct from that employed to synthesize proteins.

Of the 20 amino acids that commonly occur in proteins, glycine has no side chain and alanine has the simplest side chain, a methyl group. Side chains for all the common amino acids of proteins are collected in figure 10.1. These are grouped into three categories.

The first category is composed of the nine amino acids having nonpolar side chains, identified as those with side chains that are largely hydrocarbon in nature. The single exception is methionine, which contains a sulfur atom in its side chain. This is not a problem, since sulfur atoms are quite hydrophobic. Basically, hydrocarbons do not like water; "hydrophobic" means water-hating. Think about fats, oils, waxes.

Secondly there are six amino acids with polar but uncharged side chains. These side chains are intermediate in character between those of the first and third categories.

Finally, five amino acids whose side chains, under physiological conditions, bear charges create the third category. Three of these possess positive charges (lysine, histidine, and arginine) and the other two have negative charges (aspartic acid and glutamic acid). These side chains are very strongly hydrophilic, which means water-loving.

Each amino acid has three designations (figure 10.1). The first is its name. Some of these are rather long: phenylalanine and tryptophan, for example. When it comes to writing down the structures of proteins containing a large number of amino acids, it is inconvenient to write out the complete name. So abbreviations for the amino acids of proteins have been invented. There are two classes of these: three-letter designations and one-letter designations.[2] The former are generally the first three letters of the name of the amino acids. There are two exceptions: Gln for glutamine and Asn for asparagine. So we have Gly for glycine, Tyr for tyrosine, Met for methionine, and so on. The one-letter abbreviations are included in figure 10.1.

Amino acids with nonpolar side chains

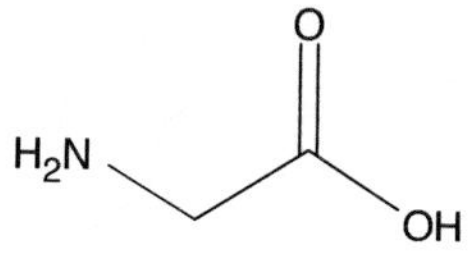

glycine; Gly, G

Glycine is in a class by itself. It is the only protein amino acid that is not chiral. It is neither hydrophilic nor hydrophobic. With the exception of proline, all other protein amino acids are derived from it by substituting various groups on the α carbon atom. Glycine is an important inhibitory neurotransmitter in the central nervous system.

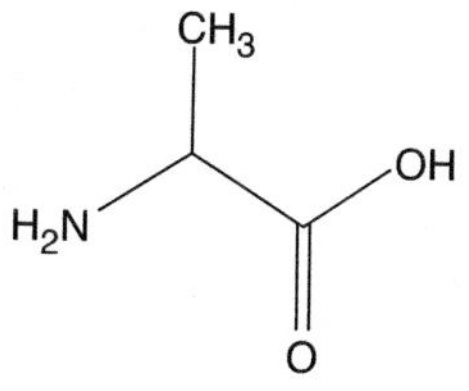

alanine; Ala, A

Alanine is the simplest nonpolar amino acid. It is derived from glycine by affixing a methyl group to the α carbon atom.

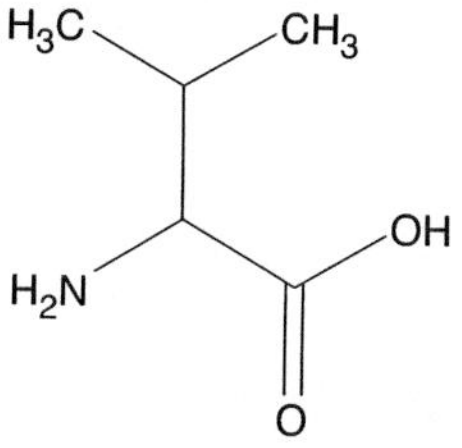

valine; Val, V

Valine is more hydrophobic than alanine: the hydrocarbon moiety is larger.

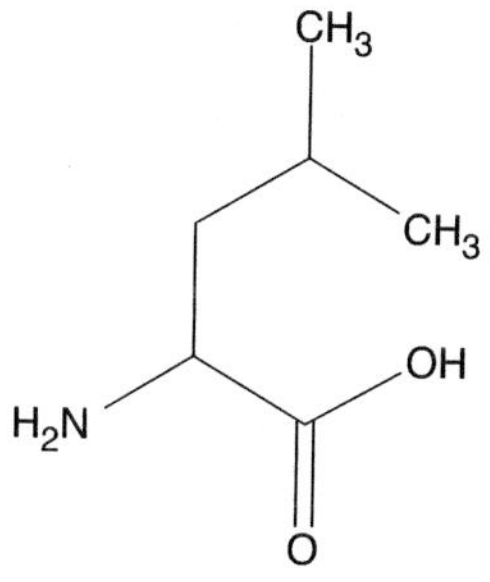

leucine; Leu, L

Figure 10.1 Structures and some properties of the amino acids that occur commonly in proteins.

Leucine is yet more hydrophobic than valine, with a larger hydrocarbon moiety.

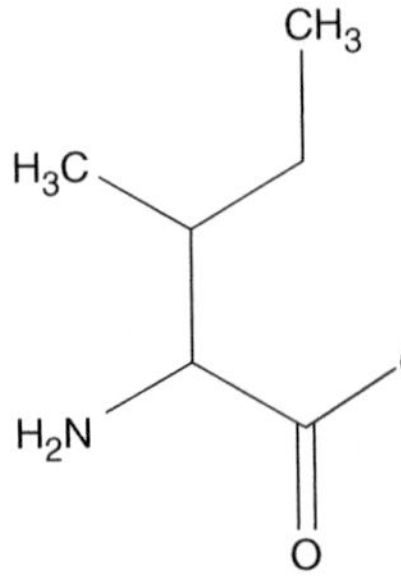

isoleucine; Ile, I

Isoleucine is a constitutional isomer of leucine. Note that isoleucine has two chiral centers.

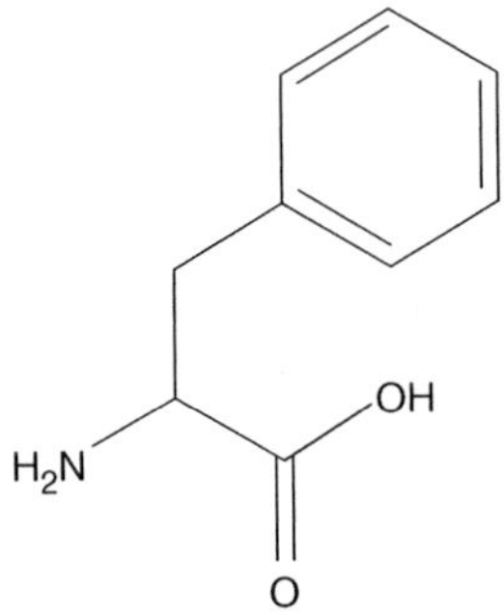

phenylalanine; Phe, F

Phenylalanine possesses a phenyl group on its side chain. It is the metabolic precursor to tyrosine.

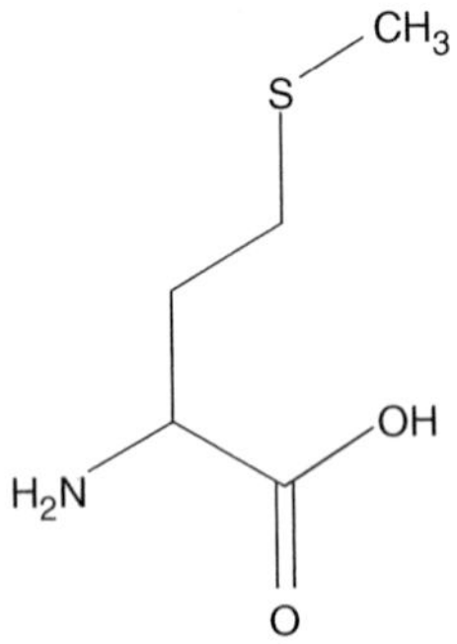

methionine; Met, M

Figure 10.1 (*Continued*)

Methionine is one of two protein amino acids that contain sulfur. It has an important metabolic role in methyl group transfer reactions.

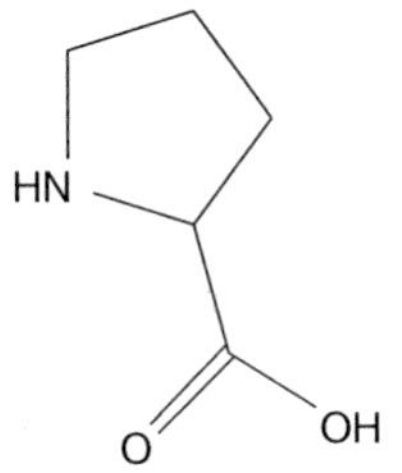

proline; Pro, P

Proline is unique among protein amino acids in that the side chain is linked back to the amino group to form a cyclic structure.

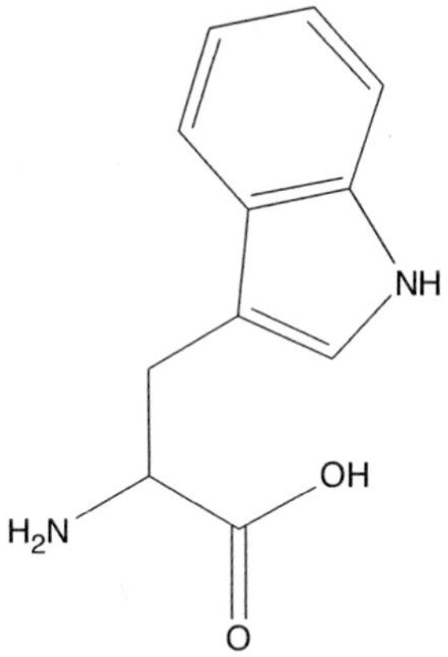

Tryptophan possesses a complex heterocyclic side chain known as indole. It is the metabolic precursor to serotonin (5-hydroxytryptamine; 5-HT), an important neurotransmitter.

Amino acids with uncharged polar side chains

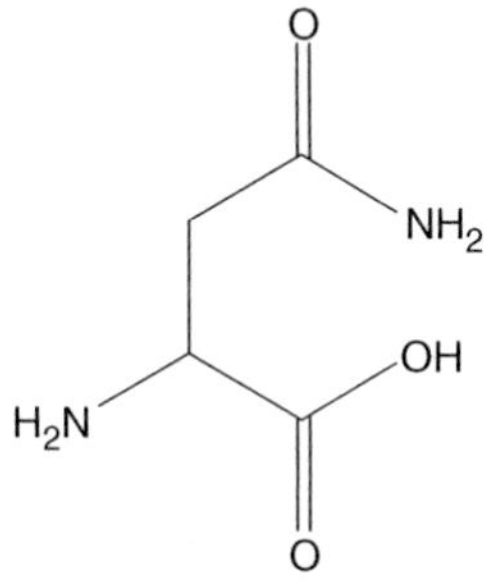

asparagine, Asn, N

Figure 10.1   (*Continued*)

Asparagine is derived from aspartic acid. Its side chain is an amide.

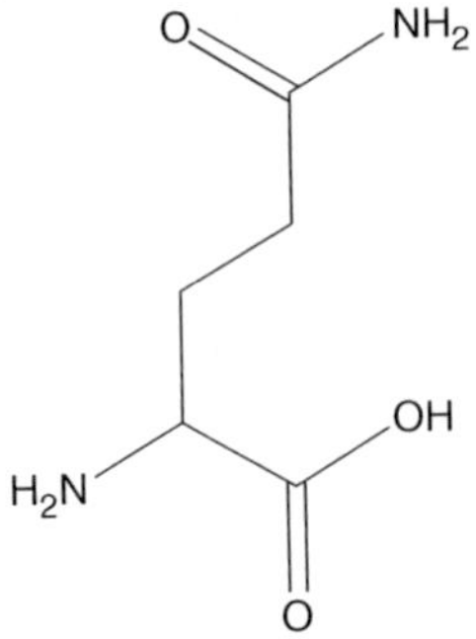

glutamine; Gln, Q

Glutamine is derived from glutamic acid. Its side chain is an amide.

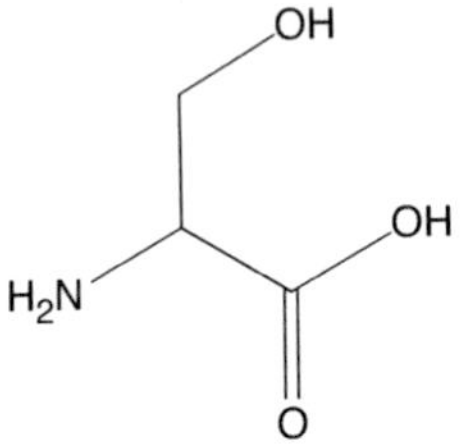

serine; Ser, S

Serine is one of two amino acids that possess a simple alcohol function.

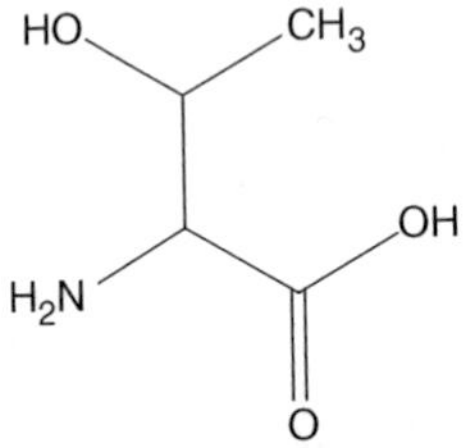

threonine; Thr, T

Threonine is related to serine. It is unusual in that it has two chiral centers, as does Ile.

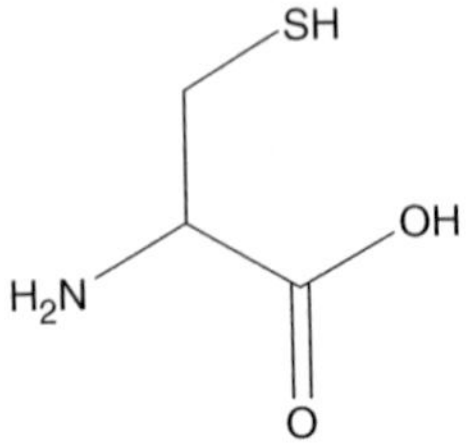

cysteine; Cys, C

Figure 10.1   (*Continued*)

Cysteine is derived from serine by replacing the hydroxyl group with the sulfhydryl group. It is the second protein amino acid that contains sulfur.

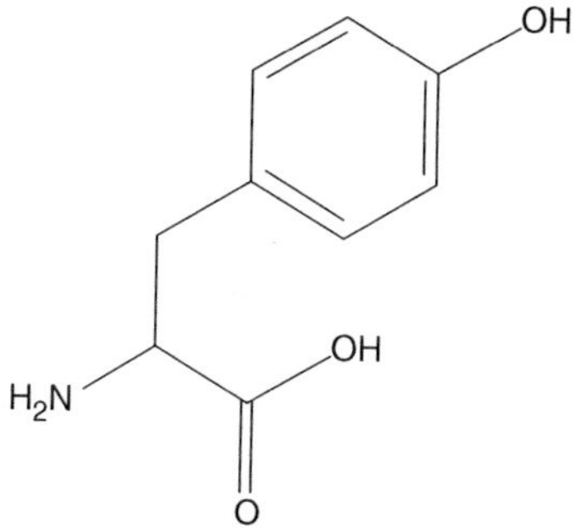

tyrosine; Tyr, Y

Tyrosine is structurally related to and derived from phenylalanine. It is the metabolic precursor to dopamine, an important neurotransmitter. Tyrosine is also the precursor to the hormones epinephrine and norepinephrine and to melanin, the pigment of skin.

Amino acids with charged polar side chains

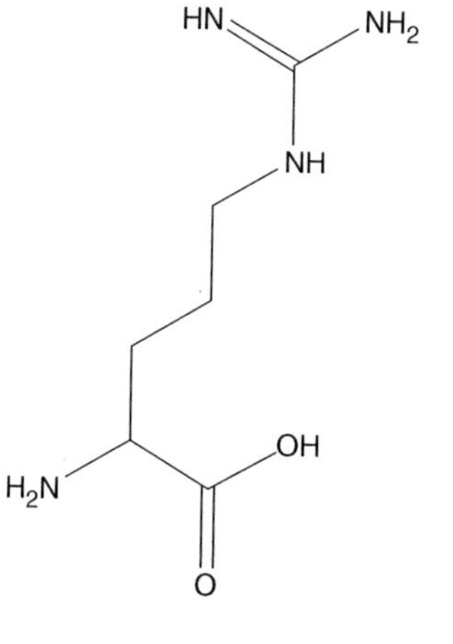

arginine; Arg, R

Arginine possesses a guanidino group on its side chain. This group is strongly basic and will add a proton and be, therefore, a cation under physiological conditions.

Figure 10.1   (*Continued*)

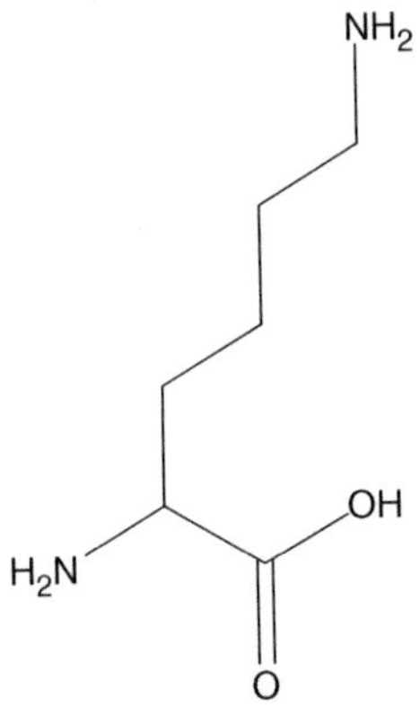

lysine; Lys, K

Lysine has a terminal amino group on its side chain that will add a proton to become a cation under physiological conditions.

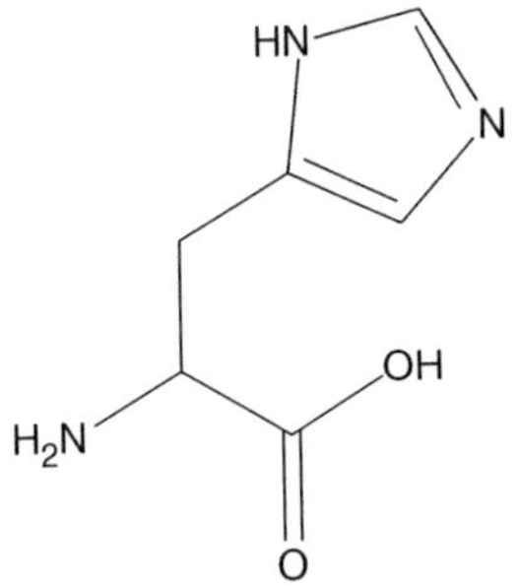

histidine; His, H

Histidine is characterized by a heterocyclic side chain known as imidazole. The imidazole group will bear a positive charge under physiological conditions. Histidine is the metabolic precursor to histamine, a potent inflammatory molecule. Antihistamines work by antagonizing the action of histamine.

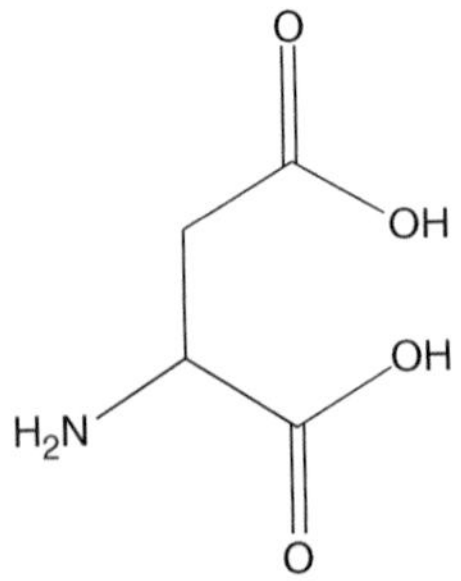

aspartic acid; Asp, D

Figure 10.1 (*Continued*)

Aspartic acid has a side chain carboxyl group that will lose a proton and become an anionic carboxylate group under physiological conditions. Aspartic acid is the metabolic precursor to gamma (γ)-aminobutyric acid (GABA), an important inhibitory neurotransmitter in the human central nervous system.

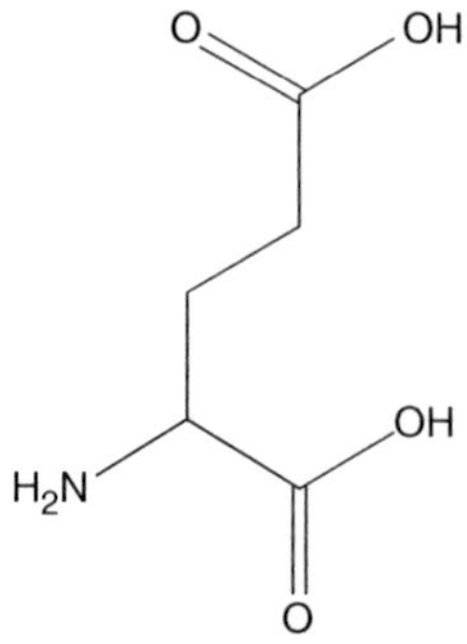

glutamic acid; Glu, E

Glutamic acid also has a side chain carboxyl group that will lose a proton and become an anionic carboxylate group under physiological conditions. Glutamic acid is the most important excitatory neurotransmitter in the human central nervous system.

Figure 10.1   (*Continued*)

Amino acids are essential components of a healthy diet

In the previous chapter, I noted that several of the amino acids are essential in the human diet. For most of these amino acids, humans are incapable of making them from simpler constituents. For a couple of these amino acids, we can make them but not in adequate amounts to sustain good health.

Here are the essential amino acids: arginine, histidine, isoleucine, leucine, lysine, methionine, phenylalanine, threonine, tryptophan, and valine. Of these, humans make some arginine and histidine and these can be omitted from the diet for short periods of time without damage. In the longer run, they are required dietary constituents. Dietary arginine is, however, required at all times for infants and growing children. The dietary requirements for the essential amino acids in young men were established in feeding experiments about 50 years ago.[3] Although the exact requirement varies some from one amino acid to another, 1–2 grams per day for each essential amino acid, other than arginine and histidine, are generally required for a growing person.

In general, proteins from meat and milk have a good balance among the essential amino acids. They are, therefore, sources of high- quality protein. Vegetable and grain proteins are, in general, more widely variable in balance among the essential amino acids. Corn, for example, is deficient in lysine and a diet in which corn is the principal source of protein would not be healthy.[4] That is not a reason not to eat corn. It is a reason to eat other sources of protein along with it. To balance relative amounts of the amino acids, vegetarians generally eat a variety of foods and do very well, as any number of healthy vegetarians will testify.

Protein deficiency has a number of manifestations. Kwashiorkor is a form of malnutrition caused by protein deficiency in a diet that includes fair-to-good calorie intake. This disease occurs largely in cases of famine or limited protein-rich food supply. It is a progressive disease in which symptoms become worse over time. Early symptoms include fatigue, irritability, and lethargy. If inadequate protein intake continues, the symptoms progress to growth failure, loss of muscle mass, edema, and decreased immunity. A large protuberant belly is common. Eventually the disease will progress to coma and death.

Amino acids in proteins are linked by peptide bonds

Now we have the structures of the 20 amino acids that constitute proteins. So the next question is how these get linked together to form the polymer. The answer is that they hook together by joining the amino group of one amino acid to the carboxyl group of another with the elimination of water:

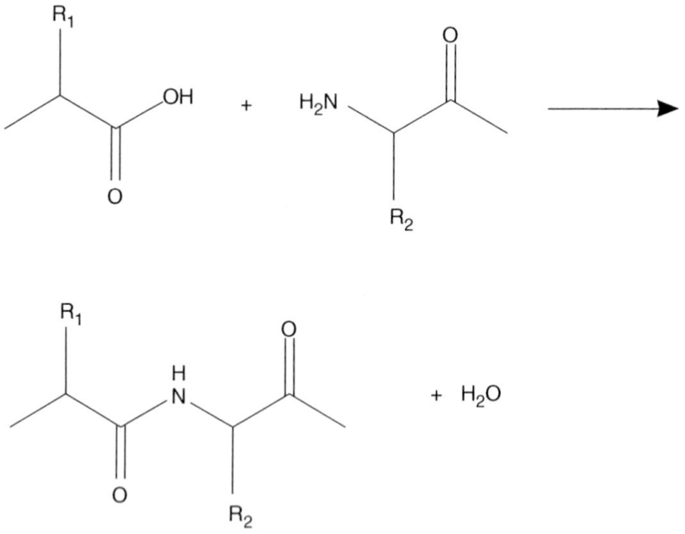

Note that I have used R_1 and R_2 to stand in for undesignated amino acid side chains. In this process, the —OH group from the carboxylic acid and a hydrogen atom from the amine form the product molecule of water. The remaining atoms form the developing protein. Each time that a molecule of an amino acid is added to a growing chain, a molecule of water is eliminated. The four-atom unit which unites the two amino acids is an amide and is termed, for historic reasons, a peptide bond.

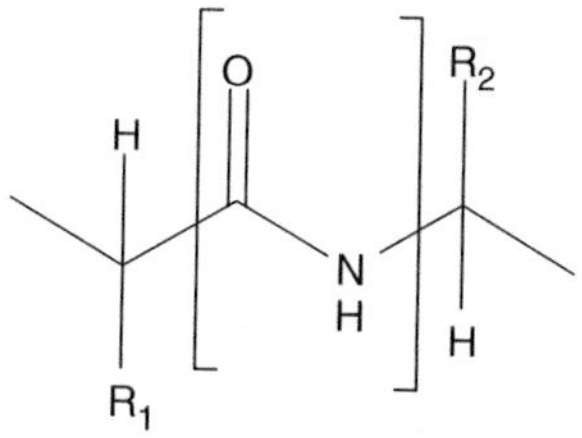

the peptide bond

Proteins are frequently referred to as polypeptides. Of course, we have dipeptides (two amino acids), tripeptides (three amino acids), oligopeptides (a few amino acids), and so forth. A typical dipeptide is Ala-Val:

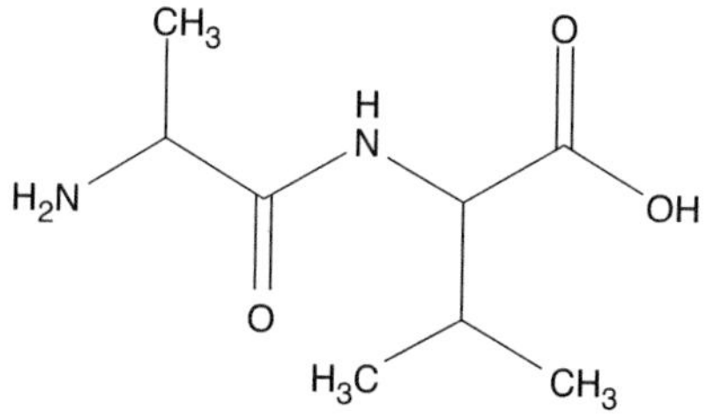

alanyl-valine

By convention, we write the free amino group to the left and the free carboxyl group to the right. How many dipeptides can we make? Well, we can have 20 different amino acids in the first position and 20 more in the second one. So for each amino acid in the first position, we can have 20 in the second: Ala-Val, Ala-Gly, Ala-Leu, Ala-Met, ... It follows that there are $20 \times 20 = 400$ dipeptides. How many tripeptides can we make out of our 20 amino acids? For each of the 400 dipeptides, we can have 20 different amino acids in the third position. So we can have $400 \times 20 = 8000$ tripeptides. There are 160,000 possible tetrapeptides and 3,200,000 pentapeptides. If the chain is n amino acids long, the number of possible structures is 20^n. The next question is how big is n? The fact is that n varies a lot. For insulin, quite a small protein, n is 51. More typically, n is greater than 100 and sometimes much greater. Even if we confine our attention to proteins the size of insulin, there are 20^4 possible structures having 51 amino acids. This is an unimaginably large number.

Therein lies the secret of the diversity of protein functions. There are so many possible protein structures that nature, through the process of evolution, has been able to pick and choose among this cornucopia of possibilities to find the cream of the cream for each function. The number of different proteins in the human body—perhaps 100,000—is an incredibly small fraction of all the proteins that one can construct using 20 natural amino acids linked in chains, say, 100 units long (1 part in 10^{125}).

The sequence of amino acids along the polypeptide chain defines the primary structure of a protein

So far, we know the structures of the 20 amino acids that appear in proteins and how they are hooked together. We also know that there is an enormous number of possible proteins.

In moving forward, it is clear that the most fundamental structure of a protein, termed its primary structure, is defined by the sequence of amino acids along the polypeptide chain. The primary structure demands that we specify which amino acid comes first in the chain, which second, which third, and so on, until we reach the end of the chain.

Bovine insulin was the very first protein to have its primary structure revealed; this was accomplished in 1955. This magnificent achievement is the work of Frederick Sanger and his colleagues at Cambridge University. Sanger won the Nobel Prize in Chemistry in 1958 in recognition for this triumph of science. The primary structure of bovine insulin is shown in figure 10.2. Note that insulin has two polypeptide chains (A and B). Two disulfide bonds hold the two chains together and there is one intrachain disulfide bond as well.

The significance of Sanger's work is immense. It proved for the first time that the structure of a protein is unique: that is, all molecules of bovine insulin, for example, possess the same sequence of amino acids along the polypeptide chains.[5] This sequence has no obvious order, but it is unique. This singular finding requires that there is a genetic code: information encoded in a molecule which specifies the sequence of amino acids in the insulin molecule and, for that matter, in all protein molecules.

We need a few words about the disulfide bonds that link the two chains of insulin.[6] The side chain of Cys is —CH$_2$—SH. The —SH group is termed the sulfhydryl group. If two of these side chains come together, the sulfhydryl groups can be linked together (oxidized) to form a disulfide bond:

$$R-SH + R-SH \rightarrow R-S-S-R$$

The formation of three disulfide bonds in bovine insulin brings parts of the chain that are distant in terms of amino acid sequence into close proximity in the three-dimensional structure of the protein.

Ribonuclease A is a small protein that has 124 amino acids in its single polypeptide chain, molecular mass 13,700. The primary structure of ribonuclease A

Figure 10.2 The primary structure of bovine insulin. This molecule possesses two polypeptide chains, labeled A and B. These are joined by two disulfide bridges between Cys amino acid residues. There is a third disulfide bridge linking two Cys residues in the A chain.

was determined in 1963 by D. G. Smyth, William Stein, and Sanford Moore.[7] It was the second protein for which the primary structure was deduced.

To provide some sense of the rate of progress in biochemistry and the rate at which information is generated, Sanger got the primary structure for insulin, a small protein, in 1955 after several years of dedicated work. We waited 8 years for the next primary structure, that for ribonuclease A, for which a Nobel Prize in Chemistry (in part) was also awarded, in 1972 to Christian Anfinsen, Sanford Moore, and William Stein. There are now tens of thousands of such structures known and the number increases substantially each and every day. The library of known primary structures is collected in several major databases and is accessible to scientists or others who have need of them.[8]

Before we take leave of primary structures for proteins, there is one last, important realization. Proteins serving the same function in different species may have different primary structures. For example, the primary structures of bovine, ovine, and human insulins are not quite the same. They are closely related but not identical. Different species have discovered different protein solutions for the same biological problem.

One might well imagine that the primary structures for proteins serving the same function in different species would most resemble each other for species that are closely related in an evolutionary sense. Such proteins are termed homologs. This turns out to be the case. Cytochrome c is a protein very widely distributed throughout nature. The primary structure for the cytochrome c of humans is very similar to that from pigs but not quite so similar to that from frogs and even more distantly related to that from insects and, at the extreme, that of wheat.

The point is that molecules such as cytochrome c act as a sort of molecular clock for evolution. By comparing the homologies among the primary structures of proteins serving the same function in different species, one can construct evolutionary trees. Assuming that we know something about the mutation rate, we can assign approximate times at which different branches on that tree diverged. Proteins serve as molecular fossils.

Extraordinarily, fragments of the protein of connective tissue known as collagen have recently been extracted from bones of *Tyrannosaurus rex*. The primary structure of these fragments strengthens the suggestion that birds are our closest extant relatives to dinosaurs. This finding is certain to set off a determined effort to recover additional protein material from ancient fossils.

The primary structure of proteins is not the whole story. To really understand how proteins work, we have got to understand them as three-dimensional objects. So on to higher dimensions in the next chapter. But first, a few paragraphs about another role for the protein amino acids: biosynthesis.

Amino acids are important metabolic precursors

Amino acids have a larger role to play in life than serving as the building blocks of proteins, itself an absolutely critical role. Specifically, several of the amino acids are key metabolic precursors to families of products. That is, these amino acids are the starting point for one or a series of chemical transformations, all catalyzed by enzymes, leading to other important molecules of life. Let's have a look at a few examples.

Phenylalanine and tyrosine provide interesting and important examples. The addition of a hydroxyl group to phenylalanine to yield tyrosine is unusually important. An enzyme known, appropriately, as phenylalanine hydroxylase catalyzes this reaction. Mutations in the gene coding for that enzyme that lead to an inactive enzyme underlie the genetic disease known as phenylketonuria, or PKU. In fact, there are about 400 known mutations in the human gene that codes for this enzyme that yield an inactive enzyme. People who have one of these mutations cannot convert phenylalanine to tyrosine. One consequence is that abnormally high concentrations of phenylalanine accumulate and this is toxic in the young. Untreated people with PKU suffer from profound mental retardation. In the United States, every child is checked for PKU at birth by assaying for a metabolite of phenylalanine excreted in the urine. Infants found to have PKU are managed by restricting the dietary intake of phenylalanine and monitoring the blood level of phenylalanine to ensure that it stays within the normal range. Phenylalanine is an essential amino acid but, in excess, it is also toxic to the central nervous system of the young. Restriction of phenylalanine intake continues until the patient is perhaps 10 years old. Phenylalanine toxicity goes away once children reach this age. PKU accounts for something that you may have noticed. On bottles or cans of diet sodas, you will frequently see the phrase "contains phenylalanine." The phenylalanine comes from metabolism of the sweetening agent aspartame. This phrase is an alert for PKU patients.

Tyrosine is a precursor of thyroid hormones as well as L-dopa. Both thyroxine and L-dopa are employed in clinical medicine: thyroxine to treat hypothyroid patients and L-dopa to treat patients with Parkinsonism. L-dopa is also the precursor to the pigment of the skin known as melanin. The enzyme that catalyzes the transformation of tyrosine into L-dopa, tyrosine hydroxylase, also catalyzes the transformation of L-dopa into melanin. Albinism is a genetic disease in which a mutation in the gene encoding tyrosine hydroxylase results in an inactive enzyme. People with albinism have no pigment in their skin, hair, or retina.

Tyrosine is also the metabolic precursor to the neurotransmitter dopamine and the catecholamine hormones norepinephrine (noradrenaline) and epinephrine (adrenaline), as well as to the alkaloids in opium, including morphine.

Tryptophan is also an important starting point for biosynthetic reactions. The decarboxylation of tryptophan yields tryptamine, a molecule found in very low concentrations in the mammalian brain where it may function as a neurotransmitter or neuromodulator. It is found in high concentrations in some cheeses.

Metabolism of tryptophan leads to an important neurotransmitter, serotonin. Adding a couple of methyl groups to the amino function of serotonin creates bufotenine, a psychoactive compound found in frog skin and some beans. In point of fact, there are a number of derivatives of tryptamine that are psychoactive.

Further chemical changes lead to psilocin, another psychoactive molecule, and finally psilocybin, a well-known hallucinogen found in "magic mushrooms."

The list of tryptophan-derived molecules could be extended almost without end; tryptophan is the parent molecule of the indole alkaloids, a large group of plant compounds.

That is enough for now. We move on to look at proteins as three-dimensional objects in the next chapter.

Key Points

1. Proteins are polymers of L-α-amino acids.

2. The essential amino acids that are required in the diet in adequate amounts for good health include Arg, His, Ile, Leu, Lys, Met, Phe, Thr, Trp, and Val.

3. In proteins, amino acids are linked together by peptide bonds, amides formed by the union of the amino group of one amino acid and the carboxyl group of the other.

4. The sequence of amino acids along the polypeptide chain defines the primary structure of a protein.

5. Proteins serving the same function in different species may have different primary structures. The extent of the differences is a measure of their relatedness on the evolutionary tree of life.

6. Several of the protein amino acids are important starting points for biosynthetic sequences, leading to important product molecules.

11

Proteins are three-dimensional objects

Under physiological conditions, nearly all proteins have at least one well-defined three-dimensional structure and the biological activity of that protein depends on that structure.

Chemists are not quite as odd as they are frequently depicted. Given that dedicating a professional life to moving around molecules one way or another may seem a bit wide of the mark to most people, chemists do have a life outside the laboratory. In my experience, chemists have more than the usual affinity for the arts: music and painting in particular.

Sometimes the science and the art come together in productive ways. I have particularly in mind a collaboration established years ago between Richard Dickerson, a protein chemist at Caltech, and Irving Geis, a talented painter. Dickerson determined structures and Geis illustrated them. One product is the coauthored book, *The Structure and Action of Proteins,* published in 1969.[1] It is full of useful and lovely depictions of protein structure. Although the modern capabilities of molecular graphics are great for presenting structures, the original paintings of Irving Geis continue to inform and intrigue. This chapter is enlivened by several examples of his work.

In the last chapter we established that the primary structure for all members of a particular protein—say ribonuclease A or insulin—in a single species is identical. We owe that key insight to Fred Sanger. This concept translates to higher dimensions: the primary structure of a protein is basically one-dimensional, a chain of symbols.

This chain can wrap itself up in space in innumerable ways. Here we learn that, for example, all ribonuclease A molecules from a particular species wrap themselves up in exactly the same way: an amazing and amazingly important fact.

Proteins have unique three-dimensional structures

Once we know the primary structure for a protein, we have the basic aspects of the structure in hand. We know the sequence of amino acids along the polypeptide backbone. Even more basically, we know which atom is bonded to which for all atoms in the protein. It is like knowing that the carbon atom is bonded to four hydrogen atoms in the structure of methane, but on a vastly larger scale.

We are now in a position to ask a more sophisticated question related to protein structure: what is the three-dimensional structure of the protein? This question assumes something that is by no means obvious: that a protein has a unique three-dimensional structure. Think of the protein ribonuclease A as 124 beads on a string. You can imagine that it could fold up in a great many ways, just like you can fold up a string of beads in a great many ways.

The following point is of great significance: *under physiological conditions, nearly all proteins have one predominant, well-defined three-dimensional structure and the biochemical activity and biological function of that protein depends on that structure.* If the three-dimensional structure of the protein is lost, so is the biological activity.

Before we get further into three-dimensional structures for proteins, I need to remind you about hydrogen bonds. We first encountered these back in chapter 6 when we invoked them to account for the remarkable properties of water. Hydrogen bonds can form when a hydrogen atom is shared between atoms that may be nitrogen, oxygen, or fluorine. Put another way, here is a model for a hydrogen bond: $X—H•••:Y$, where X and Y are N, O, or F. The hydrogen atom is shown tightly associated with X and more loosely tied to the unshared electron pair of Y in this model. Individual hydrogen bonds are rather weak interactions. However, when a substantial number of hydrogen bonds are formed, they can account for important properties of molecules or more complex substances, as in the case of water. In what follows, we shall see that the hydrogen bonds play a critical role in determining the three-dimensional structures for proteins.

Hydrogen bond formation drives protein secondary structure

An important element in the three-dimensional structure of a protein is the secondary structure. The secondary structure results from the formation of hydrogen bonds between the —N—H groups and the carbonyl ($C=O$) groups of the peptide bonds: —N—H•••:O=C. There are two basic ways to do this. We can form a helix or we can form a sheet. The great American chemist Linus Pauling won the Nobel Prize in Chemistry in 1954 for the elucidation of these structures.

The helical structure that commonly occurs in proteins is termed the α helix. A model of the α helix is provided in figure 11.1. Note that the hydrogen bonds lie roughly parallel to the long axis of the helix. The amino acid side chains bristle out from the helix. Things are just the opposite for the sheet structure, termed the β sheet. A model of the β sheet is provided in figure 11.2. Here the hydrogen bonds may be formed between different polypeptide chains or between the segments of a single polypeptide chain that has looped back on itself. Both things happen. The hydrogen bonds now are perpendicular to the long axis of the polypeptide chains. The amino acid side chains alternate between being above and below the main plane (actually a pleated sheet) of the structure.

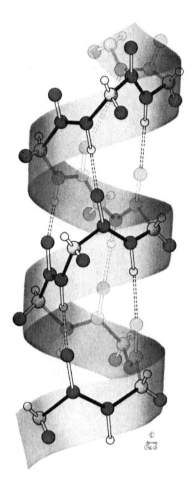

Figure 11.1 A polypeptide chain folded into a helical configuration called the α helix. The main course of the helix is shown by the shaded ribbon. Hydrogen bonds between the N—H and the C=O groups along the polypeptide chain are indicated by dashed lines. There is an amino acid every 1.5 Å along the helical axis. The distance along the axis required for one turn is 5.3 Å, giving 3.6 amino acids per turn. (Illustration, Irving Geis/Geis Archive Trust. Copyright Howard Hughes Medical Institute. Reproduced with permission.)

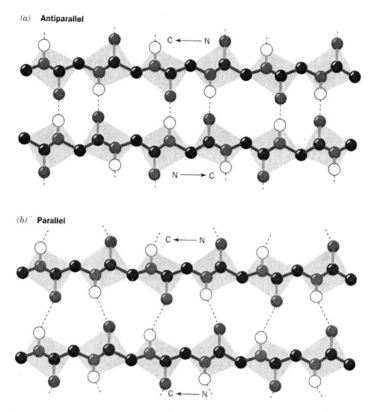

Figure 11.2 Polypeptide chains held together by hydrogen bonds, indicated by dashed lines, in a configuration called a β sheet. (a) The antiparallel β pleated sheet. (b) The parallel β pleated sheet. (Illustration, Irving Geis/Geis Archive Trust. Copyright Howard Hughes Medical Institute. Reproduced with permission.)

Helical and sheet structures in proteins may be interspersed with loops, which are neither. Individual proteins vary greatly in their content of helical and sheet structures. Some are mostly helical, some mostly sheet, some contain a good deal of both, and others contain little of either. Nature provides all possibilities for our enjoyment.

The position of each protein atom in space defines its three-dimensional structure

Consider a protein composed of a single polypeptide chain, remembering that some proteins have more than one. The tertiary structure defines the position of each atom in three-dimensional space. Thus, the tertiary structure includes secondary structural elements, helices and sheets, as well as the spatial distribution of the amino acid side chains of these elements and the positions of each atom in the loops of polypeptide chains linking secondary structural elements.

Myoglobin is the first protein for which we had knowledge of protein tertiary structure. This monumental accomplishment is the work of two crystallographers. The first is John Kendrew, who actually completed the myoglobin structure in 1959.[2] The second is Max Perutz, Kendrew's mentor. Perutz initiated work on the structure for hemoglobin a few years earlier than Kendrew got started on myoglobin.[3] However, myoglobin is the simpler molecule and Kendrew completed work a few years ahead of Perutz' work on hemoglobin. Elucidation of the structure of myoglobin and hemoglobin capped three decades of scientific effort. Kendrew and Perutz shared the Nobel Prize in Chemistry in 1962 in recognition of their accomplishments.

Myoglobin is a protein found in mammalian muscle. It is an oxygen-storing protein of middling size, containing a single chain of 153 amino acids, molecular mass about 17,500. Myoglobin contains a single atom of iron, bound in a complex structure termed heme and to which a single molecule of oxygen binds. In humans, myoglobin is found in highest concentration in the heart muscle. We might have guessed this since one wants to keep the heart supplied with a reserve of oxygen. However, the most abundant sources of myoglobin come from the muscles of diving mammals (whales and dolphins), whose myoglobin-linked stores of oxygen permit them to submerge for substantial periods of time.

Chemists have adopted several models for depicting protein structure. These include depiction of all the heavy atoms of the protein (the hydrogen atoms are deleted to keep the model simple enough to be understandable), emphasis on the backbone formed by the α carbon atoms, and ribbon diagrams that illustrate secondary structural elements. These three models for sperm whale myoglobin are collected in figure 11.3.

We should not overlook one central finding established by the structure of myoglobin. Just as the primary sequence of each myoglobin molecule is the same, so is the three-dimensional structure. Each polypeptide chain organizes itself in space in just the same way. This is highly important since, as noted above, the personality of the protein is encoded in its three-dimensional structure. If the three-dimensional structure is perturbed in some way, the biological properties of the protein are likely to be altered or perhaps lost entirely. Enzymes lose catalytic activity. Antibodies may no longer recognize the antigen that elicited them. Receptors may fail to recognize their ligand. Myoglobin may fail to bind and store oxygen, and so forth.

Here too we have an example of the rate at which scientific information is created currently. The earliest three-dimensional protein structures were products of decades of work. In the succeeding four decades, we have defined three-dimensional structures for more than 5000 proteins and complexes of proteins with small molecules. They occupy huge publicly available databases. They grow daily. This information is of great importance for several purposes. Among these is the design of novel molecules useful in human clinical medicine.

The elegant models of three-dimensional protein structures, such as those shown in figure 11.3, fail in one respect: they provide a sense of a static molecule in space. As we learned from very simple structures such as ethane, molecules are dynamic, changing conformations in space rapidly. This is surely true for proteins as well as small molecules. There is a sophisticated computational technique known as molecular dynamics that permits us to have a realistic sense of the motions of protein molecules over time. A "snapshot" of a protein is recorded at very short time intervals, typically each femtosecond (10^{-15} second), and these are played as a sort of motion picture. A protein is seen to jiggle, wobble, flex, rotate, and dance over time: a very dynamic structure indeed. The static models of proteins provide a time-averaged view of protein structure.

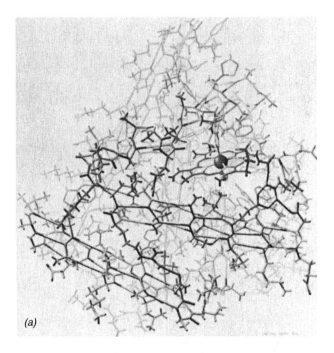

(a)

Figure 11.3 Three representations for the structure of sperm whale myoglobin. (a) The protein and its bound heme group are drawn in stick form, showing only the heavy atoms. The iron atom at the center of the heme group is large and dark, easily identifiable. It has a water molecule bound to it. In this one-of-a-kind painting of the first known protein structure, Irving Geis has employed "creative distortions" to emphasize the key structural features of myoglobin, in particular the α-helices. (b) A diagram in which the protein is represented by its chain of α carbon atoms, shown as balls. These are consecutively numbered from the N-terminus. The 153-residue polypeptide chain is folded into eight α-helices, designated A through H. The protein's bound heme group is shown in the center. Its central iron atom is large and dark. There are two side chains of His residues closely linked to the iron atom. One of the heme side chains has been displaced for the sake of clarity. (c) A computer-generated cartoon drawing in an orientation similar to that of part b, emphasizing the secondary structure of myoglobin. The heme group with its bound oxygen molecule and its two associated His side chains are shown in ball-and-stick form. (Parts a and b are based on an X-ray structure by John Kendrew, MRC Laboratory of Molecular Biology, Cambridge, UK. Illustrations, Irving Geis/Geis Archive Trust. © Howard Hughes Medical Institute. Reproduced with permission. Part c is based on an X-ray structure by Simon Phillips, University of Leeds, Leeds, UK.)

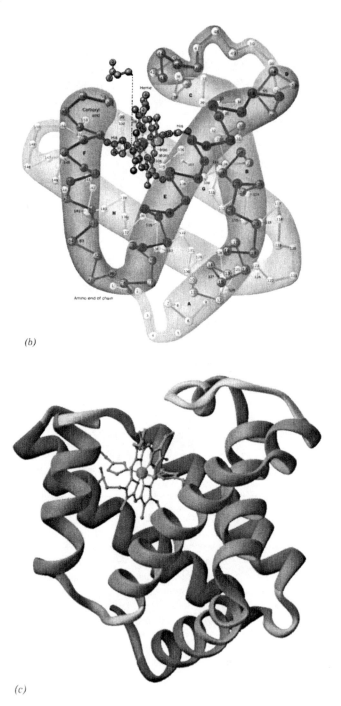

(b)

(c)

Figure 11.3   *(Continued)*

Hemoglobin provides an example of a protein possessing quaternary structure

The apex of protein structural organization is revealed in molecules such as hemoglobin that contain more than one polypeptide chain. Specifically, hemoglobin contains four such chains, two each of two types, α chains and β chains. Each polypeptide chain in hemoglobin has its own tertiary structure. Each polypeptide chain may be thought of as a subunit of hemoglobin. The hemoglobin molecule is held together by interactions among the individually folded subunits. The quaternary structure of proteins is defined by the spatial arrangement of these subunits. Hemoglobin is by no means unique in possessing quaternary structure. In fact, most proteins consist of subunits and have, therefore, quaternary structure.

The hierarchy of protein structure is illustrated in figure 11.4. Here too we have a wealth of structural information. The quaternary structures for many proteins are now known and generally available in databases. As complex as these are, this is not the end of the story. We have atom-by-atom structures for entities as complex as viruses and the ribosome, an intracellular RNA-protein complex and the site of protein synthesis. Modern structural biology continues to provide detailed insights into some of the most complex constructs of nature. We are better off for having these insights.

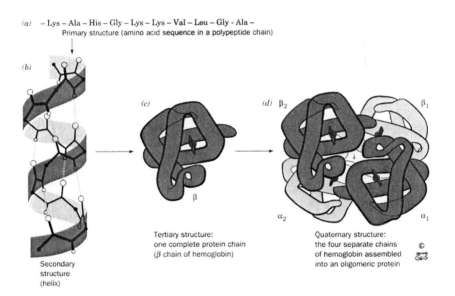

Figure 11.4 The structural hierarchy in proteins. (a) A segment of primary structure; (b) secondary structure illustrated as a segment of alpha helix; (c) tertiary structure in which helices are interspersed with coils, and (d) quaternary structure. (Illustration, Irving Geis/Geis Archive Trust. Copyright Howard Hughes Medical Institute. Reproduced with permission.)

The primary structure of a protein determines the three-dimensional structure

The biological power of proteins is coded in their three-dimensional structures: how the helices, sheets, and loops are organized in space. This organization controls the biochemical activities of the proteins. The three-dimensional structures are, in turn, coded in the primary structures: the amino acid sequences along the polypeptide chain. The essential demonstration of this fact comes from an experiment of the following sort, employing pure preparations of a single protein.[4] We have noted above that under physiological conditions proteins are folded in complex ways. Under nonphysiological conditions, they can be unfolded to yield an ensemble of floppy, unstructured polypeptide chains. In the unfolding process some or all of the secondary and tertiary structure is lost. If the physiological conditions are restored, each polypeptide chain assumes the same three-dimensional structure that it originally had. Since there is no agent present to dictate how the unfolded protein chain will refold, the refolding process must be spontaneous. Each polypeptide finds its way back to its most stable structure: helices and sheets intact. Once the primary structure of the polypeptide chain is determined, the three-dimensional structure and, hence, the biological properties are determined as well.

It follows that a change in the primary structure may result in a change in three-dimensional structure and biological properties. We know that this is not necessarily the case. Insulins from different mammalian species have somewhat different sequences but the same biological function. Cytochrome c from a great many different species has many changes in primary structure but the same biological function.

In contrast, we also know of a great many cases in which a very small change in primary structure has major implications for biological function: sickle cell anemia is a prominent example to which I return shortly. So we have a question. How are we to know when a change in primary structure is likely to alter the three-dimensional structure and biological properties of a protein? To get at that question, we need to understand more about why a protein folds up in the way that it does.

Some amino acid replacements are conservative; some are not

Let's go back to the classification of the amino acid side chains (chapter 10). Several of these are hydrocarbons: greasy, fat-like side chains. Look at the side chains of leucine, isoleucine, valine, and phenylalanine as examples. Fats and water do not mix. Fats are termed hydrophobic, water-hating. It follows that these side chains will try to do what they can to get out of contact with water. The obvious way to do this is to hide in the center of the protein structure, where they may enjoy the company of like-minded amino acid side chains and avoid that of water molecules—sort of a molecular ethnic cleansing.

At the other end of the spectrum are a number of amino acid side chains that are charged under physiological conditions: those of aspartic acid or lysine, for example. These side chains will try to insist on being in contact with water. In addition, there are a number of amino acid side chains that are polar but not charged: consider those of serine, asparagine, and threonine. These side chains too will prefer to be exposed on the surface of the protein molecule, in contact with the water environment.

The forces that cause proteins to fold up in the way that they do are complex and not fully understood. At the same time, the basic idea just introduced is clearly quite important. One finds the charged amino acid side chains almost always on the protein surface and the highly greasy ones generally in the interior of the structure. At the simplest level, one can view a protein molecule as folding up to form a tiny oil droplet surrounded by charged or otherwise polar groups on its surface. This is all that we need to know to explain a good deal about certain human diseases.

Diseases of people come in many flavors. There are infectious diseases (measles, mumps, influenza, AIDS, ...), nutritional deficiency diseases (scurvy, beriberi, kwashiorkor, ...), degenerative diseases (Alzheimer's disease, osteoporosis, ...), cancer (of the lung, breast, prostate, liver, ...), and single-gene inherited diseases or molecular diseases. In the last category, an important and instructive example is provided by sickle cell anemia. Let's consider this disease and begin to develop a sense of how we can understand it on the basis of what we now know about proteins.

We begin with the symptoms of sickle cell anemia. As the name implies, victims are frequently anemic: that is, they have a content of hemoglobin in blood less than the normal range, the result of lysis of red blood cells, the carriers of hemoglobin (hemolytic anemia). In addition, disease victims are susceptible to chronic infections, may have enlarged spleens, and suffer intermittent bouts of pain, which can be severe, in the bones, joints, and periosteum. The disease can be debilitating.

In the United States, sickle cell anemia is particularly common among African-Americans. About 8% of this population carries a gene for sickle cell anemia. There is an understandable reason for the high prevalence of sickle cell anemia in this population. Individuals harboring a sickle cell anemia gene have a built-in resistance to malaria, for reasons that are not entirely clear, a disease common in sub-Saharan Africa. Hence, carriers of the sickle cell anemia gene had a selective advantage in this region of the world. The frequency of carriers consequently increased over time through an evolutionary process. These people are the ancestors of much of the current African-American community in the United States.

Hemoglobin is a complex protein molecule whose special task in life is to act as a carrier of oxygen in humans and many other species. Hemoglobin occurs only in the blood and is confined to the red blood cells. Indeed, red blood cells are pretty much wall-to-wall hemoglobin. It is a red protein and is responsible for the color of blood. As the blood circulates through the lungs, hemoglobin picks up oxygen from the inhaled air. Typically, hemoglobin is about 96% saturated with oxygen as it leaves the lungs: that is, it is carrying about as much oxygen as it is capable of carrying. As the oxygenated blood circulates through the tissues, hemoglobin gives up some of its oxygen, required to support the life of the tissues. It returns to the lung partially depleted of oxygen, is repleted with oxygen in the lungs, and continues back to the tissues to again deliver life-sustaining oxygen. Clearly hemoglobin is a molecule essential for life in people.

Victims of sickle cell anemia have a subtle but important error in their hemoglobin molecule. This disease is, consequently, referred to as a hemoglobinopathy. There are many other genetic diseases in this category. To understand the subtlety of the molecular error in sickle cell anemia, consider that the composition of normal hemoglobin is $C_{2954}H_{4516}N_{780}O_{806}S_{12}Fe_4$ while that of sickle cell hemoglobin is $C_{2954}H_{4518}N_{780}O_{804}S_{12}Fe_4$.[5] The very modest difference in composition reflects a single, small, localized change in hemoglobin gene structure that results in substitution of a valine where a glutamic acid should be. For the moment, it is enough to understand that a very small change in a critical molecule can lead to a serious disease that afflicts a substantial number of people, here and abroad. It is worth understanding the molecular basis for sickle cell anemia in terms of the chemistry of proteins.

A nonconservative amino acid replacement underlies sickle cell anemia

We are now in a position to understand the molecular nature of sickle cell anemia. We need to remember that amino acids come in three flavors: nonpolar, charged (highly hydrophilic polar), and uncharged polar. We also need to remember that proteins are organized in a way that hides most of the nonpolar amino acid residues in the molecular interior and exposes most of the charged, in particular, and uncharged polar residues on the molecular surface.

Now we can ask what is likely to happen to the three-dimensional structure of a protein if we make a conservative replacement of one amino acid for another in the primary structure. A conservative replacement involves, for example, substitution of one nonpolar amino acid for another, or replacement of one charged amino acid for another. Intuitively, one would expect that conservative replacements would have rather little effect on three-dimensional protein structure. If an isoleucine is replaced by a valine or leucine, the structural modification is modest. The side chains of all of these amino acids are hydrophobic and will be content to sit in the molecular interior. This expectation is borne out in practice. We have noted earlier that there are many different molecules of cytochrome c in nature, all of which serve the same basic function and all of which have similar three-dimensional structures. We have also noted the species specificity of insulins among mammalian species. Here too we find a number of conservative changes in the primary structure of the hormone. Although there are exceptions, as a general rule conservative changes in the primary structure of proteins are consistent with maintenance of the three-dimensional structures of proteins and the associated biological functions.

On the other hand, we can imagine nonconservative replacements. In this case, a charged amino acid might replace a nonpolar one or the converse. Less dramatically, but important nonetheless, is the replacement of an uncharged polar residue by a nonpolar one or the converse. It is easy to imagine that a nonconservative replacement of one amino acid by another in the primary sequence of a protein might have substantial consequences for the three-dimensional protein structure and associated biological activity. After all, a residue that seeks a spot on the molecular surface replaces one that was content with a spot in the molecular interior or vice versa. Structural reorganization is likely to occur.

That is exactly what happens in sickle cell anemia. In the two β chains of hemoglobin, a unique position in the primary structure occupied by glutamic acid, a charged amino acid, is replaced by valine, a nonpolar amino acid. The consequence of this nonconservative change is that the deoxygenated form of the mutant hemoglobin molecule tends to aggregate within the red blood cell. Its ability to serve as an oxygen carrier to tissues is compromised. Sickle cell anemia is the result.

This is a beautiful example of how we can come to understand the molecular basis of an important human disease on the foundation of a few basic ideas about chemistry.

Many proteins contain nonprotein constituents

Let us return to the cases of myoglobin and hemoglobin and recognize something explicitly: these two proteins contain a nonprotein constituent. This constituent, or prosthetic group, is heme, a complex polycyclic structure, protoporphyrin IX, containing an atom of iron (as Fe (II) or Fe^{2+}) at its center.

The heme group is vital to the function of myoglobin and hemoglobin. The site of attachment of molecular oxygen to these proteins is the iron atom of the heme moiety. In fact, if the Fe^{2+} in heme is oxidized to Fe^{3+}, both myoglobin and hemoglobin lose their capacity to bind oxygen and store it or to deliver it to tissues. This is another dramatic example of the effect of small changes in molecular structure, in this case loss of a single electron, on its biological properties.

The cases of myoglobin and hemoglobin are not rare. Many enzymes are dependent for their function on the presence of a nonprotein group. For example, cytochrome c also contains a prosthetic group similar, but not identical, to heme, as do a number of other proteins. These are known generically as heme proteins. There is a family of enzymes that contain a flavin group, the flavoproteins. Another family contains pyridoxal phosphate, a derivative of vitamin B_6. There are a number of other examples.

There are three cases of proteins containing nonprotein components in addition to those just mentioned: metalloproteins, glycoproteins, and lipoproteins. More about glycoproteins and lipoproteins follows later in this book when we get around to sugars and lipids.

Metalloproteins contain a family of metal ions

Back for moment to the content of my daily vitamin and mineral pill: I noted earlier a substantial list of metals among the ingredients. Many of these metals are components of metalloproteins, establishing a rationale for their essentiality in the human diet.

Many proteins, including many enzymes, contain tightly bound metal ions. These may be intimately involved in enzyme catalysis or may serve a purely structural role. The most common tightly bound metal ions found in metalloproteins include copper (Cu^+ and Cu^{2+}), zinc (Zn^{2+}), iron (Fe^{2+} and Fe^{3+}), and manganese (Mn^{2+}). Other proteins may contain weakly bound metal ions that generally serve as modulators of enzyme activity. These include sodium (Na^+), potassium (K^+), calcium (Ca^{2+}), and magnesium (Mg^{2+}). There are also exotic cases for which enzymes may depend on nickel, selenium, molybdenum, or silicon for activity. These account for the very small requirements for these metals in the human diet.

Iron has a greater role to play in human physiology than the storage and transport of oxygen by myoglobin and hemoglobin. There are many iron metalloproteins and these are frequently involved in promoting electron transfer reactions, including those of the electron transport chain of mitochondria, responsible for the synthesis of most of the body's ATP.

Here is a variation of the theme of metalloproteins. Most enzymes that catalyze reactions involving ATP require magnesium ion, Mg^{2+}. However, in most cases, the magnesium ion is not bound to the protein but to the pyrophosphate groups of ATP. Put another way, the real substrate for these enzymes is not ATP but ATP•Mg. There are a great many reactions in this category, establishing a rationale for the need for the magnesium ion in the human diet.

That is enough for now about proteins. Let's turn our attention to the building blocks of the nucleic acids and relate these to nucleic acid structure and function.

Key Points

1. The biological activity of proteins depends critically on their three-dimensional structure.

2. Helical and sheet structures define secondary structures for proteins.

3. Proteins have unique three-dimensional structures: the position of each atom in space defines this structure, known as the tertiary structure.

4. Some proteins contain more than one polypeptide chain. These are held together by noncovalent chemical forces. Such a structure is known as a quaternary structure.

5. The primary structure of proteins determines the three-dimensional or tertiary structure.

6. Amino acid replacements in proteins may be conservative: that is, an amino acid of one class is replaced by another of the same class, e.g., alanine is replaced by valine. Conservative replacements frequently have little impact on protein structure or function.

7. Amino acid replacements in proteins may be nonconservative: that is, an amino acid of one class is replaced by one of another class, e.g., alanine is replaced by glutamic acid.

8. Many proteins have nonprotein components: these may be metal ions, carbohydrates, lipids, or complex structures such as heme.

12

Nucleotides are the building blocks of nucleic acids

The stuff of genes

DNA is the master substance of the cell, the chemical embodiment of the evolved program of life; it is the stuff of genes, the units of inheritance. DNA is composed of a complementary double helix of nucleotides, the building blocks.

Cells are the basic building blocks of living organisms. For some, the cell is the living organism: bacteria or algae for example. For others, such as humans, the organism contains a great number of cells (many trillion in the human body) of many different sorts—about 200 unique cell types in humans.[1]

There is no such thing as a typical cell. They come in a variety of sizes, shapes, and structures, each suited to a particular function or family of functions. Cells are isolated from their environment by a cell membrane that regulates what gets in and what does not. At present, we need to know just two things about cells.

First, each cell type expresses a unique biochemical personality. Liver cells are biochemically distinct from nerve cells and both are distinct from cells of a pneumococcal bacterium. The unique personalities are expressed in the spectrum of proteins that inhabit the cell and the evolution of this spectrum over the lifetime of the cell. This should not be surprising given the multitude of functions of the proteins, as described in chapter 9. But it does raise a question: what sort of cellular device determines the spectrum of proteins in a certain cell at some point in time?

Secondly, the biochemical personality of cells is preserved upon cell division: that is, the daughter cells are capable of expressing the same specific set of proteins that the parent cell did. Put more broadly, genetics works. Genetics tells us that many properties—inherited properties—are passed down from parent to progeny: think of eye color, hair color, blood type, and the like. The inheritance of these properties is not always simple, but they are inherited. This raises a second question: what cellular device is responsible for preserving and transmitting a unique biochemical personality upon cell division and organismal reproduction?

These cellular devices are molecules. In fact, both devices are DNA, deoxyribonucleic acid.

DNA is the master substance of the cell

DNA codes for its own synthesis at the time of cell division. Thus, DNA acts as the agent of inheritance. As is developed below, DNA is a double-stranded helical molecule—the famous double helix—in which the two strands are complementary. DNA is the repository of information that is expressed in synthesis of the proteins of the cell. Therefore, DNA acts as the determinant of the biochemical personality of the cell.[2]

Genetics is the science of inheritance of properties in living organisms. The units of genetic inheritance are called genes. I will get around to a formal molecular definition of a gene later. At the level of the molecule, genetics is based on the structure and function of DNA. DNA is the genetic material. Genes are made of DNA. The structure and function of DNA are unified in an amazingly elegant way. It is the central goal of this chapter to develop this structure/function relationship. I will also develop the structure and function of DNA's close relative RNA, ribonucleic acid. In fact, in some viruses (retroviruses, viruses containing single-stranded RNA as genetic material), it is RNA, rather than DNA, that is the basic repository of genetic information.

Back in the not-so-good old days, scientists speculated and experimented about the molecular nature of the genetic material—the stuff of genes—for several decades.[3] Prior to 1944, and actually for some years beyond that, most scientists believed that genes were proteins. This was a reasonable point of view given the many roles that proteins were already known to serve. But it is not so and Oswald Avery (1877–1955) is the person responsible for telling us, elegantly and persuasively, that it is not so.

Nucleic acids have long been recognized as one of the four preeminent classes of organic molecules characteristic of living organisms, along with proteins, carbohydrates, and lipids. Specifically, a Swiss scientist, Johann Friedrich Miescher, discovered DNA in 1869. He isolated a complex of DNA and protein from the nuclei of white blood cells of pus, collected from discarded surgical bandages. The next year he found a better source (salmon sperm) and isolated pure DNA from it. If pus and salmon sperm seem like exotic things to study, it turns out that many things of overwhelming significance have been discovered in places where most people would not have thought to look. Guanine, one of the constituents of DNA and RNA, was first isolated from the excreta of birds (guano), for example. Politicians are occasionally driven to ridicule science and scientists for doing what may seem like useless tasks: perhaps for studying pus or salmon sperm. Regrettably, they have no clue about how science moves forward. They make a lot of unnecessary trouble out of ignorance and a desire for publicity.

We had DNA, and later RNA, for a long time during which no one was quite sure what roles they played in life. A minority view held that DNA might be the stuff of genes but for no better reason than others held the protein view. That changed in 1944 when Oswald Avery and his collaborators, Colin MacLeod and Maclyn McCarty, working at the Rockefeller Institute, now the Rockefeller University, published the results of an elegant experiment. Working on pneumococcal bacteria, here is what they did.[4]

Pneumococcal bacteria come in a number of types, readily distinguishable by standard sorts of immunological manipulations. The various types synthesize different polysaccharides (polymers of sugars) and expose these on the cell surface, where they can be detected and identified. When a pneumococcus Type II divides, it produces more pneumococcus Type II. The same goes for pneumococcus Type III and all the other types; there are quite a bunch of them. Now, if you take a collection of Type II pneumococci and mix them with an extract of killed Type III pneumococci, some of the progeny turn out to be Type III. Some of the Type II pneumococci have been transformed into Type IIIs. Furthermore, the progeny of these Type IIIs are also Type III and this continues for as many generations as anyone is willing to study. It follows that a permanent, heritable property has been transmitted to the Type IIs by something in killed Type IIIs. What Avery, MacLeod, and McCarty did was to exhaustively purify a Type III extract and prove, beyond reasonable doubt, that the transforming substance is DNA. This is one of the defining experiments of twentieth century biology.

It was no easy task, consuming years of careful, dedicated work. In a letter to his brother, Avery notes: "To try to find in that complex mixture, the active principle!! Try to isolate and identify the particular substance that will by itself when brought into contact with the R cell derived from Type II cause it to elaborate Type III capsular polysaccharide, and to acquire all the aristocratic distinctions of the same specific type of cells as that from which the extract was prepared! Some job—full of headaches and heart breaks. But at last perhaps we have it."

They did indeed have it. The pity is that it took others several years to recognize that they had it. In 1950, six years after Avery's beautiful experiment, many scientists still held the view that protein was the genetic material. Oswald Avery died in 1955, before he could be properly honored with the Nobel Prize for his work. Nobel Prizes are not awarded posthumously.

There are two basic functions of DNA as genetic material: it codes for its own duplication and it codes for the synthesis of RNA and proteins, accounting for inheritance and cell specificity and personality. We are going to return to these functions later, but the time has come to develop the structures for DNA and RNA and relate these structures to their functions.

Nucleotides are the building blocks of the nucleic acids

Like proteins, the nucleic acids are polymers. DNA and RNA are each composed of four different monomers, just as proteins are composed of 20 different monomers. These are called *nucleotides*.

A nucleotide is composed of three parts: a nitrogen-containing base, a sugar, and phosphate. The base defines the identity of the nucleotide since all nucleotides of DNA contain 2'-deoxyribose as the sugar and all nucleotides of RNA contain ribose as the sugar. Note that numbered carbon atoms in sugars of nucleotides are indicated by a "prime." Thus, 2'-deoxyribose indicates that the 2-carbon atom of ribose has lost its associated oxygen atom. The carbon and nitrogen atoms of the purines and pyrimidines are numbered without the use of primes (see figure 12.1). All nucleotides contain phosphate. So let's have a look at the bases. Here are the structures.

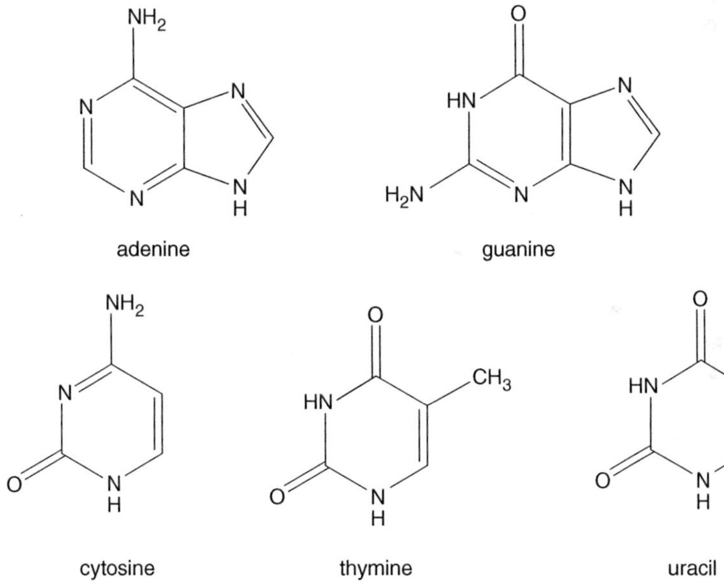

adenine   guanine

cytosine   thymine   uracil

The nitrogen-containing bases that occur in DNA and RNA fall into two structural categories: the purines and the pyrimidines. The former contain a five-membered ring fused to a six-membered ring, while the latter contain a six-membered ring only. The two purines are common to both DNA and RNA: adenine (A) and guanine (G). The pyrimidine cytosine (C) occurs in both DNA and RNA. The other pyrimidine is thymine (T) in DNA but uracil (U) in RNA.

These bases are not the only ones that occur in nucleic acids. There are many rare bases that are occasionally found in these molecules, particularly in RNA. Perhaps 100 of these are known, in addition to the five common bases. We shall encounter a few of these rare bases in transfer RNA, tRNA, molecules. It is enough to know that they exist. We need not be concerned with their structure or function.

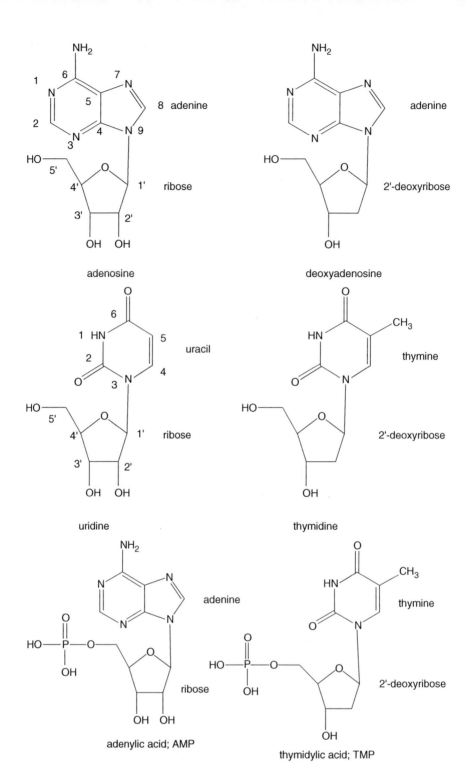

Figure 12.1 Representative structures for nucleosides and nucleotides.

The next issue is how the three components of each nucleotide are put together. The basic answer is: base–sugar–phosphate. A more nearly complete answer is provided in figure 12.1 (note in this figure how the atoms are numbered). The N9 of purines or N3 of pyrimidines is linked to C1$'$ of each sugar. These compounds are termed nucleosides. The phosphate is linked to C5$'$ of each sugar, to yield the nucleotides. So we have our four monomers for linking together to create the nucleic acids, DNA and RNA.

Now that we have the monomers, which are pretty complex in this case, it remains to define how they are joined together to create the polymer. The amino acids in proteins are linked by peptide bonds. The nucleotides in nucleic acids are linked by phosphodiester bonds, as is shown in figure 12.2. These DNA phosphodiester bonds are very stable. Indeed, samples of largely intact DNA can be recovered from organisms that have been extinct for thousands of years. This remarkable stability should not come as a surprise: the central importance of DNA in all forms of life requires that it be stable to various sorts of insults.

Figure 12.2 shows both detailed structures (a) and a convenient shorthand structure (b). The phosphate group is linked to C3$'$ of one sugar and to C5$'$ of the next one. So we get a backbone of alternating sugars and phosphates: ... –sugar–phosphate–sugar–phosphate– ... with the bases chemically bonded to the sugars. In one sense, the basic structural organization is similar to that of the proteins. For proteins, we have a backbone of peptide bonds and α carbon atoms. The amino acid side chains, which define them, dangle from the α carbon atoms. For nucleic acids, we have a backbone of phosphate–sugar from which dangle the bases, which define the nucleotides. The polymer is called a polynucleotide and nucleic acids are frequently referred to as polynucleotides.

Note that the structures of nucleic acids are written with the 5$'$ end at the left and the 3$'$ end at the right. This is a convention, just as by convention we write the amino terminus of proteins to the left and the carboxyl terminus to the right.

Final question: how big are the nucleic acids? How many nucleotide units do we typically find in DNA or RNA? The answer is enormously variable. There are small RNA molecules that contain 25 or fewer nucleotides; there are also RNA molecules that contain thousands of nucleotides. But for really, really big molecules, we turn to DNA, which may have tens of thousands of nucleotides linked together! Specific examples follow below but we need some additional insights first.

DNA is a chemical message

DNA is information.[5] We now have a picture of a DNA molecule (or an RNA molecule) as a long string of nucleotides, each bearing one of four nitrogen-containing bases. We might visualize the sequence of bases in a DNA molecule as, for example:

AGTTCAATCCGGTATAGCGGCCCTTGATCA...

or any other sequence that we might imagine. There is more here than meets the eye. This is a linear sequence of symbols. Such sequences have the capacity to store

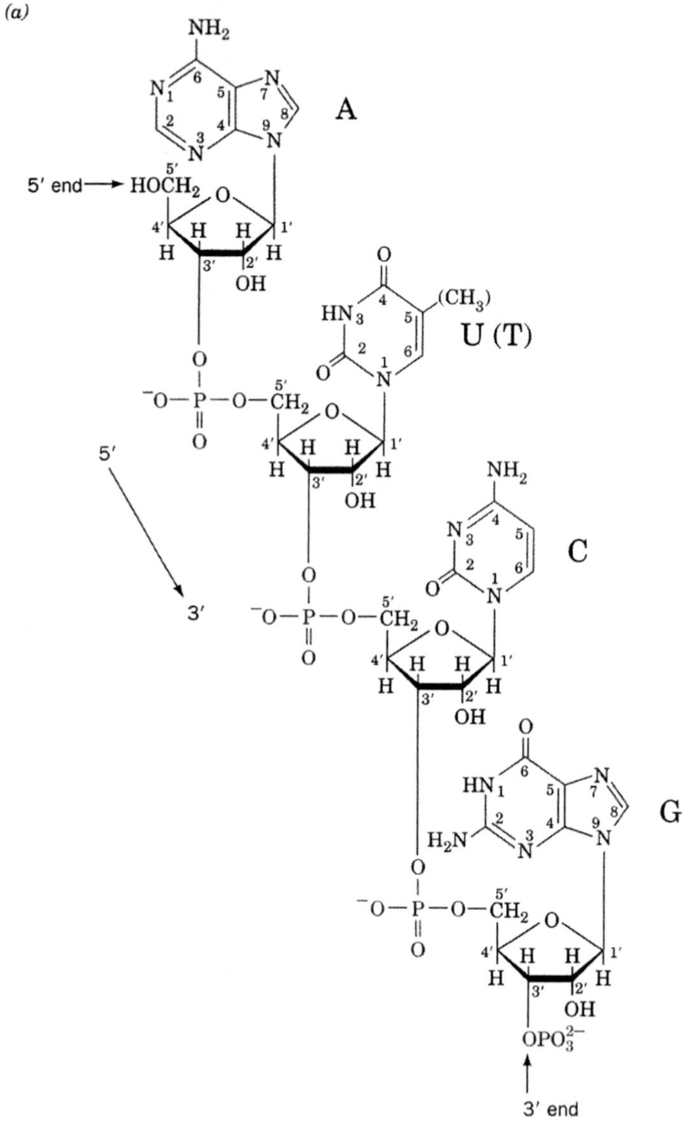

Figure 12.2 Two representations of a short polynucleotide: (a) a detailed atom-by-atom structure of a tetranucleotide showing both the 5′ and 3′ ends; (b) the same structure shown in a shorthand, schematic way. Both representations possess the same basic information. (Reproduced from D. Voet and J. G. Voet, *Biochemistry*, 3rd edn, 2004. ©2004, Donald and Judith G. Voet. Reprinted with permission of John Wiley and Sons, Inc.)

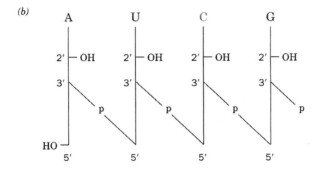

Figure 12.2   (*Continued*)

information. After all, the message that you are reading on this page is nothing more than a long, linear sequence of symbols, the 26 letters of our alphabet in this case, augmented by a number of punctuation marks; different symbols. By ordering the symbols (letters) in an appropriate way, we create words, short sequences of linear symbols. A word is the smallest structure in language that has meaning. Words on this page have meaning to those who understand the English language.

The meaning of these words is context-specific: that is, the full meaning only becomes clear in the context of other words. A favorite example is the following two statements: "Fruit flies like bananas" and "Time flies like the wind." Clearly, the word "flies" has two very different meanings in the context of the two complete statements: "flies" is the subject of the first statement but the verb of the second. Context is important but "flies" also has meaning independent of context. The meaning is simply not fully recognized in the absence of context.

DNA is a text. The language in which the DNA text is written has just four letters, A, G, C, and T. The English language, in contrast has 26. The Japanese language has a few thousand symbols. But the underlying idea is the same. Stringing symbols together creates words and stringing words together creates sentences, and so forth. The sequence of symbols creates a message.

As we will develop, the language of DNA also uses words. All the words in this language are three letters long.[6] It follows that the sequence GGT is a word in this language. We do not yet know the meaning of this word—but we will.

Here are the key points. (a) DNA is the stuff of genes. (b) DNA is a sequence of nucleotides, each of which carries one of four possible symbols. (c) DNA is organized into a sequence of genes. (d) Genes determine the sequence of amino acids in proteins. (e) The sequence of amino acids determines the three-dimensional structure of proteins, which, in turn (f) determines their biological properties. These, in turn (g) determine the nature of the cell. It follows that the sequence of bases in DNA is the ultimate repository of the information required to specify the unique biochemical personality of the cell.

Clearly, these ideas apply with equal force to RNA molecules. The only basic difference is that the RNA library has one letter different from that of DNA: U replaces T. The retroviruses employ RNA as basic genetic material. Among scientists involved in efforts to understand the origins of life on Earth, RNA is widely considered to have come before DNA and was the original repository of genetic information: the RNA World. DNA came along later and replaced RNA as the genetic material, though DNA has not replaced the other functions of RNA.

In most cells, DNA is transcribed into RNA and the latter is then translated into protein, as I develop in chapter 13. In cells infected with retroviruses, the viral RNA is first transcribed into DNA and the resulting DNA is then transcribed back into RNA, which is then translated to protein.

Now we need one more realization before we will begin to understand the language of DNA. Proteins, like DNA and the English language, also contain a linear sequence of symbols. In the protein case, the symbols identify the 20 amino acids that commonly occur along the polypeptide chain of proteins. It follows that proteins also possess information in the form of the amino acid sequence. That information is expressed in the biological properties of the protein, dependent on the three-dimensional structure of the protein. So we have two basic languages here: that of DNA and RNA and that of proteins.

If we return to natural languages for a moment, we will recognize that we can translate one language into another. The process is aided by bilingual dictionaries that provide the meaning of a word in one language in the second language. Given such a dictionary and an understanding of the syntaxes of the two languages, we can translate a message in, for example, Spanish into the same message in English or any other language.

In the simplest case, which is a good starting point—we can always complicate the picture later—a gene specifies the amino acid sequence for one protein. Now we can understand that one three-letter word in the DNA language specifies one amino acid in the protein language. For example, the sequence AAA in DNA will be translated into phenylalanine (Phe) in a protein.[7] So a stretch of DNA coded AAAAAAAAAAAA would be translated into Phe-Phe-Phe-Phe, one phenylalanine for each AAA sequence.

There are 64 ways to order four things three at a time when the order in which they are taken (permutations) matters: $n = 4^3 = 64$. So there are 64 words in the language of DNA. This is more words than we need to specify the 20 amino acids of proteins. A few of these words are used as punctuation marks—start and stop signals. Beyond that, most of the amino acids are specified by more than one word. The genetic code is provided in table 12.1.[8] Note that the code words in table 12.1 refer to those in messenger RNA, mRNA, the complement to the code words in DNA.

We have concluded that a word in the language of DNA is defined by three contiguous bases or as a triplet of bases. A long sequence of such triplets, a gene, specifies the structure of a protein. It follows that a sentence in the DNA language is a gene. The corresponding sentence in the protein language is a protein. We can push our analogy further. In all organisms, genes are organized into chromosomes, single molecules of DNA in the simplest cases (single molecules of DNA complexed with specific proteins in more complex cases but it does not matter so never mind). The genome of a virus or a bacterium is contained in a single chromosome: that is, a single molecule of DNA in each bacterial cell or virus particle. The genome of humans consists of 46 chromosomes, organized into 23 pairs. So a single chromosome contains a large number of genes: that is, a large number of sentences. Put simply, a

Table 12.1 Genetic Code: mRNA to Amino Acids

First position (5′ end)	Second position				Third position (3′ end)
	U	C	A	G	
U	UUU Phe UUC UUA Leu UUG	UCU UCC Ser UCA UCG	UAU Tyr UAC UAA Stop UAG Stop	UGU Cys UGC UGA Stop UGG Trp	U C A G
C	CUU CUC Leu CUA CUG	CCU CCC Pro CCA CCG	CAU His CAC CAA Glu CAG	CGU CGC Arg CGA CGG	U C A G
A	AUU AUC Ile AUA AUG Met	ACU ACC Thr ACA ACG	AAU Asn AAC AAA Lys AAG	AGU Ser AGC AGA Arg AGG	U C A G
G	GUU GUC Val GUA GUG	GCU GCC Ala GCA GCG	GAU Asp GAC GAA Glu GAG	GGU Gly GGC GGA GGG	U C A G

chromosome is a volume written in DNA language. If we think about volumes being organized into libraries, then the library of a virus or bacterium contains a single volume. That of a human being is larger, containing 46 volumes.

Summarizing:

- The genome is a library.
- Chromosomes are volumes in the library.
- Genes are sentences in the volumes.
- Base triplets are words in the sentences.
- *DNA is information.*

Having said all this, we need to understand how the language of DNA is translated into the language of proteins. I will get to that in the next chapter. Understanding the double helical nature of DNA first will help us understand how DNA is replicated at the time of cell division and how the translation of DNA structure into protein structure happens.

The DNA double helix is the most famous structure in chemistry

So far, I have described the primary structure of a nucleic acid. DNA is a linear polynucleotide based on 2′-deoxyribose as sugar and A, G, C, and T as bases. RNA is a linear polynucleotide based on ribose as sugar and A, G, C, and U as bases. In both

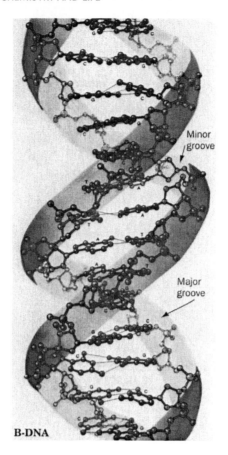

Figure 12.3 The double helix of DNA. The sugar–phosphate backbones wind about the periphery of the molecule in opposite directions. The hydrogen-bonded base pairs occupy the core of the structure and are basically flat and lie perpendicular to the long axis of the helix. (Illustration, Irving Geis/Geis Archives Trust. © Howard Hughes Medical Institute. Reproduced with permission.)

cases, individual nucleotides are linked through 3′,5′-phosphodiester bonds. Just as in the case of proteins, biological function is correlated with the three-dimensional structures of the nucleic acids. We begin with that for DNA.

The most famous structure in all chemistry is the Watson–Crick double helix for DNA (figure 12.3).[9] The discovery of this structure by James Watson and Francis Crick in 1953 was the beginning of molecular biology. An amazing number of insights about the nature of life have been derived from this structure.

The Watson–Crick double helix is the outcome of three lines of work.[10] The first is the discovery by Erwin Chargaff of Chargaff's rules.[11] Specifically, for all normal DNAs, A = T, G = C and A + G = C + T. The actual content of each base in DNA varies from species to species over a wide range. Despite this variation, the content of A is always equal to that of T and the content of G is always equal to that of C. This proved a powerful clue to the structure of DNA.

The second line of work was a series of low-resolution X-ray diffraction studies of DNA carried out by Maurice Wilkins and Rosalind Franklin. This technique reveals that certain structural features are repeated at regular, defined intervals. Without getting into technical details, the important point is that these measurements provided strong constraints on acceptable structures. Possible structures that do not possess repeating structural elements at the defined intervals are excluded.

The final line of work was model-building. The task was to find a way to account for Chargaff's rules and the X-ray diffraction data in a way consistent with the twin functions of DNA as information store and information transfer device. Model-building was aided by some basic concepts of chemistry: the need to get the highly water-loving (hydrophilic) phosphate and sugar groups in contact with solvent water and to get the water-hating (hydrophobic) bases out of such contact. Specifically, it would be good to get the sugars and phosphates on the outside of the structure and the bases on the inside. The Watson–Crick structure does just that (figure 12.3). For this pioneering work in chemistry, Watson, Crick, and Wilkins shared the Nobel Prize for Physiology or Medicine in 1962.

Here is a summary of some of the details of the Watson–Crick structure, called B-DNA:

- The molecule consists of two polynucleotide strands that are wrapped around each other in a helical fashion.
- The two strands of the double helix are antiparallel: that is, one strand runs $3' \to 5'$ while the other runs $5' \to 3'$.
- The sugar and phosphate groups are on the outside; the bases face each other on the inside. The planes of the bases are perpendicular to the long axis of the double helix.
- The bases are specifically paired: each place in one strand occupied by an A has a T on the opposing strand; each place in one strand occupied by a G has a C on the opposing strand, accounting for Chargaff's rules. This base pairing is enforced by the sizes and shapes of the bases, their capacity for hydrogen bonding to each other (depicted in figure 12.4) and the spatial constraints of the double helix itself. A purine on one strand is always opposed by a pyrimidine on the other.
- The distance between consecutive base pairs along the double helix is 3.4 Å. There are 10 base pairs per turn of the helix, accounting for the repeat distance of 34 Å found in the X-ray diffraction studies.
- Finally, the helix has one minor groove, about 12 Å across, and one major groove, about 22 Å across. Parts of the base pairs are exposed in these grooves and they are targets for a number of anticancer drugs.

The pairing of bases in the double helix provides a beautiful mechanism for the duplication of DNA and its information content at the time of cell division, a matter that we get to just below. Basically, one can simply imagine that, one way or another, the two strands of the DNA come apart and each forms a template for the synthesis of the complementary strand. Thus, the original DNA molecule has become two, each identical to the first.

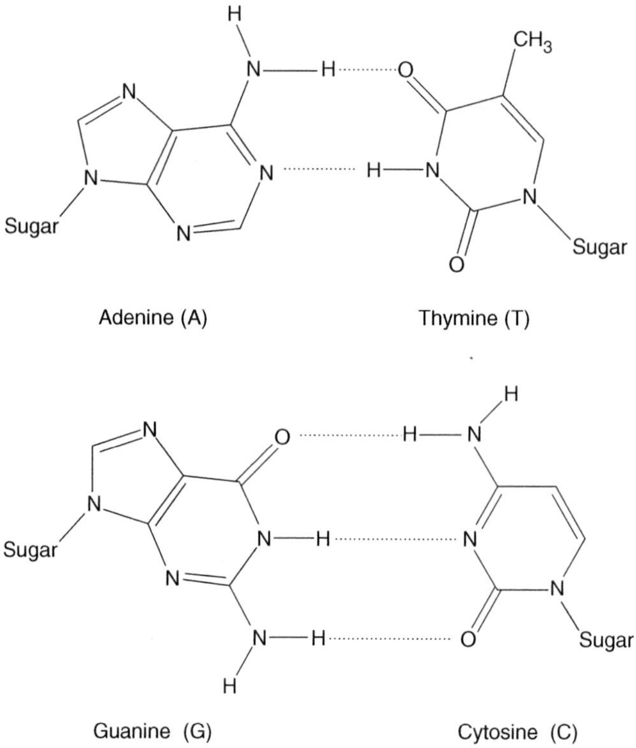

Figure 12.4   The structures of A-T and G-C base pairs in DNA.

The beauty of the Watson–Crick structure for duplication of information did not escape the discoverers. The last sentence of their ground-breaking publication in the scientific journal *Nature* reads: "It has not escaped our notice that the specific pairing we have postulated immediately suggests a possible copying mechanism for the genetic material."

The biosynthesis of DNA preserves the central cellular information

DNA is the carrier of the genetic information of all cells and many viruses. Two important points follow from this simple statement. First, DNA must be replicated in order for cell division to take place. In cell division, one cell becomes two and each must have its own store of genetic information. The genetic information is stored in chromosomes and each chromosome contains a single molecule of DNA. Secondly, the replication of DNA must be a very precisely controlled process. The goal is for each daughter cell to have exactly the same genetic information as the parent cell, which requires that each chromosomal DNA molecule replicated in the course of cell division is replicated exactly. Errors in replication may introduce mutations and many mutations are deleterious to the resulting cell.

DNA replication is semiconservative

Earlier, we did a thought experiment about one way that DNA replication could take place. In this mechanism, termed *semiconservative*, each new strand of DNA is paired with one of the old strands: that is, the two strands of the original DNA molecule are now divided between the two molecules, each of which has one old strand and one new strand. However, there is another possibility, termed *conservative*. In this case, the two new strands are paired with each other and the two old strands remain paired with each other. Matthew Meselson and W. F. Stahl provided compelling evidence that DNA replication is semiconservative. Here is how the experiment was done.[12]

Recall from our discussion in chapter 4 that all elements occur as more than one isotope. For example, most nitrogen is nitrogen-14, containing 7 protons and 7 neutrons in its nucleus. Nitrogen-15 is another isotope of nitrogen. It also contains 7 protons in the nucleus but has 8 neutrons. Chemically, nitrogen-14 and nitrogen-15 behave in much the same way. However, nitrogen-15 is heavier (one more neutron per atom) than nitrogen-14 and the two can be distinguished on that basis.

Escherichia coli bacteria were grown on a medium in which all the nitrogen present was nitrogen-15. As a consequence, all the nitrogen atoms of the DNA of these cells were nitrogen-15. Let's call this DNA **H-H** since both DNA strands are made from heavy nitrogen (nitrogen-15). Then the *E. coli* cells were transferred to a culture medium in which all the nitrogen present was nitrogen-14. It follows that DNA synthesis during the next cycle of cell division would create light (**L**, nitrogen-14) strands of DNA. Conservative DNA replication would require that half of the double-stranded DNA molecules would be **H-H** and that half would be **L-L**. Semiconservative replication would require, in contrast, that all the double-stranded DNA molecules would be **H-L.**

These possibilities can be sorted out by a technique known as density gradient centrifugation. Following one round of cell division, only **H-L** molecules were observed. This established that DNA replication in *E. coli* is semiconservative. Subsequent research established that DNA replication is semiconservative in all prokaryotic and eukaryotic cells. This is just what we would have guessed based on the Watson–Crick structure for DNA. At the same time, guesses are not good enough in science and the Meselson–Stahl experiment is a classic. Some scientists consider it to be the single most elegant experiment ever in molecular biology.

DNA replication is surprisingly fast. Let us take the process in *E. coli* as an example. We know from DNA sequencing of the entire *E. coli* genome that it contains 4,639,221 base pairs. We also know that it takes about 42 minutes (or 2520 seconds) to replicate the *E. coli* genome. It follows that the rate of synthesis is about $4,639,221/2520 = 2000$ base pairs per second. There are two growing sites (from one point of origin going both ways), so growth at each site occurs at the rate of about 1000 base pairs per second! In human cells, the rate of growth at each growing site is slower, about 100 base pairs per second. We can make a simple calculation of the minimum number of growing sites in the human genome. The number of base pairs in the human genome is about 3 billion (3×10^9; 3,000,000,000). It takes about 8 hours (28,800 seconds) to replicate the human genome. One growing site could replicate about $28,800 \times 100 = 2,880,000$ base pairs in 8 hours. It follows that we need at least 3,000,000,000/2,880,000 = 1000 growing sites. In fact, the best estimate is that there are 10,000 or more growing sites in the human genome during cell replication.

The machinery for DNA biosynthesis is complex

The key enzymes in DNA replication are the DNA polymerases. These enzymes utilize single-stranded DNA as template and catalyze the formation of a complementary strand. Each new nucleotide to be added is selected on the basis of its ability to form a Watson–Crick base pair with the base on the template strand.

Arthur Kornberg discovered the first DNA polymerase in *E. coli.* Kornberg shared the Nobel Prize in Physiology or Medicine in 1959 for his discovery.[13] This enzyme is now known as DNA polymerase I. As it turns out, DNA polymerase I is not the basic DNA replicating enzyme of *E. coli;* an enzyme known as DNA polymerase III is. However, DNA polymerase I does have a very important function. It can detect errors in the growing DNA chain, incorporation of nucleotides that violate the Watson–Crick base pairing scheme, and remove the offending nucleotides. This permits detection and correction of errors. DNA polymerase I is a proofreading enzyme.

Before we take leave of the process of DNA synthesis, there is one more, nonobvious, point that is worth making. In the foregoing, my emphasis has been on ensuring the fidelity of DNA synthesis, the goal being to pass on uncorrupted genetic information from parent cell to progeny cell. Indeed, nature has gone to considerable lengths to ensure that there is a high degree of fidelity in the replication of DNA at the time of cell division. Nonetheless, nothing is perfect and occasional errors are made. This is not all bad. Here is why.

The process of evolution requires changes in the genetic material, known as mutations. If the genetic material were not altered (mutated), from time to time there would be no opportunity for change. Species would simply perpetuate themselves over time and, being unable to respond to environmental changes, they would eventually die out. There are several mechanisms for making alterations in the structure of DNA or RNA: an error at the time of cell division is one of them. Although it is surely true that most genetic alterations are neutral or injurious, it is also true that an occasional one is a helpful advance along the road of evolutionary change.

That concludes all that we need to know about the complex process of DNA replication.

DNA can be chemically modified

The story of DNA replication is followed by one of DNA modification. The most important chemical modification of DNA is methylation: that is, the methyl group, CH_3-, is added to two of the bases of DNA, adenine (A) and cytosine (C). Three modified bases are formed.

The methyl groups attached to A and C project into the major groove of the DNA double helix. There they can interact with DNA- binding proteins.

In eukaryotes, DNA methylation is important in regulation of gene function. The predominant product of methylation in the DNA of vertebrates is 5-methylcytosine. This methylated base is found largely in CG dinucleotides in palindromic sequences. These may occur in control regions upstream of transcribed DNA sequences. There is considerable evidence to strongly suggest that DNA methylation in vertebrates turns off gene expression.

Now let's take a look at the structure of RNA and begin to develop some of its functions.

RNA is usually a single-stranded molecule

In contrast to DNA, which is quite generally found in a double-stranded structure, RNA is generally, but not always, found in nature as a single-stranded molecule. There are, as always, exceptions, one of which we note below. Double-stranded RNAs as well as hybrid DNA:RNA double-stranded molecules are found.

To say that RNA molecules are single-stranded molecules is not the same as saying that they have no higher-order structures. In fact they have several. The formation of Watson–Crick complementary base pairs is a driving force for formation of higher-order structures. These include the stem-loop and hairpin secondary structures, as well as more complex tertiary structures. Of particular note, are the complex structures for transfer RNAs, tRNAs. Examples are provided in figure 12.5 (note that there are several unusual bases in these structures; this is typical of tRNAs but not of RNA molecules in general). These structures are intimately related to the function of these molecules as adaptors in the process of protein synthesis, as developed in the next chapter.

The phenomenon of RNA interference, RNAi, involves double-stranded RNA molecules

These molecules are known as small-interfering-RNAs, siRNA. Craig Mello and Andrew Fire, who shared the 2006 Nobel Prize in Physiology or Medicine for this work, discovered RNAi in 1998.[14] RNAi is a general and ancient mechanism that organisms employ to control expression of their genes. Specifically, siRNA molecules target specific mRNA molecules, leading to their destruction and preventing their translation into protein. This, of course, ends the expression of the gene that coded for the mRNA destroyed. It is a mechanism that applies to all genes. More about how this works follows in the next chapter.

RNAi plays a key role in cellular defense against viral infection, genome rearrangement, the timing of developmental events, and brain morphogenesis.[15]

siRNA molecules are really quite small: they generally consist of perhaps 22–24 nucleotide base pairs, about one-third the size of a typical tRNA and much smaller than a typical mRNA or rRNA. Here is how RNAi works. The siRNA molecules

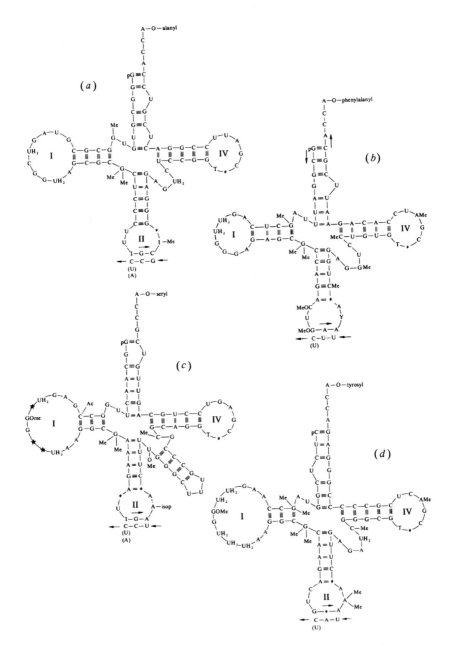

Figure 12.5 The structures for four tRNA molecules of yeast. (a) Alanyl-tRNA; (b) phenylalanyl-tRNA; (c) seryl-tRNA; (d) tyrosyl-tRNA. The single letter designations identify the sequence of bases along the single chain. Note that several of these are unusual bases, most of which are methylated (Me). Note also the ACC sequence at the 3′ terminus of each tRNA. This is the site to which amino acids are attached in the process of protein synthesis, as indicated. These tRNA molecules have a substantial amount of secondary structure created by formation of Watson–Crick base pairs. Finally, note that the anticoding triplet in the bottom loop is shown.

are synthesized in the cytoplasm of the cell from endogenous RNA molecules, either longer double-stranded molecules or RNA known as short-hairpin RNA, shRNA. In a sequence-specific manner, one of the strands of the siRNA becomes incorporated into a complex known as the RNA-induced silencing complex or RISC. Within the complex, this strand binds to and degrades complementary mRNA. This is not only an important control on cellular gene expression but also enormously useful as a research tool. Since the mechanism applies to all genes, it is possible to use RNAi to specifically shut down the expression of any gene or, sequentially, all the genes of an organism. This is a powerful tool in understanding the role of specific genes in cellular physiology.

Now we can get on with understanding the basics of the translation of DNA structure into protein structure. This will necessarily involve some thoughts about how RNA is synthesized using DNA as a template. These are topics for the next chapter.

Key Points

1. Cells are the basic building blocks of living systems.

2. Each cell type expresses a unique biochemical personality or individuality. This biochemical attribute is preserved upon cell division.

3. DNA is the agent of inheritance. DNA is the genetic material. DNA codes for its own synthesis at the time of cell division.

4. The units of genetic inheritance are genes.

5. DNA codes for the synthesis of RNA and proteins.

6. Nucleotides are the building blocks of the nucleic acids.

7. In DNA, there are four bases: adenine, guanine, cytosine, and thymine. The sugar is 2′-deoxyribose.

8. In RNA, there are four bases: adenine, guanine, cytosine, and uracil. The sugar is ribose.

9. In DNA and RNA, the nucleotides are linked together by phosphodiester bonds.

10. DNA is information. The sequence of bases along the backbone carries a message. All words in the language of DNA are three letters long. They specify amino acids in proteins. This is known as the genetic code.

11. The structure of DNA is a double helix, the Watson–Crick structure.

12. In the DNA double helix, bases along one strand are specifically paired with bases on the other strand: A is paired with T; G is paired with C.

13. DNA replication is semiconservative: that is, each new strand of DNA is paired with one of the old strands.

14. DNA polymerases catalyze the replication of DNA.

15. Certain bases in DNA may be altered chemically, by methylation for example. These alterations may have important consequences for regulation of gene function.

16. RNA molecules are usually single-stranded but structured through intramolecular base-pairing.

17. siRNA are small double-stranded RNA molecules involved in the control of protein synthesis.

13

The central dogma of molecular biology and protein synthesis

DNA directs its own replication and its transcription into RNA that, in turn, directs its translation into protein. tRNA is the adaptor molecule that mediates the interaction between mRNA and the encoded amino acid.

Suppose for a moment that you are the publisher of J. K. Rowling's series of Harry Potter novels (don't spend the money yet; we are just supposing). You have a single copy of her latest manuscript, say *Harry Potter and the Deathly Hallows*, the final book in the series. Basically, you have two tasks. The first is to replicate the manuscript in the English language several million times. The second is to translate the English-language manuscript into a number of other languages and then replicate those copies.

Now suppose that you are a cell in possession of one copy of your genome, a small collection of DNA molecules. When you sense a round of cell division about to happen, you have two related tasks. The first is to replicate your DNA molecules exactly, so as to have two copies, one for each daughter cell. Unlike the publisher of Rowling's books, you need only one new copy, not millions, but you will need to continue to replicate the DNA molecules at each round of cell division. The second task is to translate the DNA molecule language into that of RNA and then into that of proteins. A new complement of RNA and protein molecules is required for the new cell and for the process of cell division. So the task of a cell facing cell division is not unlike that of a book publisher in some basic ways.

In the last chapter, I developed the basics of how DNA is replicated with the help of DNA polymerases at the time of cell division. I have said nothing about the second task: how DNA is transcribed into RNA, and how RNA is translated into proteins. Before we get into it any further, let me emphasize the basic nomenclature. The process of making new DNA from a template of existing DNA is *replication*. The process of making RNA from a template of DNA is *transcription*. The process of making proteins from an RNA template is *translation*.

The sequence of steps along the pathway from DNA to RNA to protein is understood in intimate and revealing detail. These insights are the product of decades of determined scientific effort. My very earliest experiences in biochemical research were as an undergraduate in the laboratory of Richard Schweet at Caltech and later at the City of Hope Medical Center.[1] The Schweet laboratory had developed the first cell-free system for studying protein synthesis— specifically that of hemoglobin. Not a lot about the process was understood at the time but we were learning rapidly and it was an exciting time. A wealth of insight and detail has subsequently been gained. However, the details are not central to our understanding. Here are the essentials.

The central dogma of molecular biology was formulated by Francis Crick

Francis Crick enunciated the central dogma of molecular biology in 1958: *DNA directs its own replication and its transcription to RNA that, in turn, directs its translation to protein.*[2] This statement is frequently oversimplified to: "DNA makes RNA makes protein."

In protein synthesis, there are three classes of RNA to worry about: ribosomal RNA, rRNA; messenger RNA, mRNA; and transfer RNA, tRNA. All three classes of RNA play key roles in the final stage of the process: the biosynthesis of proteins. However, we are going to take this one step at a time. We turn attention first to the biosynthesis of RNA.

RNA polymerases catalyze the transcription of DNA into RNA

The counterpart of DNA polymerases in replication is RNA polymerases in transcription. Just as there are several DNA polymerases in vertebrate cells, so there are several RNA polymerases. To be precise, there are three of them. The different RNA polymerases are associated with three of the classes of RNA molecules found in vertebrate cells. Specifically, RNA polymerase I is responsible for the synthesis of the precursors of most rRNAs. RNA polymerase II plays the same role for the precursors of mRNA. Finally, RNA polymerase III is responsible for the synthesis of the precursors to the tRNAs as well as a few other small RNA molecules. Note here that I have specifically referred to precursors of these classes of RNA molecules. The initial products of the action of the RNA polymerases undergo further metabolism to yield the mature, functional products.

Each of these enzymes is found in the cell nucleus, as one would expect. DNA is localized in the nucleus and does not leave it. DNA is the template for the synthesis of all classes of RNA molecules. The base sequence in one of the strands of the DNA molecule is transcribed into the complementary base sequence of the product RNA molecules: A codes for U; G for C; C for G; and T for A. Watson–Crick base-pairing not only helps define the structure of DNA but also tells us how DNA is replicated and transcribed. Therefore, transcription of DNA into RNA must occur in the cell nucleus.

Unlike the DNA polymerase reaction, RNA polymerases catalyze the transcription of only one of the two DNA strands. The two DNA strands are termed the *sense* strand and the *antisense* strand. It is the antisense strand that is transcribed by the RNA polymerases. Thus, the base sequence of the newly synthesized RNA strand is identical to the sense strand of the DNA template, except of course that U replaces T.

RNA polymerases bind to promoters

The sequence of amino acids in proteins is ultimately encoded in the sequence of bases in DNA. Transcription encodes this information in mRNA molecules. Each RNA polymerase transcribes a very small part of the total DNA base sequence. It follows that RNA polymerases need specific places on DNA molecules to start and to stop transcription. Getting the amino acid sequence right is critical for protein function. It follows that getting transcription started at precisely the right place is also critical. If the mRNA base sequence is not right, the amino acid sequence in the protein will not be right.

Here is the key point. RNA polymerases bind specifically and tightly to sites of transcription initiation at DNA base sequences known as *promoters*. These promoter base sequences are recognized by one of the subunits of the complete RNA polymerase molecule. Promoters are generally about 40 base pairs long and they frequently lie on the 5′ side of the transcription start site, sometimes referred to as upstream from the transcription start site. The RNA polymerase is said to work its way downstream from the start site.

Chain initiation and elongation are catalyzed by RNA polymerases

Once we have the RNA polymerase associated with the promoter site, chain initiation can begin. The first reaction is joining two nucleoside triphosphates:

$$\text{ppp}X + \text{ppp}Y \rightarrow \text{ppp}X\text{p}Y + \text{PP}_i$$

Here I have let X and Y stand for any of the four nucleosides of RNA. Note that the first step results in a triphosphate group at the 5′ end of the RNA nucleotide chain. Each additional nucleoside triphosphate extends the chain by one unit with splitting out of inorganic pyrophosphate, PP_i. So at some point, the growing RNA molecule will look like:

$$\text{ppp}X_1\text{p}X_2\text{p}X_3\text{p}X_4\text{p}X_5 \ldots$$

30 to 50 nucleotides are added to the growing RNA chain per second.

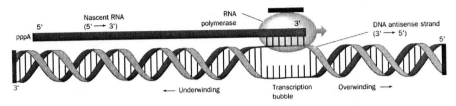

Figure 13.1 A schematic view of RNA chain elongation catalyzed by an RNA polymerase. In the region being transcribed, the DNA double helix is unwound by about a turn to permit the DNA's sense strand to form a short segment of DNA–RNA hybrid double helix. That forms the transcription bubble. Note that the DNA bases in the bubble on the antisense strand are now exposed to the enzyme and are useable as a template for chain elongation. The RNA polymerase works its way down the DNA molecule until it encounters a stop signal. (Reproduced from D. Voet and J. G. Voet, *Biochemistry*, 3rd, edn, 2004; © Donald and Judith G. Voet. Reprinted with permission of John Wiley and Sons, Inc.)

In the process of chain elongation in RNA biosynthesis, there is local unwinding of the DNA double helix with formation of a transcription bubble (figure 13.1). Local unwinding of the double helix is required to expose the sequence of bases along that strand of the DNA being transcribed. Without such exposure, DNA cannot act as a template for RNA biosynthesis. As the RNA polymerase does its job, the transcription bubble moves along the DNA molecule, exposing new DNA bases to act as template for the enzyme.

Finally, we need to terminate chain elongation at the right point. Chain termination is associated with specific base sequences in the DNA molecule. Certain sites in DNA act like periods at the end of a sentence. They indicate that a message is complete and it is time to stop with the letters and get on to the next message.

Eukaryotic cells modify RNA following transcription

Those RNA molecules destined to function as mRNA are chemically modified following transcription. The immediate product of transcription is termed pre-mRNA. There are two classes of these modifications that lead to mature, functional mRNA: alteration of the mRNA ends and splicing.

Both ends of mRNA molecules are chemically modified. The 5′ end is capped by 7-methylguanosine triphosphate. This modified guanine cap helps to protect the mRNA from attack by hydrolytic enzymes. Beyond that, the 7-methylguanosine triphosphate cap acts as a recognition element for ribosomes. The 3′ end is modified by the addition of a poly A tail. This may consist of as few as 30 or as many as 200 adenine nucleotides.

The pre-mRNA molecule contains both coding and noncoding regions (coding in the sense of containing triplet codons for amino acids). The coding regions are termed *exons*; the noncoding regions *introns*. Phillip Sharp and Richard Roberts discovered the existence of introns. These scientists shared the 1993 Nobel Prize in Physiology or Medicine for this discovery. In order to create mature mRNA, it is necessary to splice

Pre-mRNA:

5′ exon1-intron1-exon2-intron2-exon3 3′

mature mRNA:

5′ 7-methylG . . . exon1-exon2-exon3 . . . AAAAA . . . AAAA 3′

Figure 13.2 The maturation of pre-mRNA into functional mRNA. The model for pre-mRNA shown here has three exons (coding regions), separated by two introns (non-coding regions). In the maturation process, the introns are spliced out, bringing together the three exons into one contiguous coding region. In addition, the mature mRNA is capped at the 5′ end with a 7-methylguanosine group and by a chain of A's at the 3′ end.

out the introns. Short nucleotide sequences at the ends of introns identify splice sites. The actual splicing is carried out by a complex structure termed the spliceosome. The maturation of pre-mRNA is summarized in figure 13.2.

That is transcription. Now, we get on with the final step: the translation of mRNA structure into protein structure.

Protein synthesis is a multistep process

To understand what is going on in the biosynthesis of proteins, we need to keep in mind what we are trying to achieve. Here is how the logic goes.

First, the functions of proteins in cells are critical to the survival and well-being of cells. Secondly, the function of proteins is intimately dependent on the proper sequence of amino acids. So we have got to get that sequence right for each protein of the cell. Thirdly, the fundamental information specifying the amino acid sequence of proteins is encoded in the genome: that is, in DNA. Finally, we know that the information is transferred from DNA to mRNA molecules—working copies of genes—through the process of transcription. So we are left with the problem of translating the information encoded in the sequence of bases in mRNA molecules into the proper sequence of amino acids in the protein that each mRNA encodes. It follows that we need adaptor molecules, an idea first proposed by Francis Crick in 1955. The idea is that the adaptor molecules will somehow adapt each amino acid to its three-letter code word on the mRNA. The adaptor molecules are the intermediaries between the amino acids and the mRNA triplets that code for them (see table 12.1). These adaptor molecules are tRNAs. Now I need to explain how this works.

There are some basic principles for protein biosynthesis

Before we get into the biosynthetic process itself, here are some basic elements of protein synthesis. First, protein synthesis starts at the N-terminus and proceeds to the C-terminus. Secondly, the mRNA is read by the ribosome in the 5′ → 3′ direction. Thirdly, protein synthesis occurs on ribosomes attached to an mRNA molecule, polyribosomes. Basically, the ribosomes are a sequence of beads on an mRNA string. The maximum density of ribosomes on the mRNA molecule is about one ribosome per 80 mRNA nucleotides.

With these fundamentals in mind, let us have a look at the process of protein biosynthesis.

Step One is the specific attachment of an amino acid to its cognate tRNA

In order that tRNA molecules can function as adaptors between amino acids and mRNA, two things must be true. First, there must be a mechanism for attaching each amino acid specifically to its cognate tRNA. Doing so will create molecules known as aminoacyl-tRNAs. Secondly, each aminoacyl-tRNA must recognize the correct triplet code for that amino acid on mRNA. I am going to worry about the former issue first.

There is a family of enzymes that catalyze the attachment of amino acids to their cognate tRNAs, aminoacyl-tRNA synthetases. There is one or more of these enzymes for each of the 20 amino acids that occur commonly in proteins. Each of these enzymes recognizes (a) a specific amino acid and (b) its cognate tRNA. Imagine a soup of 20 amino acids and 20 tRNAs, one for each amino acid. For example, the aminoacyl-tRNA synthetase for, saline would specifically pick valine out of the soup and catalyze its attachment to the tRNA for valine, tRNA^{val}. Simply, we can write the product of the reaction as val-tRNA^{val}. This is a lovely example of the role of molecular recognition in a critical life process.

All tRNA molecules have the sequence –CCA at the 3′ end. This three base sequence is termed the acceptor stem. The aminoacyl-tRNA synthetases catalyze the formation of an ester between the carboxyl group of the amino acid and the 3′ –OH of the ribose of the terminal adenosine moiety:

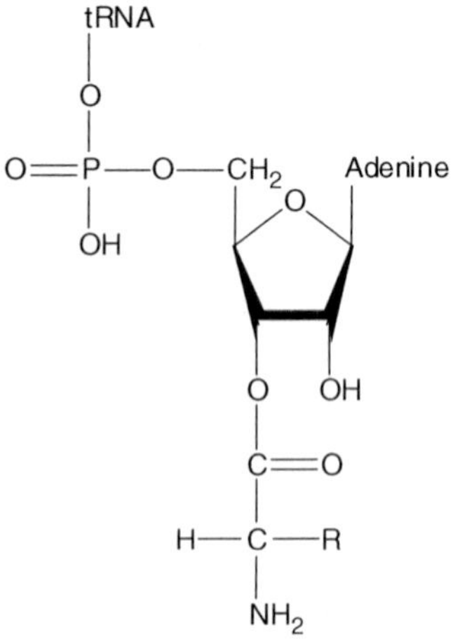

Aminoacyl-tRNA

The formation of aminoacyl-tRNA molecules is an energy-requiring process, which means that some sort of driving force is required to push the reaction to completion.

That driving force is hydrolysis of ATP. The attachment of amino acids to their cognate tRNAs is highly specific.

The anticodon on the aminoacyl-tRNA recognizes the codon on mRNA

Now that we have specifically attached each amino acid to its cognate tRNA, we need to make the interface between each aminoacyl-tRNA and the correct triplet codon on mRNA. The aminoacyl moiety has nothing to do with this: it is simply a matter of codon–anticodon recognition, which is shown schematically in figure 13.3.

We know that the anticodon is a triplet of bases occupying the anticodon loop in the tRNA structure. By Watson–Crick base-pairing, this anticodon can recognize the

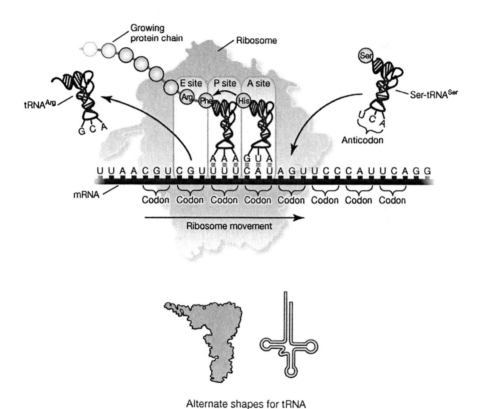

Alternate shapes for tRNA

Figure 13.3 The process of protein synthesis on the ribosome. The strand of mRNA is shown associated with the small subunit of the ribosome. The aminoacyl-tRNA molecules are shown associated with the large subunit of the ribosome and base-paired with mRNA codons. A peptide bond is in the process of formation between the two associated amino acids, extending the growing polypeptide chain by one unit. On the left, a tRNA is shown leaving the ribosome, having donated its amino acid to the growing chain. On the right, an aminoacyl-tRNA molecule is shown entering the ribosome. It is next in line to contribute its amino acid to that chain.

codon on the mRNA. This recognition is, in part, responsible for the specificity of the translation process. That is not the whole story. The smaller subunit on the ribosome (the site of protein synthesis) is actively involved in monitoring proper base-pairing between the anticodon on the aminoacyl-tRNA and the codon on the mRNA. The observed specificity is then a combination of the tendency to form Watson–Crick base pairs in preference to others and the active role of the small ribosome subunit in monitoring this process.

The ribosome is an active participant in protein synthesis

The ribosome is both the site of protein synthesis and an active participant in the process. The eukaryotic ribosome is constructed from two subunits: the smaller 40S subunit and the larger 60S subunit.[3] Basically, the 40S subunit binds the mRNA and monitors the recognition between the mRNA codon and tRNA anticodon. The 60S subunit has the binding sites for aminoacyl-tRNAs and catalyzes the formation of peptide bonds. Remarkably, the catalytic entity for peptide bond formation in the 60S subunit is the RNA component, not the protein component. Therefore, the 60S subunit acts as a ribozyme.

There are three stages in protein synthesis

The first stage in protein synthesis is initiation, in which an aminoacyl-tRNA is positioned on one of the subunits of the ribosome. The attached amino acid will be the N-terminus of the completed protein.

The second stage in protein synthesis is chain elongation. An aminoacyl-tRNA is bound to an adjacent site on the ribosome and a covalent bond is formed between the amino group of the aminoacyl-tRNA in the second site and the carboxyl group of the growing peptide chain. In order to continue chain elongation, we need to release the uncharged tRNA in one site and move the peptidyl-tRNA, together with its mRNA, to the next site.

The final step in protein biosynthesis is chain termination. Natural mRNA molecules contain termination codons: UAA, UGA, or UAG. There are no tRNAs that have anticodons which are complementary to these codons. When the growing peptide chain encounters one of these termination codons, the peptidyl-tRNA is transferred to water instead of another aminoacyl-tRNA. The peptidyl-tRNA is hydrolyzed to free the completed protein and the tRNA. Chain termination completes protein synthesis.

The genetic code is translated with remarkable accuracy

We started out this section by emphasizing the importance of getting the amino acid sequence right. After all, the sequence determines the three-dimensional structure of proteins and that, in turn, is critical for function. Through multiple mechanisms, several of which we have mentioned, translation of the genetic code is remarkably accurate. The error rate is about 1 out of 10,000. Of course, many of the errors which do occur may not be important for protein structure and function (conservative amino acid replacements).

There is post-translational modification of proteins

Just as pre-mRNA molecules are chemically modified to create functional mRNA (figure 13.2), nascent proteins may be chemically modified to create fully functional molecules. There are two general mechanisms at work here: proteolytic cleavage and covalent modification.

Many proteins are synthesized in inactive forms, termed *proproteins*. Insulin is created as an inactive single polypeptide chain and must be cleaved to create the active hormone. Many proteolytic enzymes are made as inactive precursors and must be cleaved to form enzymatically active molecules.

Proteins that are destined to be secreted or to end up as transmembrane proteins are synthesized with an N-terminal signal peptide. Signal peptides are highly hydrophobic sequences of variable length. Proteins synthesized with signal peptides are termed preproteins. For example, insulin is a secreted protein. It is therefore synthesized as a preprotein. It is also synthesized in inactive form: a proprotein. Insulin as initially synthesized is, in consequence, a preproprotein.

Finally, many proteins are covalently modified post-translation. We have already encountered examples of glycoproteins and lipoproteins. The sugar and lipid groups are attached to the side chains of certain amino acids following translation. Other examples include phosphorylation, hydroxylation, and methylation, along with more exotic modifications. In general, these post-translational modifications affect either the biological properties of these proteins, their localization within the cell, or their resistance to degradative mechanisms.

That is protein synthesis.

Key Points

1. The central dogma of molecular biology is that DNA directs its own replication and its transcription into RNA that, in turn, directs its translation into protein.

2. RNA polymerases direct the transcription of DNA into RNA. The process is complex but understood.

3. There are three classes of RNA: ribosomal RNA, messenger RNA, and transfer RNA. It is messenger RNA that codes for the structure of proteins.

4. Messenger RNA contains both coding regions (exons) and noncoding regions (introns).

5. Transfer RNA molecules are adaptors; they adapt each amino acid to its triplet codon on messenger RNA.

6. Protein synthesis occurs on ribosomes and is complex.

7. The genetic code is translated with remarkable accuracy.

14

Genomes

We have an amazing abundance of genomic data: from viruses,
bacteria, yeast, worms, plants, animals, and humans. Analysis of this
data has convincingly confirmed the essential unity of all life on Earth.

One of the most amazing things that I have seen in the life sciences is the increase in the rate of acquisition of genomic data in the form of nucleotide sequences of DNA. The Human Genome Project, a rare example of Big Science in biology, was completed in its basic form in 2001. That marked the end of an enormous international effort in multiple laboratories, lasting several years and costing millions of dollars. The completion of the Project (though work continues on unresolved details) was greeted with a major fanfare. It is rightly considered one of the major triumphs of science.

In 2007, two personal genome sequences were published: one was that for James Watson, co-discoverer of the structure of DNA, and the other was that for J. Craig Venter, a key sequencer of the human genome. These announcements were greeted with modest interest both by the scientific community and the general public. Had the donors of these genomes been less notable than Watson and Venter, my guess is that they would have been greeted with even less interest. These reports open the gate to the issue of personal human genome sequencing. Responsible scientists now talk about sequencing a human genome quickly and for the modest price of $1000. That has yet to become reality but there is little reason to think that it will not happen and it should happen in a few years. The implications of getting a complete human genome for $1000 are incredible: personalized medicine would become a reality at a level of detail far beyond what we can currently accomplish. A project has recently been initiated that will sequence 1000 human genomes from diverse ethnic backgrounds.

The accelerating collection of DNA sequence data has been accompanied by a series of revolutionary leaps in our ability to analyze and extract information from it. The enabling technology comes in the form of computational biology, a set of mathematical techniques that, when combined with the sequencing data, forms the science of genomics. So let's have a look at what has been accomplished and its implications.

We have an abundance of nucleotide sequence data for DNA and RNA

For most of the history of mankind, unraveling the nucleotide sequence of even a quite small nucleic acid was a formidable undertaking. Following 7 years of labor, Robert Holley solved the first such structure, that for an alanine tRNA from yeast, in 1961.[1] This molecule contains a linear chain of 76 nucleotides and includes some unusual bases, which actually help in base sequence determination. For this achievement, Holley shared the Nobel Prize in Physiology or Medicine in 1968.

Subsequently, the technologies for sequence determination of nucleic acids have been revolutionized (more than once and the revolution continues). The processes have been refined and automated. In laboratories doing serious sequence determination work, one finds batteries of automated, robotic machines grinding out long sequences of DNA routinely. Let's take a quick look at what has been gained so far.

I shall focus on nucleotide sequences for the genomes of organisms. The genome is the sum total of genetic information of an organism, generally coded in DNA. At present, we have knowledge of the complete genomes for a substantial and rapidly growing number of organisms.

The smallest genomes are those for viruses. These generally code for just a few viral proteins. The virus makes use of the synthetic machinery of the host cell to express its genome and make multiple copies of itself. It is not surprising that viral genomes were the first to be unraveled. An example is the genome of the bacterial virus ΨX-174 that is a single strand of DNA. Writing down the ΨX-174 genome as a sequence of single letters to represent bases requires a few pages. Were this the genome of a typical animal cell, it would require a book of half a million pages to record it in this form! In my library, a densely written book of 1000 pages is about 2 inches thick. At the same density of letters, a book of half a million pages would be about 30 yards thick.

Next on the list of genomes of increasing complexity are those for bacteria. Many genomes for bacteria are now known. These include the genome for *Escherichia coli*, a favorite experimental organism for geneticists and biochemists. We have as well those for *Haemophilus influenzae*, a flu bug, *Mycoplasma genitalium*, the species with the smallest known bacterial genome and an agent that infects the genitourinary systems causing urethritis, and *Aquifex aeolicus*, a highly thermophilic bacterium that grows on hydrogen, oxygen, carbon dioxide, and mineral salts, a very spare diet indeed. In addition, we have the complete genome for *Helicobacter pylori*, the organism responsible for gastric ulcers. The largest bacterial genomes contain 9–10 million base pairs and encode about 9000 proteins. The Archaea have not been ignored. We have the structure of the genome for *Methanococcus jannaschii*, an organism that employs molecular hydrogen, H_2, to reduce CO_2 to methane, CH_4.

We have the 120 million base pair nuclear genome for *Chlamydomonas*, a unicellular, photosynthetic green alga. This organism diverged from land plants more than one billion years ago. It is a ciliated organism, an innovation inherited from a progenitor of animals and plants, subsequently lost in land plants.

We have the genome for the spirochaete *Borrelia burgdorferi*, the pathogen for Lyme disease, coding for 853 proteins, and that for the spirochaete *Treponema pallidum*, the pathogen for syphilis coding for 1041 proteins. The fact is that we shall soon have the complete genomes for every important pathogenic organism of human beings and, therefore, have the structure of every conceivable molecular target for drug development for the diseases caused by these organisms. This is an exciting possibility in the search for more effective antibiotics, including antivirals.

Working our way up the evolutionary scale, we have the structure for the complete genome of bakers' yeast, *Saccharomyces cerevisiae*. It contains about 12,200,000 base pairs. In addition to the genome for *S. cerevisiae*, we have those for at least 20 more species of *Saccharomyces* and a total of more than 60 fungal genomes. We also have the genome for the nematode worm, *Caenorhabditis elegans*, which contains 97 million base pairs. This is an exciting experimental organism. We have a complete description of its cellular anatomy, including its complete neuronal wiring diagram, which contains a manageable 302 neural cells. This worm turns out to be a valid minimal model for the nervous systems of all animals, including the human nervous system.

The genome for the fruit fly *Drosophila melanogaster* was unraveled in 2000. The genome of this organism contains about 120 million base pairs. These encode about 13,600 genes, somewhat fewer than encoded by the smaller *C. elegans* genome. This particular fruit fly has been a favorite organism for genetic work for decades. In addition to the genome for *D. melanogaster*, we have those for 11 other species of *Drosophila*, some closely related to *D. melanogaster*, others no more closely related to *D. melanogaster* than man is to lizards.

Not only do we have many genomes for human pathogens but also we are collecting those for carriers of human pathogens. For example, the genome for the mosquito *Anopheles gambiae*, the primary vector for malaria in Africa, is known. More recently, that for another mosquito, *Anopheles aegypti*, has been sequenced. This mosquito is a vector for yellow fever, dengue fever, and the chikungunya virus fever. The genomic information will assist scientists in their search for ways to control mosquito vector diseases.

Late in the year 2000, the genome for the first plant was revealed. The genome for the weed *Arabidopsis thaliana* contains about 125 million base pairs coding for 25,500 genes. Thus, the genome is about the same size as that for the fruit flies but has roughly twice the number of protein-coding genes. Though a weed of no commercial value,

A. thaliana is a favorite experimental organism of plant biologists. The genome for the commercially important rice plant, *Oryza sativa*, is also complete. It contains about 400 million base pairs.

Closer to home, we have the genomes for the mouse, *Mus*, a key experimental animal. In addition, we have those for cats and dogs. Still closer to home are the known genomes for the Rhesus macaque, *Macaca mulatta*, an old world monkey that has been an important tool in medical research (definition of the Rh factor, for example). Finally, we have the genome of our closest evolutionary relative, the chimpanzee. Human and chimpanzee genomes are 99% identical in terms of their DNA sequences. The genomes for more than 20 vertebrates are now in hand. This is far from a complete listing of the known genomes but gives a pretty decent sampling of the diversity of organisms for which we have complete genomic sequence data.

There is a strong underlying rationale to the strategy for sequencing genomes

There are a tremendous number of known species that are available for genome sequencing studies, many of which would be attractive experimental targets. So how does one pick out those species for early attention? This is not a trivial question since resources—manpower, money, and machines—are limited for sequencing studies. Many of these efforts require creation of large collaborative groups. There are several dominant rationales for making the choices that have been made and continue to be made. Here are some of them.

First, it has proved compelling to get the genome structures for those species that have been most intensively studied in other ways over the years, including genetic studies. Abundant earlier work provides a context for getting the most understanding out of genome sequencing work. Sequencing work on the fruit fly, *Drosophila*, yeast, *Saccharomyces*, nematode, *Caenorhabditis*, a weed, *Arabidopsis*, and the mouse, *Mus*, among others, was driven by this consideration.

Secondly, getting the genomes for important human pathogens has been driven by the desire to identify every possible molecular target for the development of drugs or vaccines for the associated diseases. Among these efforts are several viral genomes and those of *Helicobacter, Treponema, Haemophilus, Borrelia*, and *Mycoplasma*, all mentioned earlier.

Thirdly, comparative genomics has been a strong motivating force for selection of species for genome sequencing. Perhaps the most compelling case can be made for sequencing efforts among mammals. In addition to the human genome, those for the mouse, rat, dog, cat, and chimpanzee are available (and that for a Neanderthal is coming, see below). It is compelling to see just how closely similar the chimpanzee and human genomes are and to tease out, eventually, the critical differences in the genome that have provided humans with, for example, our unique language ability. Comparative genomics also permits us to further unravel the long, long sequence of evolutionary events involved in getting us to our contemporary spectrum of living species.

Fourthly, ecological and economic issues have proved important factors in selection of species for genomic sequencing. The weed *Arabidopsis* is not only an important experimental species for plant biologists but also plants in general are of enormous importance to us. Getting the rice, *Oryza*, genome was driven by the commercial importance of this plant as a key component of the human diet.

Fifthly, some purely scientific issues have motivated genome sequencing efforts. For example, one hopes that getting the genome for the sea urchin, *Strongylocentrotus purpuratus*, will shed light on the gene regulatory networks for embryonic development. The sea urchin genome had some surprises for us. Among the 23,300 protein-coding genes in its 814 million base pairs were a substantial number previously thought to be vertebrate innovations. The honeybee, *Apis mellifera*, provides a second example. The honeybee provides a powerful example of a relatively simple organism with a complex and highly functional social organization. Its genome structure will provide clues as to how such social complexity is encoded.

Finally, we humans are sufficiently egocentric as to make the sequencing of our own genome compelling. So we turn to that story now.

The sequence of the human genome is very largely known

In early 2001, a substantially complete sequence for the human genome was revealed. The human genome sequence is the product of the work of two groups.[2] The first is the International Human Genome Consortium, headed by Francis Collins of the United States. Members of this Consortium were in academic or otherwise nonprofit organizations and placed the human genome data in publicly available databases as it was developed, with daily updates. This work was supported by several governmental agencies. The competing organization was Celera Genomics, a for-profit biotechnology company headed at that time by J. Craig Venter. The results from the two efforts were published simultaneously. The strategies employed by the two groups in getting the human genome sequence were markedly different.

The Consortium effort began several years before that by Celera. To its considerable advantage, Celera had continuous access to all the results of the Consortium effort. The two groups arrived at basically the same set of conclusions. Work on the human DNA sequence has continued since and refinements have been made. The story of the competition between the two groups has been beautifully described by Nicholas Wade in *Life Script*.[3]

The human genome contains about 3.2 billion base pairs. If the sequence were to be published employing standard type size, it would occupy about 75,000 pages of the *New York Times*. The human genome sequence can be accessed at http://genome.ucsc.edu.

The human genome codes for about 25,000 genes. Thus, we have about twice the number of genes of the fruit fly, a few thousand more genes than the nematode worm, and about the same number as the weed *A. thaliana*. Understanding the enormously greater complexity of human beings compared to the fruit fly, worm, or a weed on the basis of a modest increase in the number of DNA-encoded genes is in significant part understood but a full appreciation will challenge scientists for the next few years.

Now that we have the human genome sequence, what do we have? We have an enormous amount of new information of great value as we go forward but, at the same time, not as much as some would have us believe. On the positive side, we have the sequences of all the proteins coded for by the human genome. We know how many there are. Many of these genes code for proteins that were not previously recognized. Finding these proteins and assigning a function to them, a vigorous and ongoing effort, will vastly increase our understanding of human biochemistry, physiology, and pathology. The sequence reveals much new information about the control of gene expression. We know that something approaching 98–99% of our DNA codes for nothing that we can recognize. Is this DNA functional in some ways that we do not understand? Clarity on this issue is developing: see section on the ENCODE project below.

We know that humans and mice differ in genome structure by about 1.3%. We know that the differences in genome sequence from person to person are very small. The person-to-person differences are as great or greater than the race-to-race differences. In sum, unraveling the human genome is an amazing accomplishment. It is one that will yield immense benefits to humankind. Some of these will be realized in the short term; others after many years.

On the other hand, it is not true that the human genome reveals everything. We are all products of our genetics and our environment. Although the human genome has been termed *The Book of Life*, there are many aspects of human life that are not revealed by the human genome. The human genome is basically like a parts list for the human organism. Now that we have the parts list, we need to understand how the parts are fit together and work together to create human beings.

Do we carry Neanderthal genes in our human genomes?

There has been discussion for many years about the fate of the Neanderthals who occupied part of what is now Europe for a few hundred thousand years until they became extinct about 30,000 years ago. Most authorities believe that modern humans simply outcompeted the Neanderthals for food and other resources, leading to their ultimate demise. A central question remains: did modern humans and the Neanderthals interbreed? Do we carry some Neanderthal genes as a result?

Small samples of uncontaminated DNA have been successfully recovered from Neanderthal bone. It has been possible to sequence about one million base pairs of this Neanderthal DNA by a group of scientists in California and a second one in Germany.[4] So far, no evidence of interbreeding between modern humans and Neanderthals has been uncovered. There are good prospects for having the complete sequence of the Neanderthal genome sometime in 2009. That information may resolve the question of our possible genetic relationship to Neanderthals.

What does a 1% difference in genomes mean?

Above I noted that DNA sequences of the genomes of man and chimps are about 99% the same. Those sequences for the genomes of man and mouse are about 98.7% the same. What is the meaning for these apparently very modest differences at the level of the essential molecules of life? In fact, the commonly cited figure of a 1% difference between "humanness" and "chimpness" may be more misleading than useful. It does serve to indicate just how closely we really are related.

What the 1% figure overlooks is genomic differences that are not reflected by simply aligning the billions of base pairs in the two genomes. The 1% figure reflects base substitutions. It does not include the insertions or deletions of DNA segments that occur in these genomes as these cannot be aligned and compared for base differences. These insertions and deletions account for an additional 3% difference.

The 1% difference also overlooks genes that are randomly duplicated or lost. These also serve to distinguish humans and chimps. Humans have 689 extra genes not shared with chimps. How important these differences are in distinguishing the two species is not clear but the differences are there.

Control of gene expression is also clearly different in the two species and this is likely to be of profound importance. Scientists have looked at which of 4000 genes in brain are turned on at the same time, a measure of connections in gene networks. Clear differences have been revealed. For example, in the cerebral cortex, 17–18% of the revealed gene network connections are unique to humans.

Sorting out the role of all the differences in the human and chimp genomes in defining the clear differences between the two species will require time. For the present, we should remember that we are very closely related to chimps but that the 1% figure underestimates the differences in the two genomes.

The ENCODE project aims to define all functional elements of the human genome

There are follow-on efforts to the Human Genome Project. In fact, there are several others, including an effort to identify every gene involved in determining susceptibility to cancer.

Of particular note is the work of the ENCODE Project Consortium. ENCODE is the acronym for "Encyclopedia for DNA Elements." It is an effort to tease out all the cryptic functional information stored in the human genome. As mentioned earlier and developed in more detail below, most of the human genomic DNA does not code for proteins. Some of it codes for RNA molecules that are not translated to proteins and some of it is there for control of gene expression purposes. But the fact is that we do not know what function, if any, the bulk of our genomic DNA plays. The goal of ENCODE is to define all the functional elements of the human genome. Work on this project is very early at the time of this writing (about 1% of the human genome has been analyzed) but unexpected and intriguing results are already in hand.

Perhaps the most surprising result is that the human genome is pervasively transcribed. Transcription is generally associated with the process of protein synthesis or with synthesis of one of the classes of nontranslated RNA. Such transcription would involve a very small fraction of DNA bases. The fact that a majority of DNA bases are transcribed means that there is a lot of RNA floating around for which we have no knowledge as to function.

Many of these unanticipated RNA transcripts have been identified. Some of these derive from regions of the genome previously considered to be transcriptionally silent. Still others overlap protein-coding regions. The ENCODE project compares functional DNA elements from man, monkey, and mouse. Novel evolutionary relationships are certain to result. ENCODE is early but it promises great new insights into how the genome works.

The mitochondria of eukaryotic cells have their own genome

All eukaryotic cells possess subcellular organelles known as mitochondria. The central function of the mitochondria, though not the only one, is to produce ATP, the cellular energy currency, coupled to the oxidation of food-derived molecules.

The mitochondria are derived from some ancient bacteria, perhaps related to the present day *Rhodobacter*. Given the organismal background of mitochondria, it should not be surprising that they possess a DNA genome, a remnant of their original bacterial chromosome. Much of that original chromosome has been lost or transferred to the nuclear genome. There have been a few hundred transfers of mitochondrial DNA to nuclear DNA over time. What remains retains two characteristics of the chromosome from which it was derived: the mitochondrial chromosome is circular, as are most bacterial genomes, and it is not complexed with histones. Each mitochondrion has 5–10 copies of its DNA genome.[5]

Fred Sanger, a double Nobel Prize winner, sequenced the human mitochondrial genome back in 1981. This genome codes for 13 proteins and the mitochondrion possesses the genetic machinery needed to synthesize them. Thus, the mitochondria are a secondary site for protein synthesis in eukaryotic cells. It turns out that the 13 proteins coded for by the mitochondrial genome and synthesized in the mitochondria are critically important parts of the complexes of the electron transport chain, the site of ATP synthesis. The nuclear DNA codes for the remainder of the mitochondrial proteins and these are synthesized on ribosomes, and subsequently transported to the mitochondria.

Brain, muscle, and kidney cells, for example, all possess a few hundred or a few thousand mitochondria per cell. The human egg cell is remarkable in that it contains about 100,000 mitochondria. A sperm cell, in contrast, contains fewer than 100. New mitochondria are made as cells divide. The synthesis of new mitochondria requires that the proteins coded for by the nuclear genome and those coded for by the mitochondrial genome be mutually compatible to ensure optimal mitochondrial function. Since we can experience mutations in both nuclear and mitochondrial DNA, leading to alterations in mitochondrial proteins, long-term compatibility is not ensured. As I develop further in chapter 17, loss of this compatibility may underlie, in part at least, the phenomenon of aging.

Genome size does not directly reflect the complexity of organisms

We have an innate feeling that some organisms are more complex than others. Organisms that consist of a single cell are less complex than those that have many cells. Multicellular organisms that have many cell types are more complex than those that have few. Organisms that possess highly elaborated structures are more complex than those that do not. The human brain, for example, is generally considered to be the most complex organ that the process of evolution has yet devised. We rightly see ourselves as more complex organisms than, say, frogs, which are more complex than worms, which are more complex than bacteria.

In the preceding section, we have noted that the size of the genome of viruses is smaller than that of bacteria, which is smaller than that of worms, which is smaller than that of human beings. We might conclude that the cellular DNA content, or the size of the genome, correlates well with organismal complexity. While there is a trend is this direction, that statement is too simple. Let us look at some examples provided in table 14.1, in which the size of the genome is provided for several groups.[6]

The first thing that we note is that there is a lot of variation in genome content within individual phyla. Consider the birds for example. At one extreme, there are birds that have genomes containing 700 million (7×10^8) nucleotide base pairs and at the other extreme there are some that contain 2 billion (2×10^9) base pairs, a variation of about a factor of three. The variation is considerably greater for the amphibians: 600 million base pairs at one end of the spectrum to 80 billion! (8×10^{10}) at the other end.

Table 14.1 Ranges of values of base pairs per haploid genome for selected groups of plants and animals[a]

Group	Range of number of base pairs
Bacteria	7×10^5 to 1×10^7
Fungi	1×10^7 to 1×10^8
Plants	1×10^8 to 1×10^{11}
Insects	1×10^8 to 6×10^9
Molluscs	3×10^8 to 5×10^9
Cartilaginous fish	3×10^9 to 1×10^{10}
Bony fish	3×10^8 to 4×10^9
Amphibians	6×10^8 to 8×10^{10}
Reptiles	1.5×10^9 to 3×10^9
Birds	7×10^8 to 2×10^9
Mammals	2×10^9 to 3.5×10^9

[a]All figures are for a single chromosome per genome. The values provided in the table are for the purpose of general comparison. Values for a few individual species within each group may fall outside the indicated ranges. Note that 10^6 is a million and 10^9 is a billion.

The example of the amphibians immediately raises the second point. The correlation between organismal complexity and genome size is far from exact. Thus, some amphibians, members of an old group of animals in the evolutionary process, have a genome that is perhaps 25 times larger than our own, about 3.2 billion base pairs, or that of other mammals, 2–3 billion base pairs. The simple organism *Amoeba dubia* provides an extreme example: its genome includes 670 billion (!) base pairs and codes for far fewer proteins than the much smaller human genome.

At first glance, this seems quite surprising. DNA is information. The fact that it must require more information to code for something as complex as a human being than to code for a simpler organism like a frog would suggest that the genome size should be greater in humans than in frogs.

However, this finding may seem less surprising when we consider the nature of the genome in more detail. The fact is that a lot of the genome of higher organisms does not code for proteins or perhaps for anything else.

Here is a formal definition of a gene

There are alternative ways to define a gene: as a unit of inheritance in genetics, for example. However, in molecular terms, *a gene is the entire nucleic acid sequence that is necessary for the synthesis of a functional polypeptide or RNA molecule.*

This definition expands our earlier thoughts in two ways. First, we have included specification of RNA molecules as well as proteins. As noted above, many RNA molecules serve a role as message carrier from DNA to the protein-synthesizing machinery (mRNA) and are translated into protein structures. Other RNA molecules serve other functions: as components of the ribosome, rRNA, or as an interface between mRNA and amino acids in protein synthesis, tRNA. Finally, we have the very small RNA molecules known as siRNA. These species of RNA are not translated into proteins and this requires that the definition of a gene include the specification of their structure.

Secondly, our formal molecular definition makes it possible to include several classes of DNA sequence that are required either for control of the initiation of DNA, RNA, or protein synthesis.

Now we have three classes of DNA sequence: (a) sequences that code for protein structures; (b) sequences that code for RNA structures; and (c) sequences that act as part of mechanisms for control of macromolecular synthesis. There is a fourth class. Particularly in the cells of higher organisms, there is a great deal of DNA that seems to have no function at all. As noted above, as much as 98–99% of the DNA of human beings codes for neither protein nor RNA, nor is it involved in control of macromolecule synthesis in a known manner. The fact that the great bulk of the DNA of higher organisms serves no evident function makes it easier to understand the data collected in table 14.1, which show the lack of a compelling correlation between organismal complexity and genome size. Perhaps many amphibians, for example, simply have more nonfunctional DNA than do mammals.

That much of the human genome has no apparent function is not the same as saying that it has no use. In fact, we have been able to exploit the apparently nonfunctional DNA for several purposes. A good deal of this DNA consists of repeated sequences of bases. There is, for example, a more-or-less 300 base pair sequence termed Alu that is found at more than 500,000 positions in the human genome. The frequency of Alu repeats and their location in the human genome varies from person-to-person and acts as a DNA fingerprint. These and other repeats provide the basis for the use of biotechnology in forensic science.

Introns are noncoding nucleotide segments within protein-coding genes

The existence of introns, base sequences within a protein-coding gene that do not code for amino acids, marks a clear difference between the genomes of prokaryotic cells (no nucleus) and eukaryotic cells (with a nucleus such as almost all cells of mammals). Introns in prokaryotic genomes are very rare. Introns in cells from single-cell eukaryotes such as yeast are uncommon. Eukaryotic genomes from higher plants and animals are full of them. Earlier, we developed the analogy of a gene as a sentence in a chromosomal book. It follows that a gene coding for a protein in a prokaryotic cell would read like a perfectly normal sentence, making sense from beginning to end. In contrast, a gene coding for a protein in an eukaryotic cell from a higher plant or animal might well alternate between sensible words and gibberish, sequences of letters which do not make recognizable words. The exons constitute the sense and the introns the gibberish. If we want a simple coherent sentence, we would need to excise the introns and splice together what remains, a sequence of exons. Something like this happens in the process of protein synthesis.

Genes from higher organisms can be very complex indeed. Noncoding regions on the ends of the gene plus introns can account for 95% or more of the base pairs of the gene and only 5% or less may actually code for a protein. A spectacular example is the DMD gene, which is associated with the disease Duchenne muscular dystrophy. The DMD gene is the largest known human gene. It contains more than two million base pairs. The great bulk of this comprises many introns, which interrupt the coding sequence, and terminal noncoding regions. This gene is about twice as large as the entire genome of the pathogens for Lyme disease and syphilis and significantly larger than the entire genome for the causative agent of ulcers, *H. pylori*!

Gene splicing provides a path to protein diversity

Exons and introns are far more than a curiosity. In protein synthesis (see below) the introns must be eliminated in a process known as splicing: the introns are spliced out, or removed, and the exons are joined together as required to produce a coherent message for protein structure. Exons can be spliced out (removed or skipped over) as well. The upshot of all this is that one gene apparently coding for one protein can, in fact, code for several or even many. To see how this works, let's consider an mRNA molecule transcribed from DNA that contains four exons separated by three introns:

Now let us suppose that all proteins coded for by the mRNA molecule must include exon-1 and ask how many proteins we can make. The answer is seven.

If we remove everything except exon-1, then we just get a protein encoded by exon-1. If we splice out intron-1 and intron-2 and everything beyond it, then we get a protein encoded by exon-1-exon-2. If we splice out intron-1, exon-2, intron-2, intron-3, and exon-4, we get a protein encoded by exon-1-exon-3. If we remove all three introns and link the exons, we get a protein encoded by exon-1-exon-2-exon-3-exon-4, and so forth. So the set of seven proteins are encoded by:

exon-1

exon-1-exon-2

exon-1-exon-3

exon-1-exon-4

exon-1-exon-2-exon-3

exon-1-exon-3-exon-4

exon-1-exon-2-exon-3-exon-4

These are all the possibilities, beginning with exon-1. If we relax that condition, then we can clearly make several more: exon-2-exon-3, for example. These proteins are known as splice variants. It follows that a genome coding for 25,000 proteins can actually generate quite a few more than that. Complexity gets stacked on complexity.

Genes may encode polyproteins

Let's return to the basic idea of genes as genetic elements that code for proteins, without worrying about the subtleties of introns and terminal noncoding regions. The simplest view—that each coding gene codes for a single protein and does so only once in the genome—is too simple. For example, in viruses and bacteria, protein-coding genes frequently code for polyproteins, large protein molecules that contain several individual proteins linked end-to-end. These polyproteins must be subsequently cleaved to free the individual proteins.

As an example, consider the genome of the hepatitis C virus. The genome codes for a single polyprotein that contains about 3010 amino acids. The polyprotein is subsequently cleaved into at least 10 proteins, including several which are structural components of the virus particle and several which are required, in one way or another, for the process of replication of the viral RNA or cleavage of the polyprotein itself into its functional units.

In cells of higher organisms, synthesis of polyproteins is rare and individual protein-coding genes usually do code for individual proteins, or their splice variants.

In this sense, the situation is simpler. However, a complexity is that each coding (protein or RNA) gene is not necessarily confined to a single appearance in the human genome, though about 25% of such genes do, in fact, appear only once. In the three most important cases, coding genes are organized into tandemly repeated arrays: those for rRNA, tRNA, and nucleic acid-binding proteins known as histones. In these cases, one finds a sequence of identical coding genes lined up on the DNA molecule one after the other ... gene.gene.gene.gene ... These arrays may be hundreds of genes long.

Finding tandemly repeated genes is not surprising. Although RNA and protein synthesis occur rapidly, there is a limit to the rate. If each gene appeared in the genome only once, that would limit the rate of synthesis of each gene product, whether it be RNA or protein. If a given gene appears one hundred times, then the rate of synthesis of the corresponding protein or RNA may be one hundred times greater than if the gene appeared only once. Clearly, nature has provided for the rapid synthesis of rRNA, tRNA, and histones through the device of tandemly repeated gene arrays. These arrays are created by the repeated duplication of the original gene.

Gene duplication has a second important consequence: the creation of gene families and, consequently, protein families. The distinction between tandemly repeated gene arrays and gene families derives from the subsequent history of these genes. If duplication is followed by no (or very few) changes in gene structure, then we get tandemly repeated arrays of genes, all coding for an identical or nearly identical RNA or protein molecule, each serving the same function. In contrast, if gene duplication is followed by mutation (changes in the sequence of bases along the gene, for example) then the progeny gene may code for a protein similar but not identical to the parent one. Over time, small original differences may become amplified by further mutations.

The living world is strikingly unified at the molecular level

There is striking unity among all living things at the molecular level. The genetic material throughout all the diversity of nature is DNA (or RNA in the case of the retroviruses). The biochemical specificity of all living things is expressed in the spectrum of the proteins that they elaborate under the direction of DNA and which serve related purposes throughout nature. The genetic code is very nearly universal. Thus, all living organisms are related: life on Earth as we know it began once, though the beginning is vague, and the diversity of living forms reflects the ongoing process of change known as evolution.

The essentials of chemistry and physics are unchanging. They reflect the basic, static properties of matter. In contrast, life, and therefore biology, is constantly changing and evolving as different varieties of living organisms are created from the same fundamental elements. The diversity of life is a study of variations on a single chemical theme.

Perhaps the most striking manifestation of the unity of life is the commonality of genes that are responsible for some of the most basic functions of life. It is a fact that many human genes can be substituted for the corresponding yeast genes and the yeast does perfectly well. It is also true that many human genes can be substituted for fruit fly genes and the fruit fly does perfectly well. We find homologs of human genes everywhere: flies, worms, bacteria, viruses.

Let us consider one striking example: the homeotic genes. If we think about the development of the fruit fly from a single-cell embryo we will quickly see a series of major problems. How does the developing organism distinguish front from back, top from bottom? How is the body plan laid down? How does it know where the head ends and the body begins? Where the wings should be? Where the legs should be? The homeotic genes are responsible for encoding the body plan from front to back (or head to tail if you prefer), the anteroposterior axis. Mutations in these genes result in the replacement of one body part for another.

Part of the answer came in the 1970s from the work of two German scientists: Christiane Nüsslein-Volhard and Eric Wieschaus.[7] They found a series of eight homeotic genes clustered on a single fruit fly chromosome. Most surprisingly, the order of these genes on the chromosome is exactly the same as the order of the body parts that they control in the mature fruit fly. Thus, the first gene in the line affected the mouth, the second in line the face, and so on, until we reach the end of the abdomen. In the fruit fly, these became known as the *Hox* genes. It was later discovered that each of these genes has a common sequence of 180 base pairs. When translated into protein, this sequence codes for a DNA-binding domain. The protein products of the *Hox* genes control the activity of other genes: that is, they switch other genes on and off.

This remarkable discovery in the fruit fly (which was rewarded by a share of the Nobel Prize in Physiology or Medicine in 1995)[8] led others to look for corresponding genes in other organisms. They were quickly found. For example, in the mouse, there are four clusters of a total of 39 *Hox* genes. These are organized in just the same way as the *Hox* genes of the fruit fly: the first gene in the sequence affects the head, and so on down the line to the tail. Remarkably, the *Hox* genes of the mouse are quite similar to the *Hox* genes of the fruit fly. Humans have the same spectrum of *Hox* genes as the mouse and, by extension, much in common with the *Hox* genes of the fruit fly.

Finally, if we take a fruit fly and knock out one of its *Hox* genes and then replace it with the corresponding human *Hox* gene, we get a perfectly normal fruit fly. As Matt Ridley has phrased it: "Flies and people are just variations on a theme of how to build a body that was laid down in some worm-like creature in the Cambrian period."[9] This is a remarkable example of the underlying unity of living organisms.

Summary

We have moved from the relatively simple structures of the nucleotides, which form the basic structural units of the nucleic acids, through to their helical secondary structures, to the base sequences of many genomes. In doing so, we have related structure to function for both DNA and RNA, just as we did for proteins in earlier chapters.

Key Points

1. We have an abundant collection of nucleotide sequence data, including the human genome, among many others.

2. There is a family of criteria employed to prioritize investments in collection of genomic sequence data: intensively studied organisms, important human pathogens, comparative genomics, ecological and economic impacts, and key basic scientific issues.

3. Our collection of genomic data includes sequence data for many of the pathogens of humans and animals.

4. There are near-term prospects for sequencing the genomes of individuals rapidly and at reasonable cost. This ability would have profound implications for the future of individualized medicine.

5. The human genome contains about 3.2 billion base pairs and these code for about 25,000 proteins. Most of the human genome does not code for anything that we understand, though the human genome appears to be pervasively transcribed.

6. Differences in genome structure from person to person are small. These small differences are as great or greater than the race-to-race differences in genome structure.

7. For those segments of DNA that can be directly compared, the genomes of man and mouse differ by about 1.3%. However, this value somewhat underestimates the actual difference in genome structure, which is amplified by insertions, duplications, and deletions.

8. Mitochondria have their own limited genome, the remnants of the genome of the microorganism from which they are derived. Mitochondrial genes code for 13 proteins that are synthesized in the mitochondria and are critical parts of the mitochondrial electron transport chain.

9. There is no simple correlation between organismal complexity and genome size.

10. A gene is the entire nucleic acid sequence that is necessary for the synthesis of a functional polypeptide or RNA molecule.

11. The various ways in which messenger RNA molecules may be spliced provides a mechanism for diversity in protein structures derived from a single gene.

12. Genes may code for polyproteins.

13. The living world is strikingly unified at the molecular level, as revealed by the interchangeability of certain genes among widely different organisms.

15

Vitamins

Molecules of life

Many vitamins or their metabolic derivatives are essential coenzymes for multiple metabolic reactions of life; others or their metabolic derivatives are transcription factors, regulating gene expression and, therefore, protein synthesis.

Every morning I begin my day by taking a multivitamin/mineral pill. It serves me as a kind of insurance policy: whatever my diet may be on a given day, I am reasonably assured of fulfilling all my needs for vitamins. This by no means the same as saying that it does not matter what I eat once I have taken my vitamin pill; it surely does. Taking vitamin pills does not excuse one from consuming a sensible diet, rich in vegetables and fruits and sparing in high fat foods. At the same time, vitamin pills are inexpensive and are, in my mind at least, a sensible insurance policy.

The name "vitamin" comes originally from "vita" meaning life and "amine" and was originally spelled vitamine, since some early nutritionists thought that all vitamins were amines. That turns out not to be the case, as we shall see, but the name has endured anyway. The history of the discovery of the vitamins is highly interesting and I shall cite a case or two as we move forward. Physicians or nutritionists, having observed that certain diseases could be corrected by administration of vitamins, did most of the early work. These diseases, such as scurvy, pellagra, beriberi, and rickets, are vitamin deficiency diseases.

The label on the bottle of my multivitamin pills lists, in this order, the following ingredients: vitamins A, C, D, E, K, thiamine, riboflavin, niacin, vitamin B_6, folic acid, vitamin B_{12}, biotin, and pantothenic acid. There follows a substantial list of minerals, calcium, phosphorus, iodine, ..., and the list concludes with lutein and lycopene. The final two ingredients are not vitamins, and I will have nothing to say about them. The amount of each ingredient is specified on the label, both in quantitative terms and as a percentage of the reference daily intake, RDI. The RDI is the value established by the Food and Drug Administration (FDA) of the United States for use in nutrition labeling. It was initially based on (and replaces) recommended daily allowances, RDAs, established by the FDA in 1968 and updated most recently in 1989.[1] The RDAs were targeted to ensure adequacy for 97–98% of all members of a group. There are minor but significant differences between the current RDIs and the 1989 RDAs. Values for RDIs vary as a function of age, gender (occasionally), and whether or not one is pregnant or lactating. For certain nutrients, upper limit, UL, values have also been established. UL values are the maximum level of daily intake that is likely to pose no threat of adverse effects.[2] I shall cite RDIs and ULs, where available, for each vitamin I consider as we move through the list.

The great hero of the field of human nutrition is Elmer McCollum, who spent most of his distinguished career at Johns Hopkins University. McCollum did his pioneering work in the first half of the twentieth century. McCollum showed the way to determine nutritional requirements through careful use of animal experimentation. Along the way, he discovered vitamin A in 1913 while at the University of Wisconsin and vitamin D in the early 1920s. His work resulted in the fortification of such staples as milk and bread with vitamins in the 1940s and beyond, and for the increased appearance of fresh fruits and vegetables on the meal tables in American homes. In recognition of his work, *Time* magazine dubbed him Mr. Vitamin in a 1951 article. I doubt that there are many people who can lay equal claim to having contributed to human well-being.

The obvious question about vitamins is: why do we need them? What is the role of each vitamin in human health and physiology? What happens if we do not get sufficient amounts of a vitamin? The central purposes of this chapter are to identify these vitamins chemically and to provide answers to these questions. I am not going to consider all vitamins on the list but focus on those that seem most important to understand.

Vitamins or their derivatives frequently act as coenzymes

To understand the basic role served by vitamins in human health, we need to retreat to chapter 9 for a moment. We discussed there the nature of an amazing group of proteins, the catalysts of living systems, enzymes. Many enzymes are perfectly happy to work by themselves and they do work by themselves. Others, in contrast, require a partner for their catalytic functions. These partners are known by the generic name of *coenzymes*. Basically, they cooperate with enzymes in getting the catalytic job accomplished. Although his name implies that he worked alone, the Lone Ranger needed Tonto to get his job done. Think of coenzymes as Tonto.

Now here is the central understanding—the role of several vitamins is to serve as coenzymes or as metabolic precursors for coenzymes: that is, the vitamin itself may serve as coenzyme or it may be converted in the human body to a coenzyme. The other key point I suppose is obvious but I am going to state it anyway: we need vitamins in our diet because we cannot make them ourselves. In that sense, they are like essential amino acids or essential fatty acids: stuff that we need but cannot make ourselves and so must obtain from dietary sources. So let's get started in understanding these critical molecules and how they serve the needs of human beings.

Vitamin A is central to vision

Vitamin A is a bit complicated, not only in the ways that is works but also in nomenclature. We may take vitamin A to mean retinol, shown here in the all-*trans* form:

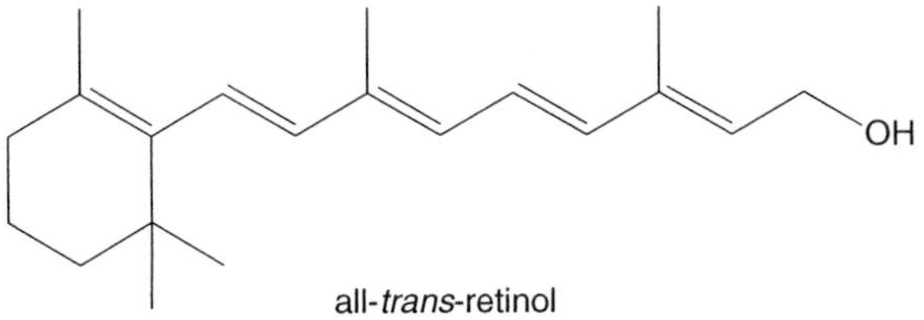

all-*trans*-retinol

plus a family of carotenoids that are converted to retinol in the human body and are collectively known as provitamins A. These include α-carotene, β-carotene, and β-cryptoxanthin. Retinol contains 20 carbon atoms; the carotenes, in contrast, contain 40. Basically, a carotene is similar to two retinol molecules hooked together tail-to-tail. The carotenes are intensely orange compounds. They are found in high concentrations in carrots, for which they are named. In vivo, the carotenes are converted, to a lesser or greater extent, into retinol. Thus, we account for the designation provitamins A.

The RDI/RDA for vitamin A for adult males is 900 micrograms/day (0.9 mg/day) and for adult females 700 micrograms/day (0.7 mg/day).[3] Children require significantly less and lactating women significantly more. There are a number of excellent sources of vitamin A: fish, dairy products, liver, leafy vegetables, and dark-colored fruits.

Here is a case in which too much of a good thing is not a good thing. Excess vitamin A is toxic, particularly to the liver. The UL is 3 mg/day for adult men and women, only three or four times greater than the RDIs. Most health food stores sell β-carotene supplements. Unless you have reason to believe that you are at risk of vitamin A deficiency, there is no reason to use these supplements.

Vitamin A has a rich associated human physiology. It is associated with vision, regulation of gene expression, reproduction, embryo development, and immune function. We cannot manage all of this but let's get started with vision.

A vitamin A derivative, 11-*cis*-retinal, is central to the visual process in humans. It is derived from all-*trans*-retinol by converting the terminal hydroxyl group to an aldehyde:[4]

$$R—CH_2OH \rightarrow RCHO$$

where "R" designates the rest of the retinol/al molecule, and changing the stereochemistry at C11 from *trans* to *cis* (see the figure below).

In photoreceptor cells, the rods and cones of the human retina, the retinal is linked to a specific protein termed opsin. The resulting pigment is known as rhodopsin. When a photon of light of the proper wavelength hits a molecule of rhodopsin, two chemical events take place. First, the 11-*cis*-retinal is converted to the all-*trans* form and, secondly, the all-*trans*-retinal is released from the rhodopsin:

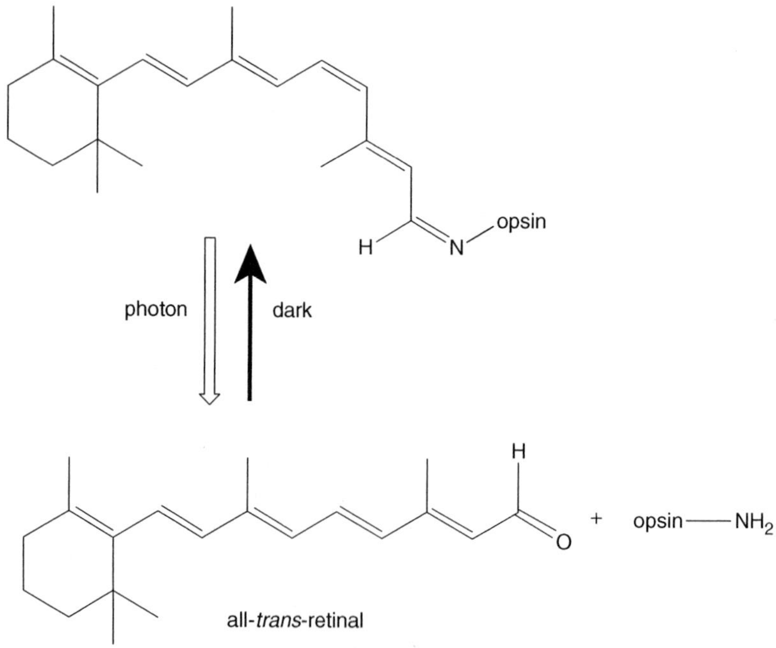

all-*trans*-retinal

It is the first of these reactions, the conversion of the retinal moiety attached to opsin from the 11-*cis* to the all-*trans* form, that generates a signal in neurons termed bipolar cells. These are neuronal cells associated with the rod and cone cells of the retina. This seemingly simple chemical reaction in which the geometry around a carbon–carbon double bond is changed from *cis* to *trans* is the fundamental molecular process underlying vision. The bipolar cells pass the signal along to ganglion cells that, with the help of yet other retinal neurons, integrate the information from all of the rhodopsin molecules of all of the photoreceptor cells. Once done, the integrated signal is sent, in the form of action potentials along the axons of the ganglion cells, to the optic nerve and, from there, to the brain. An image results. Vision happens. The released all-*trans* retinal is reconverted to the 11-*cis* form by an enzyme in a process that does not require light. It then reassociates with opsin to form rhodopsin and we are ready for the next photon. All this happens very fast, in a very small fraction of a second.

In the absence of sufficient vitamin A, one consequence is a vision defect termed night blindness or, more technically, xerophthalmia. The rod cells in the retina are largely responsible for vision in low light. If they have too little retinal, they cannot do their job and night blindness, the inability to see in low light, results. If the condition persists, it is characterized by extreme dryness of the conjunctiva and can result in permanent blindness.

Adequate vitamin A is also required for the health of the surface structures of the human body: skin, the lining of the digestive system, lungs, and mouth. These structures, particularly the skin, are an important part of the human defense system against infection. When they are compromised, the human body becomes increasingly susceptible to infection, a profound problem in areas of poor public health resources. This story, and what has been done in response, is beautifully told in Philip Hilts' book: *Rx for Survival*.[5]

Let's move on to another derivative of retinol, retinoic acid:

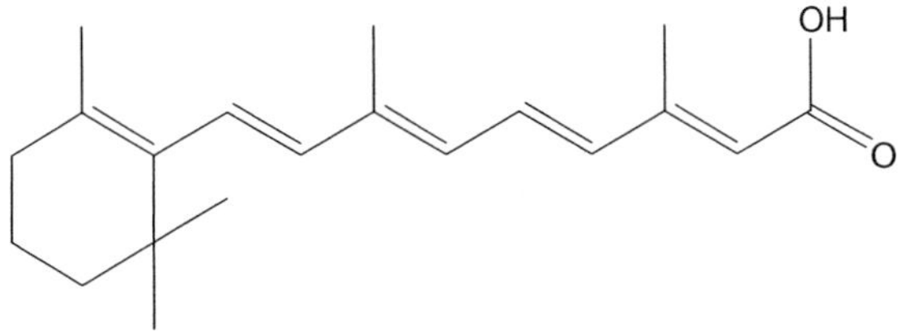

all-*trans*-retinoic acid

Here we have another simple chemical transformation, the oxidation of the aldehyde group in retinal to the carboxyl group of retinoic acid. Although the chemical change is simple, the physiology is profoundly different. Retinoic acid has nothing to do with vision but a lot to do with development and differentiation. Here we have yet another example of the sensitive interdependence of chemical structure and biological function.

We have a collection of genes known as *Hox* genes.[6] We encountered these genes earlier in chapter 14. They are intimately associated with embryonic development and are basically responsible for specifying the structural organization of the embryo. They provide the information necessary to answer questions such as: how many arms should there be? Where should they be attached? Basically, the *Hox* genes specify the nature and fates of embryonic cells. They are expressed in specific patterns and at different stages in the creation of the human embryo.[7] How this expression is controlled in time and space remains to be established. For our immediate purposes, we just need to recognize that *Hox* gene function is not something that one wishes to mess around with. Doing so is highly likely to result in getting something wrong in the development of the embryo. As it turns out, retinoic acid is a central player in the drama of differentiation and development and influences *Hox* gene expression. Since it is involved in the determination of structure, retinoic acid is known as a morphogen.

However, here, too much of a good thing is not so hot. If retinoic acid is administered systemically during embryo formation (in experimental animals to be sure), severe malformations result in the offspring. Thus, retinoic acid is a teratogen.

A derivative of all-*trans*-retinoic acid, 13-*cis*-retinoic acid (isotretinoin), is highly effective as a treatment for severe cystic acne. It is marketed under the trade name Accutane. However, remembering that retinoic acid is a teratogen, its use in women who are pregnant or who may become pregnant is contraindicated. Inappropriate use can result in infants having cranial deformations. The labeling of Accutane contains a large black box warning potential users of the teratogenic potential of this drug.

Retinoic acid works through a family of retinoic acid receptors. Activated retinoic acid receptors act as transcription factors, just as activated steroid receptors do (chapter 20), and alter the transcription of genes affecting cell division and survival. This underlies both the teratogenic potential and the therapeutic utilities of these potent molecules.

Vitamin C is the cure for scurvy

Vitamin C is commonly known as ascorbic acid:

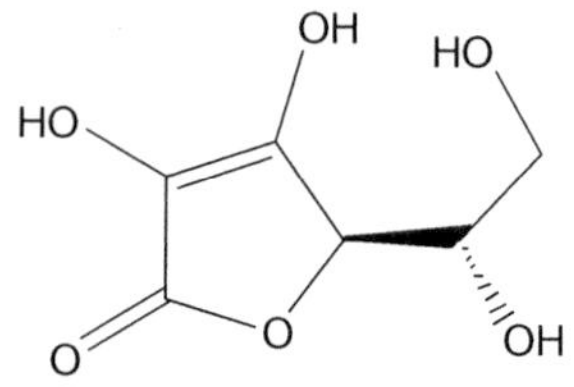

Vitamin C has both an interesting history and very wide use today. The RDI for ascorbic acid for adult males is 90 mg/day and for adult females 75 mg/day. As expected, children require less and pregnant or lactating women more. Although ascorbic acid is quite a safe molecule, the UL is 2 grams a day for both men and women. Many people in the United States get very substantially more than the RDI, largely through taking ascorbic acid as a nutritional supplement. Personally, I take a gram of this stuff each morning, along with my multivitamin pill for reasons that I get to below. There are many good food sources of vitamin C: citrus fruits, tomatoes, potatoes, cabbage, cauliflower, Brussels sprouts, spinach, and strawberries among them. Synthetic vitamin C is the pure stuff and is inexpensive. Vitamin C isolated from rose hips contains small amounts of other natural compounds and is somewhat more expensive.

A deficiency of vitamin C results in scurvy, a dreadful disease characterized by swelling of hands, feet, and gums. As the disease progresses, teeth become loose and eventually fall out, old sores and scars reopen, and eventually death results. Scurvy was unknown in Europe until the Age of Discovery, when lengthy sea voyages were undertaken without the provision of citrus fruits. The first notable outbreak of scurvy in Europe occurred in 1498 in the crew of Vasco da Gama, while exploring the west coast of Africa. In his writings, Vasco da Gama noted that his men made striking recoveries after buying oranges from Arab traders. The Arabs must have known of the disease and its cure. But the point eluded Vasco da Gama and others in spite of the clear evidence in front of them. In fact, scurvy continued as a major problem for sailors for another three hundred years.[8]

In 1746, a Scottish naval surgeon named James Lind carried out a carefully controlled study of the effect of diet on scurvy and demonstrated, beyond reasonable doubt, that oranges and lemons would cure (or prevent) scurvy. However, it was not until 1795, about three hundred years since it was known that citrus fruit would cure scurvy and about 50 years after Lind's definitive work, that the British Royal Navy insisted that sailors receive a daily dose of a citrus fruit. Opinion and prejudice outweighed scientific evidence to the detriment of many for far too long.

The isolation and structural determination of ascorbic acid proved no easy task. Only in 1932 did W. A. Waugh, C. G. King, and Albert Szent-Györgyi isolate and synthesize ascorbic acid. Szent-Györgyi, a talented and energetic Hungarian scientist, received the 1937 Nobel Prize in Physiology or Medicine largely on the basis of this work. Later, he fled the menace of Adolf Hitler and the Nazis and, at one point, was sheltered in Turkey by my former colleague at Indiana University, Felix Haurowitz. Both found their way to the United States where they, along with a host of fellow European émigré scientists fleeing the madness of the Nazi reign, spent the remainder of their productive scientific lives. Science in the United States benefited enormously.

So what does this magical molecule do? Actually, it does two things, one rather more crystalline clear than the other. The crystal clear thing that ascorbic acid does is act as coenzyme for an enzyme known as prolyl hydroxylase. This enzyme catalyzes the conversion of the amino acid proline to hydroxyproline, a major, if exotic, amino acid in the structural protein collagen:

proline   4-hydroxyproline

Collagen synthesized in the absence of ascorbic acid (i.e. without hydroxyproline) cannot form its usual stable structure. Collagen is a major component of the structural and connective tissues of the body: bone, cartilage, tendons, ligaments, teeth, and skin. Small wonder that things sort of fall apart in the absence of adequate ascorbic acid to support the activity of prolyl hydroxylase.

The less crystal clear thing that ascorbic acid does is act as an antioxidant. Without delving into detail, suffice it to say that there are various lines of evidence that strongly suggest that oxidative reactions cause cellular and tissue damage and, in part, underlie the process of aging. These oxidation reactions are mediated mainly by superoxide, O_2^-, a species formed by the addition of an electron to molecular oxygen, hydrogen peroxide, H_2O_2, the hydroxyl radical, $OH\bullet$, and singlet oxygen (we met these species back in chapter 7). These species are formed in the course of normal cellular metabolic activity and the human body has various, if not completely adequate, defenses against them, including ascorbic acid and vitamin E. Such molecules react with these oxidizing species and neutralize them. They are generically known as antioxidants.

Many people, including myself, take substantial amounts of ascorbic acid, since it is cheap and quite safe below the UL level, in the hope, unsupported by much scientific evidence, that its antioxidant properties will contribute to health and longevity.

Vitamin D is critical to bone health: rickets and osteomalacia

Vitamin D is really a small family of closely related molecules that prevent the bone disease rickets in children and osteomalacia in adults. In both cases, inadequate mineralization of bone results in bone deformation and weakness. Calcium, Ca^{2+}, homeostasis is one goal of vitamin D activity, a goal it shares with parathyroid hormone and calcitonin. Calcium is intimately involved in bone mineralization and disturbances of calcium levels in the blood can result in inadequate bone mineralization or excessive calcification of other tissues.

The RDI for vitamin D is 5 micrograms/day for men and women, adults and children, a value independent of state of pregnancy or lactation. Middle-aged adults, ages 50–70 years old, require 10 micrograms/day and for those over 70 years old the value goes up to 15 micrograms/day. Fish oils, the flesh of fatty fish, and fortified milk and cereals are excellent sources of vitamin D. Of course, vitamin D is also present in most multivitamin pills.

Vitamin D is a lipid-soluble vitamin, as is vitamin A. Lipid-soluble vitamins are stored in the body, in contrast to water-soluble vitamins such as vitamin C. Excess consumption of lipid-soluble vitamins can result in excess storage and resultant toxicity. The UL for vitamin D is 50 micrograms/day. Excess consumption may raise the blood level of calcium to the extent that calcification of organs (particularly the kidneys), occurs, and formation of kidney stones may follow.

Vitamin D is commonly found in two forms, both derivatives of steroids. These are known as cholecalciferol (vitamin D_3) and ergocalciferol (vitamin D_2). Cholecalciferol is formed directly from the steroid known as 7-dehydrocholesterol by the action of sunlight, or other source of ultraviolet (UV) light, in the skin of humans and other animals. When parents encourage their children to go out and play in the sun, there is good reason for doing so (but avoid sunburn).[9]

Ergocalciferol is formed from the plant sterol ergosterol by the action of UV light. It is ergocalciferol that is commonly found in vitamin pills and vitamin D-fortified foods. The biological activity of cholecalciferol and ergocalciferol is basically the same.

In the human body, cholecalciferol and ergocalciferol undergo two metabolic transformations to yield the active vitamin D molecule. These are additions of hydroxyl groups, first in the liver to produce 25-hydroxyvitamin D and then in the kidney. The final product has the unwieldy name 1α, 25-dihydroxycholecalciferol, and is more commonly known by its simpler name 1,25-dihydroxy vitamin D or, even more simply, $1,25(OH)_2D$.

The principal circulating form of vitamin D is 25-hydroxyvitamin D and it is this form of the vitamin that is measured in efforts to determine your vitamin D status.

Here are some insights into how $1,25(OH)_2D$ works. Like steroid hormones and retinoic acid, $1,25(OH)_2D$ binds to and activates a cytosolic receptor present in most cells of the human body. The activated receptor migrates to the cell nucleus, binds to a specific nucleotide sequence in the nuclear DNA, and acts as a transcription factor. Directly or indirectly, the expression of some 200 genes is affected as a result.

Much discussion of vitamin D focuses on bone health, though this is by no means the only focus on vitamin D action. One result of $1,25(OH)_2D$ action is the upregulation of the synthesis of a calcium-binding protein whose function is to transport dietary calcium across the intestinal mucosa and into the systemic circulation. Phosphate accompanies the calcium. This has the effect of increasing the fraction of dietary calcium that is actually absorbed and is, therefore, potentially useful for bone formation. In addition, $1,25(OH_2)D$ has the effect of mobilizing calcium from bone. Both actions tend to raise the extracellular level of calcium.

A clear lack of adequate vitamin D in childhood results in the disease rickets, characterized by deformed bones and stunted growth. Extracellular calcium levels are too low to permit normal bone mineralization. A related disease, osteomalacia, occurs in adults deficient in vitamin D. Osteomalacia is characterized by weakened bones and bone pain as a result of insufficient mineralization.

Osteoporosis, a condition in which bone becomes porous and weak (potentially leading to fractures), is a far more prevalent disease than osteomalacia. While modest levels of serum 25-hydroxyvitamin D will prevent osteomalacia, these levels may not be sufficient to minimize the risk of osteoporosis. Clinical studies have demonstrated that bone mineral density is directly related to serum 25-hydroxyvitamin D levels up to 40 ng/ml. It has also been demonstrated that in elderly women given unusually high doses of calcium and vitamin D_3 the risk of both hip and vertebral fractures is substantially reduced. Optimizing bone health in both young and old may require higher levels of vitamin D activity than are typically achieved at recommended doses. This story will play out over time.

Bone health is not the only issue. Muscle strength has been shown to improve with increasing serum 25-hydroxyvitamin D levels up to 40 ng/ml. Low levels of vitamin D activity are associated with increased risk of colon, prostate, and breast cancers. The incidence of multiple sclerosis decreases with increasing serum 25-hydroxyvitamin D levels above 24 ng/ml. These are a few of several examples suggesting that many of us have lower levels of vitamin D activity than are optimal for health.[10]

Vitamin B₁ is a precursor to thiamine pyrophosphate, a coenzyme

There is a disease known as beriberi that causes weakness, fatigue, psychosis, gastrointestinal problems, peripheral nerve damage, brain damage, and ultimately death. Many years ago, this disease was particularly prevalent in rice-eating Asian societies that polished their rice. The polishings contain an essential human nutrient thiamine, also known as vitamin B_1. It is a deficiency of thiamine that causes beriberi. Thiamine deficiency in the United States is rare, being largely confined to alcoholics, some of whom drink in preference to eating. In addition, absorption of dietary thiamine is antagonized by alcohol.

The thiamine molecule is the stuff that is in your multivitamin pills. The active form in human physiology is thiamine pyrophosphate, in which a pyrophosphate group is added from ATP.

The RDA for thiamine is 1.2 mg/day for adult males and 1.1 mg/day for adult females. Children need less and pregnant or lactating women need more. Whole-grain products, including breads and cereals, are good sources of thiamine, particularly if fortified or enriched. There are no known toxic effects from thiamine ingestion and there is no specified UL.

Thiamine pyrophosphate is a coenzyme for several enzymes involved in carbohydrate metabolism. These enzymes either catalyze the decarboxylation of α-keto acids or the rearrangement of the carbon skeletons of certain sugars. A particularly important example is provided by the conversion of pyruvic acid, an α-keto acid, to acetic acid. The pyruvate dehydrogenase complex catalyzes this reaction. This is the key reaction that links the degradation of sugars to the citric acid cycle and fatty acid synthesis (chapters 16 and 18):

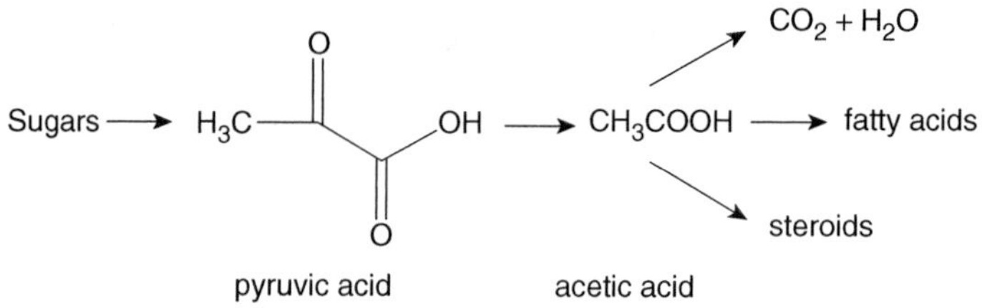

This thiamine pyrophosphate-linked reaction is, thus, a metabolic reaction of central importance.

Riboflavin is converted into the flavin coenzymes

Riboflavin is also known as vitamin B_2. It contains a complex isoalloxazine ring that humans are unable to synthesize. The complex ring is hooked onto a five-carbon sugar derivative, ribitol, closely related to the ribose that occurs in RNA. The RDA for adult males is 1.3 mg/day and for adult females 1.1 mg/day. Values decrease with increasing age but increase in pregnancy and lactation. Organ meats, milk, bread products, and fortified cereals are substantial sources of riboflavin.

Like thiamine, there is no known toxicity associated with ingestion of riboflavin and there is no established UL. At the same time, there is no known benefit to ingesting industrial quantities of riboflavin.

A deficiency of riboflavin is rare in the United States. When it does occur, it presents as pallor and maceration of the mucosa of the mouth and vermillion surfaces of the lips. The tongue may become magenta. Cutaneous lesions can occur at various sites in the body and eye problems may develop.

So what does riboflavin do? As such riboflavin does nothing. Like thiamine, riboflavin must undergo metabolic change to become effective as a coenzyme. It fact, it undergoes two reactions. The first converts riboflavin to riboflavin-5-phosphate (commonly known as flavin adenine mononucleotide, FMN), about which we will say no more, and the second converts it to flavin adenine dinucleotide, FAD. The flavins are a class of redox agents of very general importance in biochemistry. FAD is the oxidized form and $FADH_2$ is the reduced form.[11]

FAD is a coenzyme for a large number of oxidation reactions, largely of carbohydrates. Correspondingly, $FADH_2$ is a coenzyme for a number of reduction reactions. Certain of the reactions of FAD and $FADH_2$ are involved in the electron transport chain in mitochondria, associated with the synthesis of ATP. We shall see examples in chapter 17.

Niacin is converted into nicotinamide adenine dinucleotide coenzymes

Niacin is a generic name for a small family of molecules having niacin biological activity. The most common structures that fall into this category are nicotinic acid and nicotinamide:

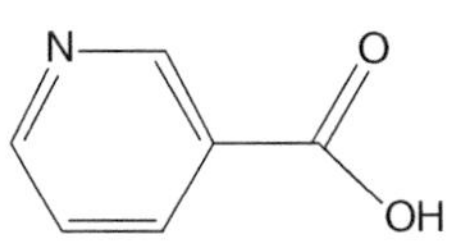

nicotinic acid 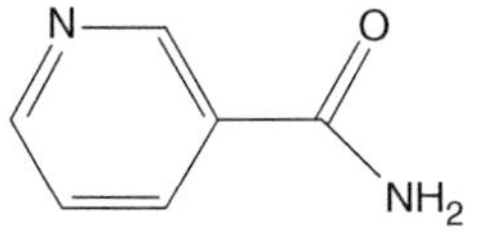   nicotinamide

In contrast to most of the vitamins encountered so far, here we have simple structures. Humans are able to synthesize these molecules from the amino acid tryptophan but not in quantities adequate to meet physiological needs. Consequently, we need to find adequate amounts in our diet. The UL for niacin is 35 mg/day for adult men and women.

The RDA for adult males is 16 mg/day and for adult females 14 mg/day. As usual, children require less and pregnant or lactating women a bit more. Niacin is not hard to come by in your diet: good sources include enriched and whole-grain bread and bread products, fortified cereals, meat, fish, and poultry.

Although the structures for molecules having niacin activity are simple, the forms in which they act in human biochemistry are not so simple. Nicotinic acid and nicotinamide are precursors for three complex coenzymes in multiple oxidation/reduction (redox) reactions: nicotinamide mononucleotide, NMN; nicotinamide adenine dinucleotide, NAD^+; and nicotinamide adenine dinucleotide phosphate, $NADP^+$. I shall use NAD^+ as representative of the class. NADH is the corresponding reduced form.[12]

The oxidation/reduction reactions that require one of the nicotinamide coenzymes are everywhere in metabolism: in the glycolytic pathway, the citric acid cycle, the synthesis and degradation of fatty acids, the synthesis of steroids, and so on. Certain of these reactions are coupled to the electron transport chain of mitochondria and are, therefore, closely involved with cellular energy generation.

A deficiency of niacin in the diet results in the disease known as pellagra, characterized by the four D's: diarrhea, dermatitis, dementia, and death. In the early years of the twentieth century in the United States, pellagra was common among poor tenant farmers and mill workers in the rural South. The diet there at that time was rich in corn that contained little niacin and little available tryptophan from which to synthesize it.

At the time, the medical community widely believed that pellagra was an infectious disease in spite of clear evidence that this was not the case. The hero in this story is Dr. Joseph Goldberger, a physician in the U.S. Government's Hygenic Laboratory, the forerunner of the National Institutes of Health. Goldberger was correctly convinced that pellagra is a dietary deficiency disease. He made a heroic effort to persuade critics that pellagra was not an infectious disease:[13]

> On April 26, 1916 he injected five cubic centimeters of a pellagrin's blood into the arm of his assistant, Dr. George Wheeler. Wheeler then shot six cubic centimeters of such blood into Goldberger. They then swabbed out the secretions of a pellagrin's nose and throat and rubbed them into their own noses and throats. They swallowed capsules containing scabs of pellagrin's rashes. Others joined what Goldberger called his "filth parties," including Mary Goldberger. None of the volunteers got pellagra.

This dramatic demonstration convinced some, but not all, physicians of the essential point. Goldberger went on throughout the 1920s demonstrating that pellagrins who consumed diets including fresh meat, milk, and vegetables recovered from their disease and that those who did not have the disease did not contract it on such diets. It was a long struggle, as much political as scientific. Many Southerners at the time felt that identification of pellagra as a dietary deficiency disease would reflect badly on their region, as indeed it did. Happily, political considerations do not determine the truth. In the end good science wins.

Pellagra in the United States is now very rare, the consequence of the good work of Joseph Goldberger and others and of more balanced diets and the vitamin fortification of many foods.

Vitamin B₆ is converted to pyridoxal phosphate, a coenzyme

Vitamin B_6 is again a small family of related compounds having the same biological activity. These include pyridoxine, pyridoxal, and pyridoxamine. In humans, these molecules are readily interconverted, accounting for their equivalence as vitamins. The stuff in your vitamin pill is likely to be pyridoxine. The actual molecule that functions as a coenzyme in metabolism is pyridoxal phosphate, in which a phosphate group has been added to pyridoxal in an ATP-dependent reaction.

Depending on age, adult men require 1.3–1.7 mg/day and adult women 1.3–1.5 mg/day. Women who are pregnant or lactating require 1.9–2.0 mg/day. Deficiency of vitamin B_6 in the developed world is rare as many foods are good sources of this vitamin, specifically organ meats, fortified cereals, and soybean-based foods.

The molecules that constitute vitamin B_6 are quite safe and there is no established UL. However, the ingestion of industrial doses of pyridoxine, 2–6 g/day for 2–40 months, is known to have caused sensory ataxia and sensory peripheral neuropathy. The senses of touch, temperature, and pain may be altered. Recovery may be slow and incomplete.

Pyridoxal phosphate is a required coenzyme for many enzyme-catalyzed reactions. Most of these reactions are associated with the metabolism of amino acids, including the decarboxylation reactions involved in the synthesis of the neurotransmitters dopamine and serotonin. In addition, pyridoxal phosphate is required for a key step in the synthesis of porphyrins, including the heme group that is an essential player in the transport of molecular oxygen by hemoglobin. Finally, pyridoxal phosphate-dependent reactions link amino acid metabolism to the citric acid cycle (chapter 16).

Folic acid prevents spina bifida

Folic acid is a complex molecule centrally involved in the metabolism of one-carbon molecular fragments: the methyl, hydroxymethyl, and formyl groups.

Adult men and women require 400 micrograms/day of folic acid. Substantial amounts are found in leafy green vegetables, enriched cereal grains, and whole-grain bread and bread products. Folic acid gets special attention in pregnant women since an inadequate intake of this vitamin is associated with a severe neural tube defect in the fetus known as spina bifida. Although it is rare for health authorities to recommend supplemental vitamins, this case is an exception. All women who are pregnant or capable of becoming pregnant are recommended to take 400 micrograms of folic acid a day from supplements or fortified foods to eliminate the potential for spina bifida.

Enzymes dependent on folic acid as coenzyme include participants in the synthesis of thymine, an essential component of DNA, and methionine, a common amino acid in proteins, among other important metabolites. A deficiency of folic acid results in the disease megaloblastic anemia.

Vitamin B_{12} is a cure for pernicious anemia

Of all the vitamins, vitamin B_{12} is unique both in the sense of its structural complexity and the presence of a cobalt atom. The cobalt-containing structure is a highly substituted tetrapyrrole related but hardly identical to those found in myoglobin and hemoglobin. An adenosine moiety attached to cobalt creates an active coenzyme derived from the vitamin.

Nothing about vitamin B_{12} has come easily. Vitamin B_{12} activity was initially discovered in 1926 when it was observed that liver contains a nutrient factor required for the cure of pernicious anemia. This observation was a boon to victims of the disease, but the boon came with a caveat: cooking liver destroys the vitamin B_{12} activity. Consequently, pernicious anemia victims had the dubious distinction of eating raw liver. Worse, liver contains very little vitamin B_{12} activity, so that they had to eat a lot of it: that was preferably to dying perhaps, but maybe not by much.

Two things hampered efforts to isolate and purify vitamin B_{12} from liver: first, liver contains little of this vitamin and, secondly, the assay for the activity employed the response of pernicious anemia patients. The work required processing factory quantities of liver and it is not surprising that the preparation of pure crystals of vitamin B_{12} took until 1948 and was accomplished nearly simultaneously in two industrial laboratories, one in the United States and one in England.

The isolation of the pure material was not the end of the story. Deducing the structure of vitamin B_{12} was a saga itself. Dorothy Crowfoot Hodgkin, an English scientist, won a Nobel Prize in Chemistry in 1964 in part for getting this structure. This was one of the early triumphs of the technique of X-ray diffraction. She was a amazingly productive structural chemist, getting the structures for cholesterol and insulin as well.

The number of vitamin B_{12}-dependent reactions is not large. Most of these involve rearrangements of the carbon skeletons of metabolites. Such reactions are important in linking some aspects of fatty acid metabolism to the citric acid cycle. In another form, a vitamin B_{12}-derived coenzyme is involved, along with folic acid coenzymes, in the metabolism of one-carbon fragments, including the biosynthesis of methionine.

Among the aged, deficiency of vitamin B_{12} is rather common. Upon aging, many people become unable to absorb adequate amounts of vitamin B_{12} from the diet. Functional deficits are the result. These are easily overcome by the monthly injection of vitamin B_{12} supplements.

Pantothenic acid is converted to coenzyme A

Finally, we come to the last of the vitamins that appear on the contents list of my multivitamin pill—pantothenic acid. This water-soluble vitamin serves a single purpose in physiology and biochemistry: it is a precursor to a far more complex molecule known as coenzyme A or, simply, CoASH.

Once the human body has a supply of pantothenic acid, it can add the remaining parts to create the intact molecule. The business part of coenzyme A is a terminal sulfhydryl ($-SH$) group. It is here that acyl (e.g., the acetyl group, $-COCH_3$) groups are attached in the process of acyl transfer reactions. There are a large number of such reactions in human metabolism and they are concerned with all aspects of metabolism: carbohydrates, lipids, amino acids, proteins, steroids, ... We shall see specific examples in chapters 16–19.

The RDA for pantothenic acid in adult men and women is 5 mg/day. Pregnant and lactating women need, respectively, 6 and 7 mg/day. As usual, children need less. Organ meats, milk, bread products, and fortified cereals are excellent sources of this vitamin. There are no reports of pantothenic acid toxicity and there is no established

UL. A deficiency of pantothenic acid results in malaise, abdominal discomfort, and a burning sensation in the feet. Such deficiency is rare, however.

Before taking final leave of the vitamins, let's note that, for vitamin after vitamin, good sources have been green vegetables, fruits, whole-grain products, milk, and meat. Mother was right: eat a balanced diet and include plenty of veggies and fruits.

Key Points

1. The role of several vitamins in human physiology is to act as coenzymes or as metabolic precursors for coenzymes.

2. We need vitamins in our diet because we cannot make them ourselves.

3. The family of related compounds known as vitamin A is essential for vision and is an important regulator of gene expression, reproduction, and immune function.

4. The molecular event that triggers the visual process is the light-induced transformation of the 11-*cis* form of retinal in rhodopsin to the all-*trans* form.

5. Scurvy is a vitamin C deficiency disease.

6. Vitamin C, ascorbic acid, is a coenzyme for the enzyme prolyl hydroxylase. The action of this enzyme is critical for the formation of normal collagen, a key component of structural and connective tissues.

7. The small family of molecules known as vitamin D is essential for bone health and other aspects of human well-being. Rickets and osteomalacia are vitamin D deficiency diseases.

8. Vitamin B_1, or thiamine, is a precursor to thiamine pyrophosphate, an essential coenzyme for several enzymes. Beriberi is a vitamin B_1 deficiency disease.

9. Vitamin B_2, or riboflavin, is the metabolic precursor to two flavin coenzymes essential for the integrity of a spectrum of redox reactions.

10. Niacin, which refers to nicotinic acid and nicotinamide, is the metabolic precursor to three nicotinamide coenzymes. These are essential for the activity of a large number of enzymes catalyzing redox reactions. Pellagra is a niacin deficiency disease.

11. The vitamin B_6 family of molecules are metabolic precursors to pyridoxal phosphate, an essential coenzyme for multiple enzymes involved in amino acid metabolism.

12. Folic acid is a coenzyme involved in the metabolism of one-carbon fragments: the methyl, hydroxymethyl, and formyl groups. An inadequate intake of folic acid during pregnancy can result in spina bifida.

13. Vitamin B_{12} is a very complex molecule that contains an atom of cobalt. Pernicious anemia is a vitamin B_{12} deficiency disease. Vitamin B_{12} is a

metabolic precursor to a coenzyme frequently involved in reactions in which carbon skeletons are rearranged.

14. Pantothenic acid is a metabolic precursor to coenzyme A, which is involved in a very large number of reactions that occur in all phases of metabolism.

16

Carbohydrates

Sweetness and life

The basic units of carbohydrates are the monosaccharides or simple sugars. These may be assembled into more complex structures such as disaccharides, polysaccharides, glycoproteins, and glycolipids. Carbohydrates serve as energy sources, energy reservoirs, structural components, and for cellular communication.

For most people, the mention of carbohydrates, more commonly known simply as "carbs," rightly brings to mind bread, pasta, potatoes, rice, and suchlike—all carbohydrate-rich foods, as are fruits and fruit juices. Proponents of certain diets, most notably the Atkins diet, have vilified carbohydrates: eat the hamburger but throw away the bun. Throughout 2003 and much of 2004, supermarket shelves were loaded with a host of new products touting their low-carbohydrate content. Seemingly, everyone was eager to cash in on the faddish popularity of the Atkins diet. It went so far as to include low-carb beers, as if "lite" beers were not enough insult to the taste buds of aficionados of this blue-collar staple. Happily, the Atkins diet fad has run its course as all fad diets do, sooner or later. The simple, compelling fact is that carbohydrates are essential in the diet of healthy human beings.

Carbohydrates may also bring to mind one of the most widely consumed compounds in the world: sucrose, simple table sugar. Personally, I add it to my morning breakfast cereal, usually either raisin bran or shredded wheat, along with fresh fruit. Of course, you can buy the sugar built-in, usually in quite high amounts, in any number of sweetened breakfast cereals. These seem to be particularly directed at and appreciated by children. Many people add sugar to their coffee or tea. Sugar is found in our soft drinks, candy bars, baby foods, and so on. Sucrose is pretty much everywhere that we are.

Sucrose is a specific term for a specific sugar molecule. The term "sugar" refers to a collection of small molecules that are the subject of this chapter. The only sugar that most of us come in daily contact with is sucrose, so that the terms "sugar" and "sucrose" are frequently used interchangeably, as in: "please pass the sugar." What we are really asking for is sucrose. There are a great many other sugars and we met two them in chapter 12: ribose, a constituent of RNA, and 2-deoxyribose, a constituent of DNA. There is at least one other sugar that finds rather wide use as a sweetening agent: honey is basically a syrup of the sugar fructose flavored by extracts of various blossoms, enthusiastically collected by bees.

Sucrose and other sugars pack a pretty decent punch in terms of calories. Each gram of pure carbohydrate eaten contributes about 6 calories to your diet. That corresponds to about 175 calories per ounce. Many people watching their weight make an effort to avoid the calories through the use of artificial sweeteners. The most commonly used artificial sweeteners in the United States include aspartame, a methyl ester of a dipeptide (NutraSweet, Equal), saccharine, and sucralose, a chlorinated derivative of sucrose (Splenda). There are a number of other known compounds that are exceptionally sweet. One example is stevioside, a natural product of complex structure isolated in 1955 from the leaves of a Paraguayan plant, *Stevia rebaudiana*. This substance is also known as yerba dulce. Artificial sweeteners save calories because they are very much sweeter, calorie for calorie, than sucrose. Sucralose, for example, is 600 times sweeter than sucrose. In addition, the human body does not metabolize it. Aspartame is about 300 times sweeter than sucrose. However, most people can tell the difference between the sweetness of sucrose and the sweetness of artificial sweeteners. There is a price to pay in exchange for fewer calories.

Carbohydrates are essential to life

Carbohydrates play an indispensable role on the stage of life. These molecules provide for several critical functions in living systems. Among these functions are maintenance of structure, energy storage and supply, and intercellular signaling.

Let's start by recognizing where the term "carbohydrate" comes from. "Carbo" refers to carbon and "hydrate" to water. Therefore, carbohydrate suggests a combination of carbon with water. In fact, this combination is reflected both in the composition and structural organization of carbohydrates.

Most carbohydrates have the composition $(C \cdot H_2O)_n$, where n is three or greater. The relationship of one carbon atom to one molecule of water is recognized in the name "carbohydrate." We can also write this composition in the more usual way: $C_nH_{2n}O_n$. Example: if we let n equal 6, the corresponding composition is $(C \cdot H_2O)_6$, or, more usually, $C_6H_{12}O_6$. This is the composition of the common sugars glucose and fructose, among many others. Simple sugars, known as monosaccharides, containing six carbon atoms are known as hexoses. Thus, glucose and fructose are examples of hexoses. You will note as we move forward that all sugar names end in "ose" (glucose, fructose, ribose, etc.). If it ends in "ose" it is very likely a sugar. This occasionally comes in handy in crossword puzzles.

Which brings to mind a wonderful story recounted by the famous Hungarian biochemist Albert Szent-Györgyi in a memoir he wrote years ago for *Annual Reviews of Biochemistry*. Szent-Györgyi was a discoverer of vitamin C, which he thought was a sugar, but not knowing the structure he elected to term it "ignose" (reflecting his ignorance of its structure but his conviction that it was a sugar) in a manuscript submitted for publication in a respected biochemistry journal. The editor of that journal took serious objection to this effort at cleverness and rejected the manuscript, demanding that a more suitable term for this molecule be employed. Szent-Györgyi complied by renaming it "Godnose." That did not get published either. In the end, Szent-Györgyi had the last laugh: he won the Nobel Prize in Physiology or Medicine in 1937, in significant part for his work on vitamin C.

Ribose has the composition $C_5H_{10}O_5$. Simple sugars containing five carbon atoms are known as pentoses. Most of the common simple sugars are either pentoses or hexoses. Several of the carbon atoms in glucose, fructose, and ribose are linked with one atom of hydrogen and the hydroxyl group, the elements of water: H—C—OH.

The basic structural unit of carbohydrates is the monosaccharide. Molecules in this class contain just one sugar moiety: a hexose, pentose, or whatever. Monosaccharides are the building blocks of more complex carbohydrates in much the same sense that amino acids are the building blocks for proteins and nucleotides are the building blocks for nucleic acids.

The structural chemistry of carbohydrates, including the monosaccharides, is complex. We are going to ignore most of the complexity. Let's get started with the essentials by looking at a model of glucose:

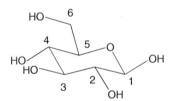

In this model we can pick out the chain of six carbon atoms, numbered 1–6, of which five carry OH groups. So each of these carbon atoms carries the functional group characteristic of alcohols. Note that five of the carbon atoms are chiral. Specifically, carbons 1–5 all possess four distinct substituents and are, therefore, chiral. Finally, note that carbon-1 is linked to two oxygen atoms, one in the hydroxyl group and the one in the ring.

The disposition of the substituents in space on carbon atoms 2–5 defines D-glucose. The mirror image of D-glucose is L-glucose. Since we have a chiral center at carbon-1, we must have two D-glucose molecules. The one shown in the model above is known as α-D-glucose. If the hydroxyl group at carbon-1 were found below the plane of the ring, we would have β-D-glucose.

Since monosaccharides containing six carbon atoms, such as D-glucose, have five chiral centers, we have a total of $2^5 = 32$ possible hexoses. For the monosaccharides containing five carbon atoms, there are four chiral centers, for a total of

$2^4 = 16$ possible pentoses. Many of these occur naturally in living systems but most have rather modest importance and we shall not worry about them further.

As one would expect for molecules having several hydroxyl groups, monosaccharides are quite hydrophilic and have high solubilities in water.[1] Concentrated solutions of simple sugars in water are known as syrups. As noted earlier, honey is basically a thick, flavored syrup of fructose.

Sucrose is a disaccharide

"Oligo" means "a few" or "some." It follows that oligosaccharides contain a few monosaccharide units. The simplest oligosaccharides are the disaccharides, which contain two monosaccharide units. These are followed by the trisaccharides, which contain three; and so forth. The most common disaccharide is sucrose, common table sugar.

Sucrose contains one unit of glucose and one unit of fructose. Sucrose is made by linking one hydroxyl group of glucose to one hydroxyl group of fructose with the elimination of a molecule of water. Sucrose has nine chiral centers. You get a lot of stereochemistry here for a dollar or so a pound!

In addition to linking up monosaccharides with one another to create more complex sugars, monosaccharides also link up with molecules in other structural classes to form structures called glycosides. For example, vanillin-α-D-glucoside is the natural source of vanilla flavor. Digitoxigenin-β-D-glucoside is a molecule having the power to stimulate heart muscle contraction and is used clinically for congestive heart failure and some cardiac arrhythmias. Sinigrin is a glycoside constituent of that pungent root horseradish.

Polysaccharides include glycogen, starch, and cellulose

The bulk of all carbohydrates in nature exists in the form of polysaccharides. These are very large molecules formed by linking together long chains of monosaccharide units. These chains may be linear, like polypeptides or polynucleotides, or branched. They may contain a single type of monosaccharide unit, similar to polyglycine or polyA for example, or two or more types of monosaccharide, like nucleic acids (four types of nucleotides) or proteins (20 types of amino acids). However, polysaccharides that contain more than two types of monosaccharide are rare in nature.

In terms of function, polysaccharides fall into one of two groups: structural and nutritional. For example, cellulose is a principal structural component of plants. Glycogen and starch, in contrast, are nutritional reservoirs for animals and plants, respectively. Monosaccharides may be mobilized from storage reservoirs such as glycogen and starch and then be metabolized to generate energy.

The most abundant structural polysaccharide of higher plants, and indeed the most abundant organic substance on Earth, is cellulose. Plants synthesize about 10^{11} tons of cellulose each year. Cellulose is a linear array of D-glucose units linked together into very long chains.

The molecular mass of cellulose preparations varies from about 50,000 to more than one million. Cellulose is organized into structures, microfibrils, which contain many molecules parallel to each other. These are localized in the cell walls of higher plants, providing strength. It is the strength of these cell walls that basically creates the structural stability of plants: witness the strength of trees.

Chitin is a polysaccharide structurally and functionally related to cellulose. The structure is derived from that of cellulose by replacing one of the hydroxyl groups on each monosaccharide unit by an acetamido group, $-NHCOCH_3$. Chitin is the structural polysaccharide of lower plants, such as fungi, and of invertebrates, particularly arthropods. It is the second most abundant organic substance on Earth.

The basic nutritional reservoir of carbohydrates in plants is starch. The simplest starches, the amyloses, consist of long, linear chains of glucose units. This structure is identical to that of cellulose except for the way in which the glucose units are linked together.

More complex starches are termed amylopectins and contain branched chains. Starches can be very large molecules indeed, with molecular masses ranging up into the millions.

In animals, glycogen functions much like starch does in plants. It is an important reservoir of carbohydrate that can be mobilized to meet energy demands. Structurally, glycogen is related to amylopectin, being a highly branched structure derived entirely from glucose. Molecular masses for glycogen molecules also range into the millions.

More complex polysaccharides play important roles in connective tissues and elsewhere. For example, hyaluronic acid is universally present in connective tissues of animals, as well as in their vitreous and synovial fluids. It helps to provide the fluids present in joints with shock-absorbing and lubricating properties. Unlike cellulose, chitin, starch, and glycogen, hyaluronic acid contains two different monomers: glucose and N-acetylglucosamine alternate in the structure. Thus, hyaluronic acid is a regular alternating copolymer ABABABA . . . .

Carbohydrates are hydrophilic molecules

We encountered the properties of hydrophilic and hydrophobic molecules in our thoughts about driving forces for formation of three-dimensional protein structures. Specifically, proteins fold in a way that puts most of the hydrophobic amino acid side chains into the molecular interior, where they can enjoy each other's company and avoid the dreaded aqueous environment. At the same time, they fold to get the hydrophilic amino acid side chains onto the molecular surface, where they happily interact with that environment. The same ideas are important for the double-stranded helical structure of DNA. The hydrophobic bases are localized within the double helix, where they interact with each other, and the strongly hydrophilic sugar and phosphate groups are exposed on the exterior of the double helix to the water environment. Now, we need to understand something more about structural features that control these properties.

The processes that constitute life take place in water (there may be a few exceptions but never mind). There are many substances characteristic of life, both simple molecules and highly complex ones, which interact readily with water. Common examples include sucrose, table salt, ethanol, rubbing alcohol (isopropyl alcohol or 2-propanol), and cotton, for example. Such substances are hydrophilic. As we shall see, most hydrophilic substances have a structural element in common with water. This fact is frequently captured in the idea: "like likes like."

At the other end of the spectrum are substances that do not interact well with water: oils, fats, waxes, and Teflon provide four examples. Oils are liquids that create films on the surface of water; many are hydrocarbons. Fats, waxes, and Teflon, a fluorocarbon polymer, are solids upon which water beads. Think about what the waxed hood of your car looks like after rain. Substances in this class are hydrophobic. We have a spectrum extending from very hydrophilic substances, on the one hand, to very hydrophobic ones, on the other hand. There is a comfortable middle ground and many substances are balanced in their hydrophobic/hydrophilic character.

It is worthwhile to emphasize one feature of hydrophilic molecules: their ability to form hydrogen bonds with water molecules. Carbohydrates, with their richness of hydroxyl groups, are abundantly able to do this. Lipids generally lack this capability and are, in consequence, quite hydrophobic.

Many proteins contain a carbohydrate component: glycoproteins

Carbohydrates are certainly important in their own right: as key intermediates in cellular metabolism; as an essential structural feature of the nucleic acids; as sources of energy; as energy reserves; for maintenance of structure; for lubrication of joints; and for other purposes. However, carbohydrates are neither limited to these functions nor confined to the structures we have developed thus far. Indeed, many of their roles, and much of the excitement associated with carbohydrate chemistry, derives from the properties of their conjugates with other classes of molecules, notably those with lipids and those with proteins.

The glycoproteins are of great importance. These are structures derived from adding carbohydrate moieties to proteins. At the outset, it is well worth noting that there are many glycoproteins in living systems. Indeed, the glycoproteins may constitute the majority of all proteins. The carbohydrate part of glycoproteins may contribute less than 1% or more than 90% of the structure by weight. Diversity is a central feature of glycoproteins.

The carbohydrate moiety of glycoproteins is added to a preformed protein molecule. Thus, the structure of the protein part is determined by the corresponding DNA structure, but the structure of the carbohydrate part is determined by the specificity of the enzymes (the structure of which is ultimately encoded in DNA as well) that catalyze the decoration of the protein once formed. The carbohydrate structures are generally quite complex and their formation is the product of several enzymes acting with several substrates.

We need to think about two issues. The first is the nature of the carbohydrate that gets added to proteins and the second is how it is attached. In general, carbohydrates covalently attached to proteins are oligosaccharides. A typical oligosaccharide linked to a protein contains 5–12 monosaccharide units.

Glycoproteins in which one of the sugar components is linked to the amide nitrogen atom of asparagine form one key group. The other key group has the sugar components hooked to an oxygen atom of the side chains of either serine or threonine.

Regardless of whether the carbohydrate is linked to the protein via asparagine, serine, or threonine, there is abundant evidence to establish that the oligosaccharides extend out from the surface of the protein into the aqueous environment, as we would expect. Thus, in the general case, we have a picture of a more-or-less globular protein with a collection of oligosaccharides protruding from its surface. In some cases, the carbohydrate units may pretty much cover the surface of the protein and form an interface with the external aqueous environment. The carbohydrate component of glycoproteins can frequently be removed without altering the structure of the protein in a detectable way, which is consistent with a picture of oligosaccharides forming whiskers on the protein surface.

The carbohydrate component of glycoproteins has the effect of increasing their water solubility and increasing the resistance of the protein component to degradation by proteases and peptidases. The latter phenomenon has led to a novel technology: the linking of long polyethylene glycol (PEG) chains to the surface of proteins:

$$\text{Protein}-[O-CH_2-CH_2]_n-OCH_3$$

Several of these PEG chains, each having perhaps 50 $-O-CH_2-CH_2-$ units may be attached to an individual protein molecule. Thus, the PEG chains may effectively cover much of the protein surface. Such proteins are said to be "pegylated." The pegylation of proteins has two key effects. First, as in the case of the glycoproteins, these proteins are protected from degradation by proteases and peptidases. Thus, their duration in human blood may be increased from a few minutes or hours to a week or more. Secondly, the addition of the PEG groups makes the proteins basically invisible to the immune system. The consequence is that pegylated proteins may be injected into patients without eliciting an immune response. The pharmaceutical industry takes advantage of these facts to the benefit of patients. Several therapeutic proteins are administered as the pegylated derivatives. Here are two examples.

A common side effect of many chemotherapeutic agents for cancer is neutropenia, a lack of adequate numbers of white blood cells, compromising the ability of cancer patients to ward off infections. To potentiate the ability of the bone marrow to produce white blood cells, human recombinant granulocyte colony-stimulating factor (G-CSF or filgrastim), a protein, is often administered to such patients. Filgrastim is marketed as Neupogen. To increase the duration of action of filgrastim, a pegylated derivative, pegfilgrastim, has been developed and approved by the FDA. It is marketed as Neulasta. Human recombinant interferon alfa-2a (Roferon) has long been used in the treatment of chronic hepatitis C infections and for renal cell carcinoma (though not very successfully in either case). A pegylated form, PEG-INF or Pegasys, has been developed and approved by the FDA. Here, too, duration of action is extended, a convenience for patients, since these protein drugs must be given by injection. Getting an injection once a week is a lot more convenient than getting one each day. Now, back to the glycoproteins themselves.

Many enzymes are glycoproteins, as are many receptors, transport proteins, and hormones. They form an integral part of the membranes of mammalian cells.

Nature has not gone to the trouble of creating these complex oligosaccharides on the surface of proteins for no reason. One common role of these carbohydrates is to act as recognition markers for several biological processes. Thus, the oligosaccharide part of glycoproteins may be recognized by transport mechanisms and direct the glycoprotein to a specific location within the cell. For example, mannose-6-phosphate components of glycoproteins guide them to lysosomes, intracellular vesicles that house hydrolytic enzymes. Glycoproteins exposed on the surface of a cell may be recognized by receptors on other cells. In this way, they may be key players in cell–cell interactions, including cell–cell communication. Alteration of carbohydrate structures on glycoproteins may be a signal to cellular mechanisms to degrade that protein. The central idea of their use as recognition markers is important though not the sole role of the carbohydrate component of glycoproteins.

An exotic function of glycoproteins is to act as antifreezes. Specifically, a number of Antarctic fish live in water cooled to about $-1.9°C$, a temperature below the freezing point of water and below that where the blood, mostly water, of these fish is expected to freeze. Clearly, this would be a disaster for these fish. They are saved from this fate by antifreeze glycoproteins. These proteins contain about 50 repeats of the tripeptide Ala-Ala-Thr. To each of these threonine residues is hooked a specific disaccharide.

Sugars react with proteins: hemoglobin A_{1c}

The construction of glycoproteins is catalyzed and guided by enzymes, as we have noted. However, this is not the only way that sugars get linked up with proteins.

Sugars are reactive molecules. Their reaction with amines, $R-NH_2$, is of particular importance to us. Proteins possess a number of amino groups, one at the N-terminus of the protein and one for each lysine residue in the protein. There is a lot of sugar floating around in the human body. So it should not surprise us that there is nonspecific reaction of these sugars with proteins. These reactions are generally slow but they do occur sufficiently fast to be important.

As one concrete example, consider the reaction in the blood between glucose and the protein hemoglobin. Glucose adds to hemoglobin, Hb, to form a product known as Hb A_{1c}. The rate of that chemical reaction is linearly related to the concentration of glucose in the blood. It follows that the more glucose you have in your blood, the more Hb A_{1c} you will expect to find. Hb A_{1c} is a long-lived molecule, as is Hb itself. Put simply, the greater the amount of Hb A_{1c}, the greater your average blood glucose concentration has been over the past few weeks. This turns out to be quite useful for diabetics. Here is the story.

The hallmark of diabetes is inappropriately high blood glucose concentrations. Too much blood glucose is linked to the sequelae of diabetes. These include blindness (retinopathy), kidney failure (nephropathy), pain and tingling sensations (neuropathy), and amputations (small vessel damage).

Diabetics often measure their blood glucose concentration several times each day in an effort to ensure that their glucose level is under acceptable control. These measurements give results applicable to the blood glucose levels at the moment that the measurement is made. Levels of Hb A_{1c} provide a different measure of blood glucose control, basically providing an indication of the overall, as opposed to the moment-by-moment, control of blood glucose levels. The point is that the blood levels of glucose change minute to minute. In contrast, Hb A_{1c} is long-lived in the blood—a matter of months. The level of Hb A_{1c} in the blood of diabetics is an important measure. It is established that the better the control of blood glucose levels, the higher the probability of avoiding certain pathological consequences of diabetes.

Finally, the accumulation of glycosylated proteins in tissue may be a cause as well as an indication of aging. These modified proteins are termed advanced glycosylation end products, AGEs. The role of AGEs in the aging process is not well-understood but then the aging process itself is not well-understood. Efforts have been made to slow the accumulation of AGEs in an effort to slow the aging process. It is one approach to the key issue of aging, which is a matter of interest to all of us.

Carbohydrates are important sources of metabolic energy

So why are simple sugars such as glucose and fructose important for life? There are many answers, but a fundamental one has to do with their function as energy sources. Almost everyone knows that consuming sugars provides a source of energy. A lot of carbohydrate-rich products are marketed specifically as energy sources, with names like GatorAde, POWERade, and CytoMax. So the next question is how do carbohydrates generate energy for life? The answer is both complex and well-understood. As usual, we are going to ignore all of the more arcane details and focus on the principal points.

In the human body, sugars are metabolized to many end products, some quite simple and others quite complex. What happens depends on the situation. In times of plentiful food, sugars may be converted to glycogen, a storage form of carbohydrate, or to fat, a storage form of lipid. These energy reservoirs may be called on later at a time when food is not plentiful (e.g., famine) or when food intake is voluntarily limited, as when dieting.

For energy generation, sugars are metabolized to simpler molecules such as small carboxylic acids and, ultimately, to water and carbon dioxide. In the process, energy is created in the form of the molecule adenosine triphosphate, ATP. Let's begin to see how this works.

First, we need a definition and some generalities: the metabolism of a cell or multicellular organism is the totality of the chemical processes that it is capable of performing. Metabolism is not time-invariant. Specifically, only part of the total metabolic potential of an organism is expressed at any given point in time. Furthermore, the fraction of the metabolic potential that is expressed changes as a function of time. Genes are turned off and on as required to meet the physiological needs of the organism as determined by the environment. I shall frequently use the term "metabolism" to refer to the subset of all potential chemical processes in a living organism actually expressed at some point in time. For example, as noted above, the expressed metabolism of a human being in a time of plentiful food is not the same as it is in a state of starvation. The expressed metabolism of a pregnant woman is not the same as that of the same woman when she is not carrying a child. The expressed metabolism of a trained athlete during a long-distance race is not the same as that of the same athlete relaxing in front of the TV set.

These simple facts directly raise the issue of metabolic control or regulation. What factors control those aspects of our metabolism that are actually expressed at some point in time and under some set of conditions? This is an enormously important question and we will see an example later. For the moment, we note that regulation is performed by a delicate set of interrelated checks and balances. These have both intrinsic (i.e. genetic) and extrinsic (i.e. physiological, environmental) components.

Metabolism is organized into connected metabolic pathways

All metabolism forms an interconnected whole in time and space. It follows that regulatory controls pervade and influence metabolism as a whole. Thus, actions at a local level have global outcomes. All metabolism and its control must be the consequence of, and explicable in terms of, the enzymatic make-up, including ribozymes, of the organism.

The metabolism of an organism is highly complex. The degree of complexity varies a good deal from organism to organism. For example, the metabolism of parasites is generally less complex than that of nonparasitic organisms. Parasites are parasitic precisely because they lack one or more metabolic capabilities required to maintain themselves as free-living organisms. The host provides the parasite with the metabolic capabilities that it lacks. But even very simple organisms, including parasitic ones, possess quite a complicated metabolism, a set of interconnected chemical reactions that are responsible for its life, whether free-living or parasitic. To gain some sense of the complexity and of the interconnectedness of metabolism, consider the scheme provided in figure 16.1. This entirely schematic view shows compounds as dots and chemical reactions as lines linking them. No structures and no names are included. The exact nature of this scheme will vary some from organism to organism. Beyond that, there are metabolic reactions that cannot be portrayed in a scheme of this sort. Nonetheless, the scheme captures most of the essential elements of the metabolism of a typical organism. The level of complexity should be clear.

As one can perhaps glean from figure 16.1, metabolism can be organized into a series of interconnected metabolic pathways. For example, the linear set of reactions shown in boldface running down the figure is the glycolytic pathway to which I turn in

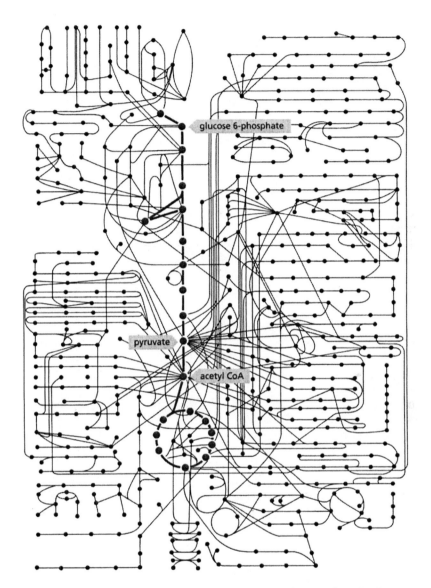

Figure 16.1 A schematic representation illustrating the chemical reactions that interconvert small molecules in cells. Each dot represents a compound and the lines that connect them represent reactions. The heavy dots and lines down the center of the maze reflect the glycolytic pathway and the heavy circle near the bottom of the maze reflects the citric acid cycle. (© 2007 From Molecular Biology of the Cell, 5th Edition, by Alberts et al. Reproduced by permission of Garland Science/Taylor & Francis, LLC.)

the next chapter. The circular reactions shown in boldface near the bottom of the figure constitute the citric acid cycle, also considered in the next chapter. These are arrays of enzyme-catalyzed chemical reactions that bring about major transformations of organic compounds of importance to the life of an organism. The compounds that are encountered along these metabolic pathways are termed metabolites.

To summarize these general considerations: the metabolism of a cell or organism is a complex and carefully controlled set of interconnected chemical reactions, each under the control of a specific enzyme, that are organized into pathways. We begin the next chapter with a description of a catabolic pathway that degrades glucose and generates energy in the form of ATP.

Not all sugars are sweet; not all sweets are sugars

One of the consistent messages of this book is that biological properties of molecules are a sensitive function of molecular structure. Although molecular properties in biological systems generally change with structure in an orderly way, small structural changes can result in major property changes. So it is with sugars. When we think of sugars, we usually think of the property of sweetness. We have sucrose on the table to add sweetness to coffee, breakfast cereal, or whatever. The sweetness of honey comes from its high content of the sugar fructose. However, not all sugars are sweet. And there are molecules that are very sweet which are not sugars. The molecular basis of taste is developed in chapter 25.

Let's look at several examples of the sensitivity of taste to molecular structure. D-glucose is quite sweet; in contrast, its enantiomer (nonsuperimposable mirror image) L-glucose is salty. The stereochemistry of D-glucose and L-glucose is different at four carbon atoms. Even more striking is the observation that α-D-mannose is sweet but β-D-mannose is bitter. These two isomers differ in stereochemistry at only one carbon atom. This simple, single change has altered their affinity for the receptors for sweet and bitter.

Similarly, D-isoleucine is sweet but L-isoleucine is bitter. Both D-leucine and D-valine are sweet but their L isomers are tasteless to bitter.

We generally associate saltiness with salts. But not all salts are salty. For example, most salts of lead are sweet. The Romans were accustomed to adding lead salts to their wine to add sweetness. Unhappily, lead salts are also toxic. To what extent lead poisoning from their wines contributed to the downfall of the Roman Empire is not known. But it certainly did nothing positive for the health of the Romans of the time.

Finally, many compounds that are neither sugars nor salts are sweet. Saccharin provides one example. Gram for gram, saccharin is 306 times as sweet as sucrose, which explains why very small amounts of saccharin are sufficient to sweeten a cup of coffee. Addition of a single methyl group to saccharin renders it tasteless (see figure 16.2). The addition of the methyl group presumably makes the modified saccharin molecule too large to fit into the effective site on the sweet receptor. Several additional examples of the relationship between molecular structure and taste are provided in figure 16.2. Suffice it to say here that small changes in structure can elicit major changes in taste.

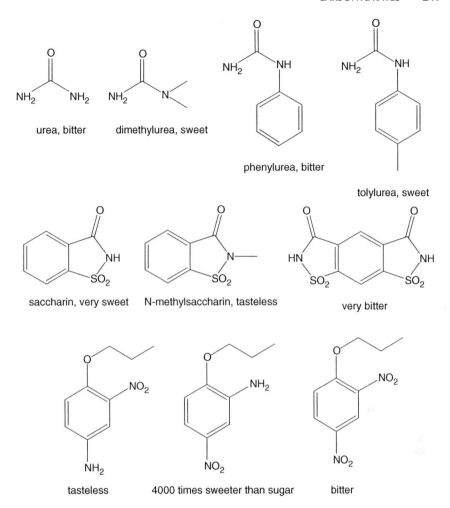

Figure 16.2   Examples of the relationship between molecular structure and taste.

Key Points

1. Carbohydrates are essential to life: for energy storage and supply, for cellular communication, and for structural purposes.

2. The basic structural units of carbohydrates are the monosaccharides.

3. Monosaccharides generally have multiple chiral centers.

4. Carbohydrates are generally very hydrophilic and have high solubility in water.

5. For energy generation, sugars are metabolized to simpler molecules, ultimately to water and carbon dioxide, with the generation of ATP.

6. For energy storage, sugars are metabolized to polysaccharides such as glycogen and starch and are stored as such.

7. Polysaccharides such as cellulose and chitin play key structural roles.

8. Glycoproteins are proteins that have been modified by the addition of one or more carbohydrate groups. Glycoproteins include blood group substances, among many others.

9. Sugars can react with proteins to form, among others, hemoglobin A_{1c}, a measure of blood glucose control over time.

10. "Metabolism" refers to the subset of all potential processes in a living organism actually expressed at some point in time.

11. Metabolism is organized into connected metabolic pathways, such as glycolysis and the citric acid cycle.

17

Generating energy from catabolism

The metabolism of a cell or organism is a complex and carefully controlled set of interconnected chemical reactions, each under the control of a specific enzyme. These are organized into pathways. The transport of electrons generated in catabolic pathways down the electron transport chain of mitochondria results in the production of essential quantities of ATP.

In the preceding chapter, I emphasized the importance of carbohydrates as sources of metabolic energy. I also introduced the idea of metabolic pathways. Now it is time to pull those two themes together and understand how the pathways for metabolism of carbohydrates yield useful metabolic energy and how these processes are controlled. On the way, we will learn how a number of important drugs for human medicine work their therapeutic magic.

Catabolism generates useful metabolic energy

Catabolic pathways are degradative. Molecules of varying degrees of complexity are broken down to simpler cellular constituents or excretory products. These pathways are oxidative in nature: that is, those molecules that provide the starting points for catabolic pathways are converted into other molecules at higher states of oxidation.[1] Electrons are pulled out of them. In human beings and many other living organisms, the ultimate oxidizing agent is oxygen, which is the final acceptor of electrons in the stepwise process of oxidizing (e.g., withdrawing electrons from) molecules. We get our oxygen from the atmosphere through the process of breathing. Catabolic pathways generally have clearly defined starting points, with less clearly defined end points. Catabolic pathways frequently begin with substrates provided by food, say the sugar in your morning coffee.

The other key point about catabolism is that it generates chemical energy, generally in the form of ATP. The organism can use this energy to provide for its maintenance, repair, growth, and reproduction. This chemical energy can also be employed for doing work, generating heat, and creating electrical impulses.

Suppose we start with a starch-rich meal, say one containing a lot of pasta or bread. The digestion of starches begins in the mouth. Saliva contains an enzyme, salivary amylase (aka ptyalin), which catalyzes the conversion of starch to simple sugars such as glucose. This process is completed in the small intestine under the influence of other enzymes in the amylase class. This completes the first phase of carbohydrate catabolism: the conversion of complex, polymeric carbohydrates (e.g., starches) to their simple monomeric units, the sugars.

We now turn to the metabolic fate of these sugars. In most cells, the metabolic pathway responsible is termed *glycolysis* or the glycolytic pathway. The starting point for glycolysis is D-glucose. The end products of glycolysis

in human tissues include pyruvic acid, $H_3C-\overset{\overset{O}{\parallel}}{C}-COOH$ and lactic acid, $H_3C-\overset{\overset{OH}{|}}{\underset{H}{C}}-COOH$. These acids are readily interconverted. Under physiolog-

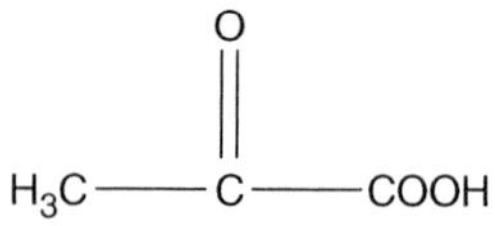

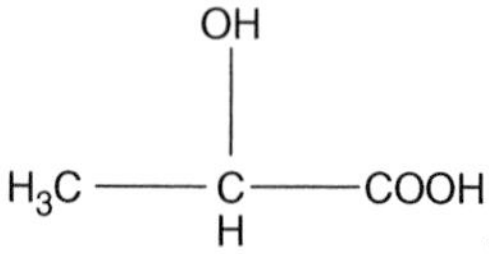

ical conditions, they exist as their anions, pyruvate, $H_3C-\overset{\overset{O}{\parallel}}{C}-COO^-$ and lactate, $H_3C-\overset{\overset{OH}{|}}{\underset{H}{C}}-COO^-$. I shall generally refer to these anionic species.

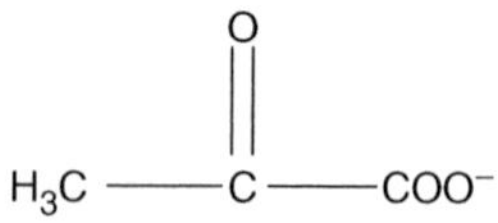

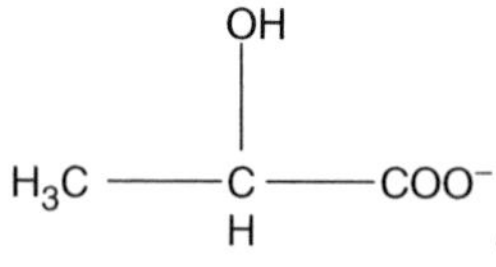

A central consequence of glycolysis is the production of limited quantities of ATP, the energy currency of the cell. Specifically, the sum total of the glycolytic pathway from D-glucose to lactate is:

$$\text{D-glucose} + 2\,P_i + 2\,ADP \rightarrow 2\,\text{lactate} + 2\,ATP$$

In this equation, I have used several common abbreviations. P_i stands for inorganic phosphate, which exists as $H_2PO_4^-$ and HPO_4^{2-} under physiological conditions. ADP stands for adenosine diphosphate and ATP for adenosine triphosphate:

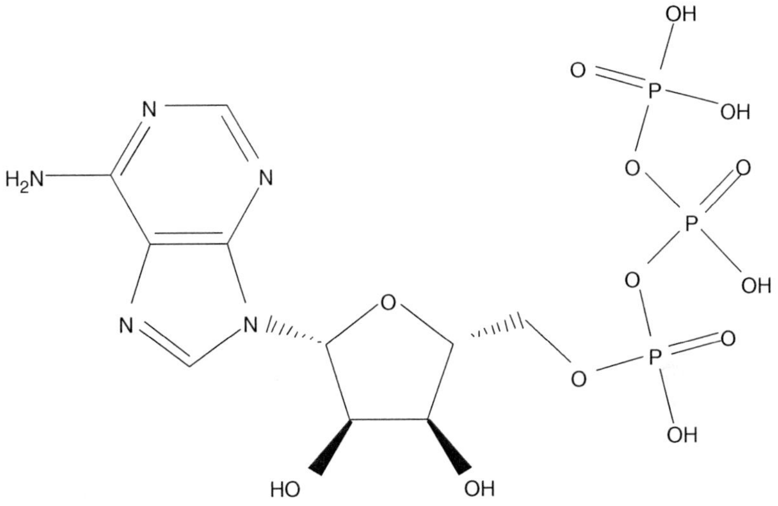

adenosine triphosphate

This equation tells us that the conversion of one molecule of D-glucose, which contains six carbon atoms, to two molecules of lactate, which contain three carbon atoms each, is accompanied by the generation of two molecules of ATP from its constituents ADP and P_i. Note that the structure of ADP is identical to that of ATP with the exception that the terminal phosphate group is missing.

A historical note: the glycolytic pathway was the very first metabolic pathway to be worked out in detail. This pathway is very well understood, as is each enzyme involved. We shall be concerned with the principal features of the pathway but not with the wealth of detail provided by research in biochemistry over several decades.

Hydrolysis of ATP can be coupled to energy-requiring processes

We have used up a molecule of glucose to generate a couple of molecules of ATP. So what is it about ATP that makes this molecule suitable as a cellular energy currency? The key point is that hydrolysis of the pyrophosphate bonds of ATP to adenosine diphosphate, ADP, plus inorganic phosphate, P_i, or adenosine monophosphate, AMP, plus inorganic pyrophosphate, PP_i, is accompanied by the release of energy. These reactions have large, negative free energy changes: ΔG is negative.[2] This energy is available to drive energy-requiring reactions and processes such as muscle contraction, nerve conduction, discharge of electric organs (if you were an electric eel or firefly this would be important to you), and the biosynthesis of proteins, nucleic acids, complex carbohydrates and lipids, among many others.

Although the detailed mechanisms that permit the energy coupling of chemical reactions or physiological processes are complex, the underlying concept is not: energy available from one chemical reaction can be used to drive another process, including other chemical reactions. We can add up the energy changes.

As Christian de Duve has put it: "All biological expenditures are, in the last analysis, supported by the hydrolysis of ATP to ADP and inorganic phosphate, one of the most noteworthy singularities offered by life today."[3]

We return to the generation of ATP later but it is time to consider the regulation of metabolism.

Metabolism is tightly regulated

It is crucial that the flux of metabolites through metabolic pathways be finely controlled to meet the needs of the organism at all points in time and under a variety of physiological and environmental conditions. Since enzymes catalyze basically all of the reactions in metabolic pathways, it will come as no surprise to learn that control is often exerted at the level of the enzymes. There are two basic ways to do that: the first way is to control the amount of an enzyme that is present, either by controlling its rate of synthesis or its rate of degradation, or both; the second way is to control the activity of the enzyme. This can happen in a number of ways, frequently by interaction of an enzyme with a small molecule.

Feedback inhibition is an important metabolic control mechanism

Anabolic metabolic pathways are the flip side of catabolic ones. Anabolic reactions are biosynthetic: that is, they create complex molecules out of simpler ones. Anabolic pathways are reductive in nature and consume energy. In all these ways, anabolic pathways stand in contrast to catabolic ones. It is frequently the case that the end product of an anabolic pathway will inhibit the first enzyme in the same pathway. This makes a good deal of sense. Anabolic pathways require energy and if there is enough end product available there is little reason to keep making more of it. So an excess of the end product simply turns off the pathway by inhibiting the first enzyme:

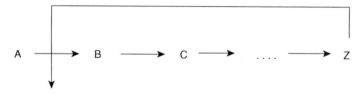

Product Z binds to the first enzyme and forms a complex that is catalytically inactive. That shuts down the biosynthetic pathway. Should the amount of product Z diminish, the inhibition will be relieved and the pathway will again become active. This is an example of feedback inhibition of a metabolic pathway and is a very important control mechanism for metabolism. All this is very neat and quite general.

Let's get back to glucose as energy source and consider a far more complex example of metabolic control. As noted above, glucose can be oxidized with the release of energy when it is required or stored as glycogen in liver or muscle when it is not. So what determines what happens?

Glycogen synthesis and degradation are regulated in a complex way

Glycogen is a polymer of glucose, as described in chapter 16. It is stored in muscle and liver. Muscle and liver glycogen stores are mobilized when required to meet physiological needs. The fates of muscle and liver glycogen are distinct. Muscle mobilizes glycogen to meet the need for ATP. Liver mobilizes glycogen in response to low blood glucose (hypoglycemia). Liver glycogen is converted to D-glucose that enters the bloodstream to elevate the blood glucose level.

Now let's think about a couple of situations in which it may be necessary to mobilize glycogen to meet the needs of the human body. To begin with, the brain has a voracious appetite for glucose. Unlike most tissues that can utilize either sugars or fats to meet energy needs, the brain is dependent on glucose alone. The brain uses about 50% of the total glucose consumed by the human body in the course of a day despite constituting only about 1–2% of the weight of the body. It follows that restriction of caloric intake will quickly create a situation in which liver glycogen must be mobilized to meet the needs of the brain for glucose. There must be a signal generated in the brain and transmitted to the liver requesting that blood glucose be elevated.

A second scenario is provided by vigorous physical activity that exercises the major skeletal muscles of the body, perhaps participating in a 10-Km road race or engaging in demanding physical labor. Such activity will rapidly consume muscle ATP. Replenishing that store requires that glycogen is mobilized in muscle and that ATP is generated through glycolysis and subsequent processes. Note that the stores of glycogen in muscle can be exhausted if vigorous physical activity is sustained for long periods. Marathoners typically exhaust their muscle glycogen stores around the 20-mile mark—hitting the wall.

A key control enzyme is glycogen phosphorylase. Glycogen phosphorylase activity controls the overall rate of glycogen catabolism and therefore the rate at which glucose from liver glycogen enters the blood or the rate at which muscle glycogen is utilized to generate ATP. A phosphorylase is an enzyme that catalyzes the cleavage of a chemical bond with transfer of a group to inorganic phosphate. Specifically, glycogen phosphorylase catalyzes a reaction in which one glucose unit is clipped off the end of one of the branches of the glycogen molecule in a form suitable for metabolism. The glycogen molecule now has one glucose unit fewer. This process can be repeated over and over again, generating metabolically active glucose and shorter glycogen molecules.

Since glycogen phosphorylase controls the rate of glucose production, the question is how signals from the brain or muscle are relayed to this enzyme. The signaling pathway for getting this done is summarized in figure 17.1. It is complex but the essentials are not difficult to grasp.

Glucagon signals the liver to produce glucose in response to a low blood sugar level. Glucagon is a peptide hormone consisting of a single chain of 29 amino acids secreted by the α cells of the islets of Langerhans of the pancreas. The pancreatic α cells have a glucose sensor. When the sensor relays the information that blood glucose is low, the pancreas secretes glucagon into the bloodstream. From the pancreas it is carried to the liver where it binds to and activates glucagon receptors. The activated receptors then stimulate adenylate cyclase activity, a key enzyme responsible for catalyzing the synthesis of the second messenger known as cyclic AMP (or cAMP) from ATP.

The situation in muscle is different. Here the responsible hormones are adrenaline (epinephrine), and noradrenaline (norepinephrine):

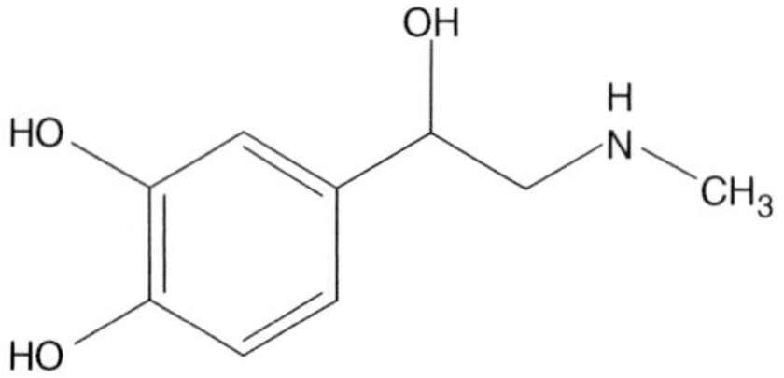

Adrenaline 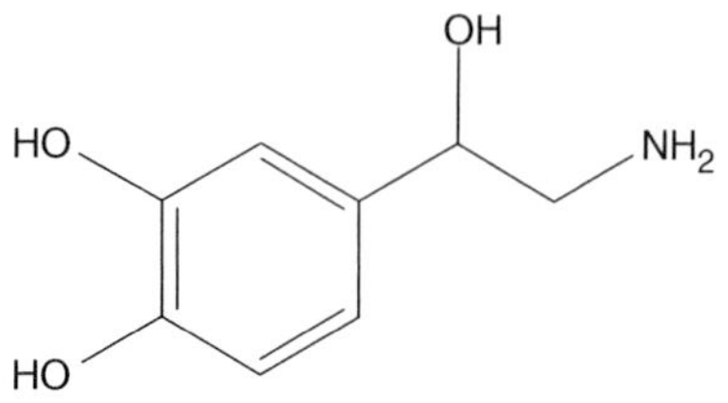   Noradrenaline

Note that these structures are related to that for the amino acid tyrosine, from which they are derived. The adrenal glands, small pieces of tissue that ride on top of the kidneys, secrete these hormones. When they activate adrenergic receptors on the surface of muscle cells, adenylate cyclase is activated, increased cAMP results, and the cascade of events in muscle cells is started (figure 17.1).

The next key point is to realize that each enzyme in the pathway exists in both active and inactive forms. cAMP initiates a cascade of reactions by activating protein kinase A (PK-A),[4] the active form of which activates the next enzyme in the sequence, and so on. At the end of the day, glycogen phosphorylase is activated and glucose or ATP is produced. This signaling pathway is a marvelous amplification system. A few molecules of glucagon or adrenaline may induce formation of many molecules of cAMP, which may activate many of PK-A, and so on. The catalytic power of enzymes is magnified in cascades of this sort.

The system just described is a *signaling pathway*. Glucagon and adrenaline are signaling molecules. Their interaction with their receptors initiates a series of enzymatic reactions, the pathway, leading to the degradation of glycogen in the liver or muscles and the production of glucose and ATP. These signaling pathways are quite common in human physiology and we shall encounter others as we move forward.

An interlude: many drugs work at adrenergic receptors

Before we leave adrenaline and noradrenaline, I want to take advantage of the opportunity to develop the mechanism of action of some of our most important and widely employed drugs in human medicine.

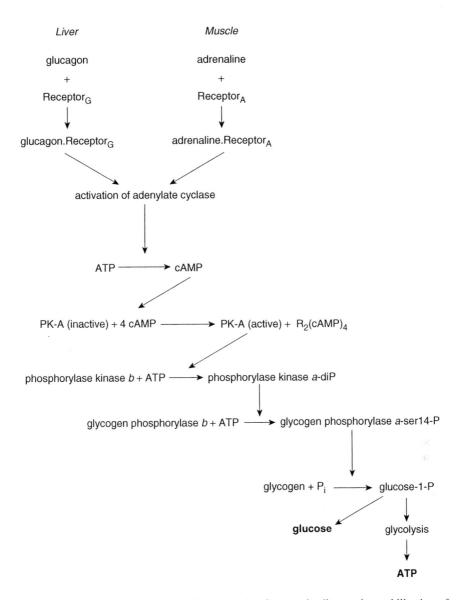

Figure 17.1 A schematic diagram of the cascade of events leading to the mobilization of glucose from stores of glycogen in liver and muscle. In liver, the cascade is initiated by the peptide hormone glucagon in response to low blood sugar levels. Through interaction of glucagon with the glucagon receptor, adenylate cyclase is activated. In muscle, the cascade is initiated by adrenaline (epinephrine) or noradrenaline (norepinephrine) that activates adenylate cyclase through interaction with the adrenergic receptor. Adenylate cyclase catalyzes the conversion of ATP into cAMP, which in turn, activates protein kinase A (PK-A). PK-A then catalytically activates phosphorylase kinase, which catalytically activates glycogen phosphorylase, which catalyzes the degradation of glycogen to glucose-1-phosphate (glucose-1-P). The net result is an increase in blood glucose in response to glucagon or a net increase in muscle ATP from metabolism of glucose-1-P via glycolysis and the citric acid cycle.

Back in 1913, adrenaline was found to cause either constriction or relaxation of blood vessels, depending on the anatomical site. Scientists correctly concluded that there must be two classes of adrenergic receptors. These are termed alpha and beta adrenergic receptors. When adrenaline hits the alpha-type receptor, blood vessel constriction results; at the beta-type receptor, blood vessel relaxation results. Thus, adrenaline and noradrenaline are nonspecific agonists (activators) at adrenergic receptors.

Many years later, Sir James Black identified the beta adrenergic receptor in heart muscle. Adrenaline increases heart rate by activating the beta adrenergic receptor in this organ. Within a decade, Black and his colleagues had synthesized a molecule known as propranolol, marketed as Inderal.

Propranolol is a beta adrenergic receptor blocker. Molecules in this category are usually known simply as beta blockers. Unlike adrenaline, propranolol is specific for beta-type adrenergic receptors. Unlike adrenaline, propranolol is an antagonist (blocker) at this receptor, not an agonist. Put another way, propranolol blocks the action of adrenaline at beta receptors. It does so by physically occupying the adrenaline binding site on the beta adrenergic receptor.

It turns out that blocking beta receptors has a number of constructive uses in human health. In fact, there are many beta blockers in medical practice and they are among our most widely employed drugs: for control of high blood pressure, heart failure, angina, certain cardiac arrhythmias, glaucoma, and stage fright. Black shared the Nobel Prize for Physiology or Medicine in 1988 for his seminal contributions.

This is not the end of the story about beta blockers. Subsequent research demonstrated that there are two subclasses of beta receptors, termed beta-1 and beta-2.[5] Both are activated by adrenaline. Both are blocked by propranolol. Beta-1 receptors are found mostly in heart muscle but not much in the lungs, whereas beta-2 receptors are found mostly in the lungs but not much in the heart. These facts provided the opportunity for better drugs. Here is the argument.

Adrenaline, which turns on both receptors, increases heart rate via the beta-1 receptor and dilates airways by the beta-2 receptors. Propranolol, which is a nonspecific blocker of both beta-1 and beta-2 receptors, decreases heart rate and lowers blood pressure, beta-1 receptor action, but can constrict airways in the lung, beta-2 receptor action, and that is not so good.

Now consider atenolol (marketed as Tenormin, among other names), a newer beta blocker that is specific for beta-1 receptors. Atenolol maintains all the good properties and therapeutic uses of propranolol but does not have the same potential to compromise lung function, a clear therapeutic advantage. There are a number of beta-1 specific beta blockers in medical practice. The point is to find a molecule that does just one thing and does it very well.

Beta agonists also find medical use: their ability to dilate the airways in the lung is useful for treatment of asthma and other situations in which bronchoconstriction is a problem such as chronic bronchitis and emphysema. Earlier, adrenaline itself and other nonspecific beta agonists, such as isoproterenol, Isuprel, were widely employed to terminate asthma attacks. These agents work but have a downside: they are potent at increasing heart rate, which is not so good. I am an asthmatic and remember childhood experiences of adrenaline injections to terminate an asthmatic attack. They worked quickly and effectively but pushed my heart rate up to perhaps 200 beats per minute. Later I used sublingual isoproterenol instead. Avoiding the necessity of finding a qualified person to do the injections was a step ahead but the heart rate problem endured. Albuterol, marketed as Ventolin, is a beta agonist that, in contrast, is specific for beta-2 adrenergic receptors. Due to its specificity, albuterol is effective as a bronchodilator but has at most a modest affect on the heart, another clear therapeutic advantage (for which I am personally grateful) as a result of our ability to target receptor subtypes specifically. There are several beta-2 specific agonists in medical practice.

Signaling pathways are as important as metabolic pathways

Signaling pathways begin with a molecule that carries a message—an agonist at some receptor. They end with the consequences of the delivery of that message: generally, the regulation of some metabolic pathway; specifically, including the synthesis of proteins. As discussed, one such example is the signaling pathway involving cAMP as second messenger, triggered by glucagon in the liver or adrenaline or noradrenaline in the muscle and eliciting the release of glucose or production of ATP.

The type of signaling pathway triggered by glucagon or adrenaline is very general. This pathway is also triggered by light in the visual process, a host of odorants, and a long list of neurotransmitters and neuropeptides, including acetylcholine, dopamine, histamine, glutamate, serotonin, and opioids, in addition to glucagon and adrenaline.

Although details will vary, in each case an agonist at its receptor activates adenylate cyclase and the second messenger cAMP is produced from ATP. cAMP activates protein kinase A and a cascade of reactions may follow. These may be metabolic reactions, as in the cases just described, or activation of a cAMP response-element protein, CREB. CREB is a transcription factor with affinity for specific sites on DNA. Control of protein synthesis follows.

One final point: like all signals in biological systems, there must be an off switch. Turning a signaling pathway off may not be as simple as it appears. The clearest thing that can happen is the dissociation of the ligand from its receptor, the reverse of the event that started the whole process in the first place. Over time, that works. However, the initial interaction may have generated a lot of cAMP and the pathway would continue to operate. A second off switch is an enzyme that degrades cAMP to AMP. In the absence of the second messenger, the process halts.

In certain tissues, the key cAMP-degrading enzyme enzyme is phosphodiesterase 5, PDE5. Inhibitors of this enzyme, sildenafil (Viagra), tadalafil (Cialis), and vardenafil (Levitra), are approved for use for erectile dysfunction. By blocking the degradation of cAMP, its metabolic effects are prolonged. In the case of PDE5 inhibitors, the net effect is to relax smooth muscle of the vascular system, facilitating the development of an erection. Note that these drugs are not aphrodisiacs. They may be necessary but are not in themselves sufficient.

Many metabolic pathways are conserved among species: the citric acid cycle

One of the great unifying features of life is the similarity in metabolic patterns. As diverse as life forms are, their patterns of metabolic activity—how molecules are formed and degraded—are remarkably closely related. That is not to say that they are identical. They are not. Indeed, identity in metabolic pattern would imply identity in structure and physiology, which is certainly not the case. Nonetheless, the similarities are striking. Variations on a unified central metabolic theme give rise to the diversity of life forms. Nowhere is this fundamental fact more clearly evident than in the central metabolic pathway known as the citric acid cycle.

The citric acid cycle is at the heart of aerobic cellular metabolism, or respiration. This is true of both prokaryotic and eukaryotic organisms, of plants and animals, of organisms large and small. Here is the main point. On the one hand, the small molecule products of catabolism of carbohydrates, lipids, and amino acids feed into the citric acid cycle. There they are converted to the ultimate end products of catabolism, carbon dioxide and water. On the other hand, the molecules of the citric acid cycle are intermediates for carbohydrate, lipid, and amino acid synthesis. Thus, the citric acid cycle is said to be amphibolic, involved in both catabolism and anabolism. It is a sink for the products of degradation of carbohydrates, lipids, and proteins and a source of building blocks for them as well.

In sum, the citric acid cycle (a) completely oxidizes the small molecule degradation products derived from amino acids, sugars, and lipids to generate useful metabolic energy in the form of ATP and (b) is the source of molecular building blocks for multiple complex molecules of life. The balance between these two roles that the citric acid cycle plays is strongly organ-dependent. The overall role of the citric acid cycle as it relates to catabolism of proteins, carbohydrates, and fats is provided in figure 17.2.

A total of eight enzyme-catalyzed reactions are involved in the citric acid cycle. We are not going to be concerned with the details or individual reactions. One complete turn of the cycle carries out the following overall reaction:

$$\text{Acetyl-SCoA} + 3\,\text{NAD}^+ + \text{FAD} + \text{GDP} + \text{P}_i$$
$$\rightarrow 2\,\text{CO}_2 + 3\,\text{NADH} + \text{FADH}_2 + \text{GTP} + \text{CoASH}$$

Note that this overall reaction requires three coenzymes that we encountered as metabolites of vitamins in chapter 15: NAD^+, derived from nicotinic acid or nicotinamide; FAD, derived from riboflavin; and coenzyme A (CoASH), derived from pantothenic acid. In the overall process, acetyl-SCoA is oxidized to two molecules of carbon dioxide with the release of CoASH. Both NAD^+ and FAD are reduced to, respectively, NADH and FADH_2. Note that one molecule of guanosine triphosphate, GTP, functionally equivalent to ATP, is generated in the process.

As noted earlier, coenzymes are frequently altered structurally in the course of an enzymatic reaction. However, they are usually reconverted to their original structure in a subsequent reaction, as opposed to being further metabolized. One turn of the citric acid cycle converts NAD^+ into NADH, FAD into FADH_2, and acetyl-SCoA into CoASH. Coenzyme A is consumed in the metabolism of pyruvate (see below) but regenerated in the citric acid cycle. Both NADH and FADH_2 are reconverted into NAD^+ and FAD by the electron transport chain.

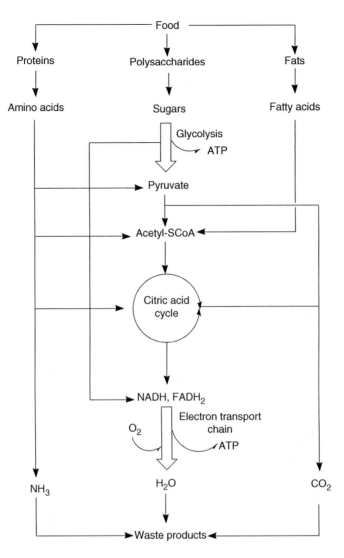

Figure 17.2 A schematic diagram of the catabolism of dietary nutrients to their metabolic end products. The basic nutrients are proteins, carbohydrates (in the form of polysaccharides), and fats. In the first stage of catabolism, these are broken down into their basic building blocks: amino acids, sugars, and fatty acids. In the second stage of catabolism, these building blocks are converted to acetyl-SCoA by glycolysis to pyruvate and then on to acetyl-SCoA in the case of sugars, by fatty acid oxidation, or by various routes in the case of the amino acids. These transformations yield modest amounts of energy in the form of ATP and reducing power in the form of NADH. In the third stage of catabolism, acetyl-SCoA is completely oxidized into CO_2 and water via the citric acid cycle. In this process abundant reducing equivalents are generated in the form of NADH and $FADH_2$. These are available for use in biosynthetic processes or they may be reoxidized by the electron transport chain, with the consumption of oxygen, to yield major amounts of energy in the form of ATP. Note that the catabolism of amino acids may yield pyruvate, acetyl-SCoA or intermediates of the citric acid cycle, depending on the amino acid. These coenzymes cycle between structural states and, therefore, are continuously available for more turns of the cycles or reactions that consume and regenerate them.

We know that anaerobic glycolysis of glucose yields pyruvate and/or lactate, interconvertable metabolites. Pyruvate is converted into acetyl-SCoA in the following reaction, catalyzed by the pyruvate dehydrogenase complex:

$$CH_3COCOOH + CoASH + NAD^+ \rightarrow CH_3CO\text{-}SCoA + CO_2 + NADH$$

In this reaction, pyruvic acid is oxidized to carbon dioxide with formation of acetyl-SCoA and NAD^+ is reduced to NADH. As noted in chapter 15, this reaction requires the participation of thiamine pyrophosphate as coenzyme. Here too the NADH formed is converted back to NAD^+ by the electron transport chain. As noted above, the acetyl-SCoA is consumed by the citric acid cycle and CoASH is regenerated.

The citric acid cycle has a long and distinguished history. Important early work by several scientists in the 1930s, including that of the Hungarian biochemist Albert Szent-Györgyi, who, as noted above, won a Nobel Prize for his work on vitamin C and other contributions, revealed several novel enzymatic reactions. These turned out to constitute a part of the citric acid cycle. Employing this background and based in very significant part on his own investigations, the citric acid cycle was first proposed in much of its current form by Hans Krebs in 1937. This cycle is also known as the *Krebs cycle,* to recognize his seminal contribution to biochemistry. It is also sometimes known as the tricarboxylic acid cycle, citric acid having three carboxyl groups. Some important details of the citric acid cycle were not established for some years following the work of Krebs. For example, Nathan Kaplan and Fritz Lipmann discovered coenzyme A in 1945, which is 8 years following the formulation of the cycle by Krebs.

We should recognize that the citric acid cycle is catalytic. Specifically, the molecule that is required to condense with acetyl-SCoA in the first step of the cycle is regenerated in the last step. In principle therefore, the cycle is capable of consuming acetyl-SCoA and producing carbon dioxide endlessly as it turns. Here is a schematic view of the idea:

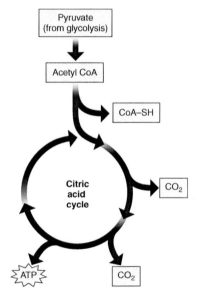

As I emphasized earlier, one of the principal reasons for (and outcomes of) catabolism is the production of chemical energy in the form of ATP.

In summary, I have provided two examples of catabolic metabolic pathways linked to production of ATP: glycolysis, in which glucose is converted to lactate and pyruvate; and the citric acid cycle, in which acetate (derived from pyruvate) is converted to carbon dioxide and water. In fact, these and other catabolic pathways generate more molecules of ATP than I have so far let on. Now we need to do two things: quantitate the actual yields of ATP and say something about how they are created. We begin by directing attention to the mitochondria.

Mitochondria are energy-generating machines

Subcellular organelles termed mitochondria are the energy-generating machines of the cell.[6] The mitochondria possess a series of multimolecular complexes that, taken together, are known as the electron transport chain. The components of the electron transport chain are localized in the inner mitochondrial membrane. The various complexes catalyze the transport of electrons from reduced coenzymes, NADH and $FADH_2$, generated in the metabolic oxidation of substrates, to molecular oxygen with the generation of ATP.

To get the basic idea, think about a river that flows over a series of waterfalls. At each waterfall, the potential energy of water at a high elevation is converted to the kinetic energy of the falling water. The kinetic energy can turn turbines to generate electricity, capturing the kinetic energy in a useful form. That electricity can be stored and used later.

The electron transport chain in mitochondria is like that in a certain way. There are multiple enzyme-catalyzed electron transfer steps along the pathway. The electrons flow from a position of higher energy to one of lower energy. Three of these steps generate useful energy in the form of a proton concentration gradient across the inner mitochondrial membrane. The idea is captured in figure 17.3.

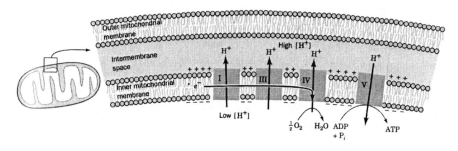

Figure 17.3 The electron transport chain of mitochondria and the coupling of electron transfer reactions to the creation of a proton concentration gradient across the inner mitochondrial membrane. This proton concentration gradient is ultimately employed to drive the synthesis of ATP by ATP synthase, noted here as complex V. (Reproduced from D. Voet and J. G. Voet, *Biochemistry*, 3rd edn, 2004; © 2004, Donald and Judith G. Voet. Reprinted with permission of John Wiley and Sons, Inc.)

To summarize: as food is oxidized, say by glycolysis or in the citric acid cycle, the extracted electrons reduce NAD^+ to NADH and FAD to $FADH_2$. The electrons in NADH and $FADH_2$ are at a high energy level. These reduced coenzymes contribute their electrons to the electron transport chain and are reconverted to NAD^+ and FAD in the process. As the electrons are passed from carrier to carrier, the energy released is captured in the form of a proton concentration gradient. Finally, the electrons are transferred to molecular oxygen with the formation of water.

What do I mean by a proton concentration gradient? Simply, there is a higher concentration of protons in the space between the inner and outer membranes of the mitochondrion than in the mitochondrial interior. The gradient is formed from the energy released in the transfer of electrons down the electron transport chain. Put another way, the released energy is employed to pump protons across the inner mitochondrial membrane into the intermembrane space.

Concentration gradients possess energy: things flow down concentration gradients, not up them. An enzyme known as ATP synthase, located in the inner mitochondrial membrane, couples the proton concentration gradient to the synthesis of ATP. For every three protons that flow back down the concentration gradient across the inner mitochondrial membrane, one molecule of ATP is formed. The central idea behind the mechanism of coupling substrate oxidation to phosphorylation, leading to ATP, was that of Peter Mitchell. Mitchell was awarded the Nobel Prize in Chemistry in 1978 for his contribution.

The importance of Mitchell's work cannot be overestimated. Good scientists worked on the mechanism of oxidative phosphorylation for decades without notable success. The reigning idea was chemical coupling of oxidation to phosphorylation: that is, it was thought that the transfer of electrons would generate some reactive, high-energy chemical intermediate and the energy available would drive the synthesis of ATP. From time to time, someone would report the identification of such an intermediate to great excitement. None of these findings held up. Mitchell's idea of a proton concentration gradient rather than a reactive intermediate was entirely novel and, finally, settled the issue. His theory is generally known as the chemiosmotic hypothesis.

The machinery for doing all this is highly complex but also very well understood. As usual, the details need not concern us. What we do need to know are the following two facts: first, the transfer of a pair of electrons from NADH to oxygen is accompanied by the production of three molecules of ATP; second, the transfer of a pair of electrons from $FADH_2$ to oxygen is accompanied by the production of two molecules of ATP. NADH and $FADH_2$ enter the electron transport chain at different places, $FADH_2$ entering later than NADH and missing the first site of proton transfer. Now we can get back to our energy calculation.

To begin with, let us return to the aerobic catabolism of simple sugars such as glucose to yield two molecules of pyruvate + two molecules of ATP + two molecules of NADH. We noted just above that coupling the oxidation of the two molecules of NADH to the electron transport chain yields an additional six molecules of ATP, three for each molecule of NADH, for a total of eight. Now let's ask what happens when we further metabolize the two molecules of pyruvate via the pyruvate dehydrogenase complex and the citric acid cycle.

First, the conversion of pyruvate to acetyl-SCoA produces one molecule of NADH. Coupling the oxidation of the NADH to the electron transport chain will generate three molecules of ATP per molecule of pyruvate or six ATP molecules per glucose molecule. Now we are up to a total of $8 + 6 = 14$ ATP molecules per molecule of glucose.

We still need to consider the outcome of converting the two molecules of acetyl-SCoA to carbon dioxide and water via the citric acid cycle. Each of these will yield a total 12 ATP molecules: one from a specific reaction in the cycle, nine from the oxidation of the three molecules of NADH, and two from the oxidation of one molecule of FADH$_2$. So the two molecules of acetyl-SCoA will yield a total of 24 ATP molecules from two turns of the citric acid cycle. That brings the grand total to $14 + 24 = 38$ ATP molecules per molecule of glucose converted to carbon dioxide and water:

$$C_6H_{12}O_6 + 6\,O_2 + 38\,ADP + 38\,P_i \rightarrow 6\,CO_2 + 6\,H_2O + 38\,ATP$$

The central point is that we get a lot of ATP molecules from the complete oxidation of one molecule of glucose via glycolysis, the pyruvate dehydrogenase reaction, and the citric acid cycle. This equation explains, in chemical terms, why sugar is such a good energy source: lots of ATP for each molecule of glucose completely metabolized to CO$_2$ and water. Incidentally, as noted above, sucrose is a disaccharide: that is, it contains two hexoses units. The metabolism of each of hexoses to CO$_2$ and water will yield 38 molecules of ATP, for a total of 76. You get a bunch of calories from a bit of common table sugar.

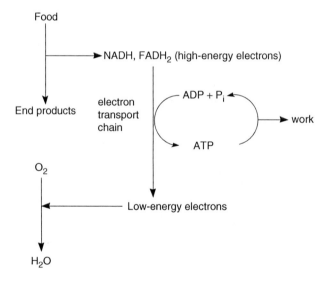

Figure 17.4 The energetics of mammalian metabolism. The food that we ingest is oxidized to a family of metabolic end products, including carbon dioxide and urea. In the process, NAD$^+$ and FAD are reduced to NADH and FADH$_2$, sources of high-energy electrons. Electrons derived from these reduced coenzymes are passed along a series of electron carriers in the electron transfer chain of mitochondria. At several steps, the electrons fall to a lower energy level and the energy released is captured in the form of ATP. At the end of the electron transport chain, the low-energy electrons are transferred to molecular oxygen (the terminal electron acceptor) with the formation of water. (Modified from a drawing in C. de Duve, *Singularities: Landmarks on the Pathways of Life.*)

However, you get even more calories from the metabolism of fats and oils, a matter to which we turn our attention in the following two chapters.

Before we move on to the next chapters, figure 17.4 provides a concise summary of all the key points.

Key Points

1. Glucose is metabolized to pyruvate and lactate with generation of ATP by the sequence of reactions known as glycolysis.

2. Metabolism is tightly regulated by a number of mechanisms: feedback inhibition, compartmentalization, covalent modification of enzymes (e.g., phosphorylation), and hormone action, among others.

3. Cyclic AMP (cAMP) is an important second messenger in signaling pathways.

4. Signaling pathways are important for metabolic control: they begin with a signaling molecule, a hormone for example, that acts through a receptor molecule with the ultimate consequence of some change in metabolism. A number of clinically useful drugs target adrenergic receptors.

5. The citric acid cycle is a catabolic, energy-generating metabolic pathway that is found enormously commonly in living nature.

6. Mitochondria are subcellular organelles that are energy-generating machines. The electron transport chain is key to ATP generation in mitochondria.

7. The complete oxidation of one molecule of glucose to carbon dioxide and water via glycolysis and the citric acid cycle generates 38 molecules of ATP.

18

Fatty acids

The building blocks of lipids

The fatty acids are a family of long-chain carboxylic acids that may be saturated, monounsaturated, or polyunsaturated. Several fatty acids are essential components of the human diet. Dietary intake of fats has important implications for health and well-being.

On January 23, 2006, a scientific advisory panel to the Food and Drug Administration, FDA, in the United States recommended that a weight-loss drug previously available by prescription only be approved for sale over the counter (OTC). The drug in question is known by the generic name orlistat and it has been marketed in the United States by Hoffmann–La Roche since 1999 as Xenical. The OTC product is known as Alli and is marketed by GlaxoSmithKline.

Orlistat is a structurally complex natural product, isolated from the fermentation broth of *Streptomyces toxytricini* in 1985. Orlistat works for weight loss, though the results are far from dramatic.[1] Use according to the recommended regimen for 6 months generally results in a drug-dependent body weight loss of about 5%. So if you start out at 200 lb and follow the directions, you can anticipate weighing in at 190 lb after 6 months of daily orlistat use. The loss of 10 lb is worthwhile but not striking. One point is that getting really significant weight loss with orlistat also involves dietary control and exercise. The other point is that significant weight loss can be achieved by diet and exercise alone.

It is known how orlistat works. The mechanism is plausible, both in terms of effecting weight loss and causing annoying side effects. Digestion of dietary fat begins in the small intestine under the action of an enzyme known as pancreatic lipase. This enzyme converts fats and oils into their constituent fatty acids and glycerol. These products are then absorbed across the intestinal wall and further metabolized. Orlistat is an inhibitor of pancreatic lipase and thus inhibits the digestion of fat. The calories that count for body weight are not the calories consumed but the calories absorbed by the body. Orlistat has the net effect of inhibiting the absorption of dietary fat and thus reducing the calories absorbed.

However, dietary fat consumed has got to go somewhere. If it is not absorbed, then it must traverse the gut and be excreted in the feces. Here the net effect is to increase the level of fat and oil in the feces and the result is not difficult to guess. About half of the people on orlistat have a problem with fecal incontinence and underwear soiling. There is a price to pay for orlistat-assisted weight loss.

However, the point here is not to detail all the plusses and minuses of orlistat but to focus attention on the national problem of overweight and obesity in the United States.

The unhappy fact is that the United States is the midst of an epidemic of weight gain.[2] So is much of the rest of the developed world. About two-thirds of all Americans are overweight and perhaps one-quarter are frankly obese. This is not good for multiple reasons such as health, self-image, mobility, and attractiveness to others. There is no doubt, from many studies, that overweight people suffer from more health problems than those whose weight falls into the ranges regarded as normal. That is not the same as saying that all heavy people are unhealthy. Adverse health effects in overweight people may derive in part from a poor diet and in part from a lack of exercise, independent of weight itself. At the same time, it is clear that excess body fat disposes the individual to the development of type 2 diabetes, which has reached epidemic proportions in the United States.

Most overweight people would prefer to be slimmer. This fact accounts for the business success of Weight Watchers, Jenny Craig, and others as well as the popularity of fad diets. I sometimes do some grocery shopping for my wife and myself. There is always a collection of popular magazines available at the checkout counter. It is generally the case that one or more has a new diet prominently displayed on the cover. If these diets actually worked, the obesity problem would be getting better, not worse.

The body-mass index is a useful measure

Obesity is defined as an excess of fat over that required to maintain health. The most reliable method to estimate obesity is the body mass index, BMI. The BMI is defined as body weight in kilograms divided by the square of the height in meters: $BMI = weight/height^2$ (kg/m^2). Suppose that we do a couple of examples to see how this works out. Consider a man 5 feet and 10 inches tall weighing 200 lb. His height in inches is 70 and that translates into about 1.77 meters, since 1 inch = 0.0253 meters. The 200 lb translates into about 91 kilograms, since 1 lb = 0.45 kg. So the BMI for our hypothetical (though rather typical) man is $91/(1.77)^2 = 29$. Let's consider one more example, this time for a woman standing 5 feet 5 inches and weighing 135 lbs.

Proceeding as above, her height is 1.64 meters and her weight is 61 kilograms. It follows that her BMI is $61/(1.64)^2 = 23$.

The ideal BMI for men and women is about 21 or 22. For men, a BMI of ≥ 28 is considered overweight; for women, the standard is 27. People with BMI of ≥ 30 are considered obese. So our hypothetical man is clearly overweight and on the verge of being obese, whereas our imaginary woman is just slightly over her ideal body weight.

As noted above, obesity is a health problem. It is associated with both elevated mortality and morbidity. More specifically, obesity is a risk factor for cardiovascular disease, including heart attack and stroke, and for high blood pressure (hypertension), diabetes, and hyperlipidemia (elevated levels of lipids in the blood, a risk factor for atherosclerosis and its sequelae), and for cancer.

The fat content of the US diet has increased very substantially over the past few decades. Currently more than 40% of dietary calories are consumed in the form of fat. This can easily lead to weight gain. The energy required for conversion of dietary fat to fat in adipose tissue involves the loss of only 3% of the calories in the dietary fat. In contrast, the energy requirement for conversion of dietary carbohydrate to fat in adipose tissue involves the loss of 23% of the calories in the dietary carbohydrate. Clearly, dietary fat is a great source of body fat.

The control of body weight is both complicated and not fully understood. Here is the core of how it works.

The control of body weight is complex

Earlier, I stated the First Law of Thermodynamics: energy is conserved. If we apply the First Law to the case of body weight control, we have the following equation:

$$\text{Energy intake} = \text{Energy expenditure} + \text{Energy stored}$$

This relationship needs to be understood in the following manner. Energy intake means net food calories absorbed. As noted above, not all food eaten is absorbed. Some is indigestible or not completely digested and passes directly through the gut and out in the feces. However, in general, the more you eat, the more food is absorbed.

Energy expenditure is the total of heat produced by the body to maintain body temperature plus the work done by the body on the environment. It is easier to maintain body weight in winter months, when the body uses substantial energy to maintain body temperature, than in the summer when it does not. Active people doing a lot of work on the environment (exercise) are better able to maintain ideal body weight than are couch potatoes. Finally, energy stored should be read as body fat, the primary form of stored energy in the body. Simply put, if your energy intake in the form of calories absorbed is greater than the energy you expend by maintenance of body temperature and exercise, you are going to gain weight in the form of stored fat. There is just no way to get around that fact.

In body weight control, the First Law of Thermodynamics wins every time. There is no diet that will result in weight loss that does not limit the number of calories absorbed from the diet to fewer than those consumed by energy expenditure.

You should believe that and nothing to the contrary. The notion that you can eat all that you want and still lose weight holds true only if the diet is so unpalatable that you don't want much to eat or you exercise more than most people are able to manage.

It is generally true that energy intake and energy expenditure are closely matched over long periods of time: that is, body weight does not usually change much over short periods of time. This provides a nice example of homeostasis: the balance of control mechanisms to maintain a physiological state constant over time. At the same time, a very small imbalance between energy intake and energy expenditure (say 0.5% per day) can lead to substantial weight gain in the long run. For example, suppose an individual consumes 2000 cal per day in his or her diet but expends only 99.5% of those calories. The difference amounts to $2000 \times (1-0.995) = 2000 \times 0.005 = 10$ cal per day. In 1 year, this will amount to $365 \times 10 = 3650$ cal. One pound of fat is worth about 4000 cal. It follows that this minute difference in energy intake and energy expenditure will cause a weight increase of 1 lb in about $4000/3650 = 1.1$ years. In 10 years this will amount to a weight gain of 9 lb, all stored as fat.

Life is not fair. It is simply true that some people are genetically disposed to be slender and others are genetically disposed toward obesity. Some of the genetic factors that act to control food intake and energy expenditure are well known. Others are not. Multiple genes are clearly involved. At the moment, we know a good deal. Some of this comes from the study of mutations affecting body weight in model organisms, generally rodents. Let us start by having a look at two of these.

One of the classical obesity mutations in mice is termed *ob*, for obesity. Mice that are homozygous for the *ob* mutation (*ob/ob* mice) are grossly obese. Jeffrey Friedman at the Rockefeller University elucidated the defect in *ob/ob* mice in 1992. Specifically, the *ob* gene encodes a protein termed leptin.Leptin is produced in fat tissue and acts on the central nervous system at the hypothalamus. Basically, it reports nutritional information to this control center.

Leptin is a homeostatic hormone. It inhibits food intake and promotes energy expenditure. Therefore, in the face of a loss in body fat, the levels of leptin decrease and food intake is stimulated. In contrast, if body fat stores increase, the levels of leptin increase as well and food intake is decreased and energy expenditure increased. Therefore, leptin tends to maintain body weight within fairly close limits.[3] Since *ob/ob* mice are genetically deficient in leptin, it follows that administration of leptin should tend to restore normal body weight. That is exactly what happens. Injection of leptin into *ob/ob* mice reduces their food intake dramatically and they tend toward the body weights of their normal littermates. In these mice at least, leptin is a powerful force for treatment of obesity.

The leptin story has been augmented by a second mouse genetic defect leading to obesity. These mice are known as *db/db*; they are very similar to *ob/ob* mice. However, these mice have normal levels of leptin. Scientists at Millennium Pharmaceuticals identified the molecular defect in *db/db* mice. They lack the normal leptin receptor. Therefore, we have both sides of the coin: *ob/ob* mice cannot make leptin, eat too much, and are therefore obese; *db/db* mice make leptin, cannot respond to it for lack of the leptin receptor, eat too much, and are obese. Administration of leptin to *ob/ob* mice normalizes their body weight but administration of leptin to *db/db* mice has no effect. If you do not have a functional receptor for leptin, it does not matter how much leptin you have.

The results with *ob/ob* mice stimulated enormous interest in the potential use of leptin for the treatment of human obesity. Amgen, a prominent biotechnology company, gained the rights to develop leptin for human use. In fact, it is known that human mutations leading to leptin deficiency are associated with massive obesity in humans. However, these mutations are very rare and account for a minute part of the human obesity problem.

The role of leptin in human physiology is not restricted to body weight control through regulation of food intake and energy expenditure. In addition, leptin levels are also known to affect bone metabolism, the immune system, and reproduction.

Adipose tissue is an endocrine organ

Adipose tissue, fat, is usually thought of as a metabolically sluggish energy reservoir and mechanical and thermal insulator. It has proved to be much more than that. Adipose tissue influences the body weight, the immune response, the control of blood pressure, hemostasis, bone mass, and the functions of thyroid and reproductive glands. It does these things largely on the basis of synthesis and release of a family of adipocyte peptide hormones.

I have already mentioned leptin. There are several others, notably including adiponectin, a protein present in high concentrations in blood. Adiponectin is an antihyperglycemic agent, acting to increase insulin sensitivity. There is a close relationship between obesity, insulin action, and development of type 2 diabetes. Without listing additional peptides secreted by adipose tissue, it is important to understand that fat is an important player in human physiology.

There is complex regulation of food intake

Leptin is not the only player in regulation of food intake. There are several other players, including insulin and a number of hormones secreted by the gut. Insulin plays a key role. The blood levels of insulin are proportional to the amount of body fat. Insulin acts in the brain to decrease food intake. Therefore, insulin also acts to maintain body weight constant in addition to its multiple effects on glucose homeostasis.

The gut secretes a family of about 20 physiologically active peptide hormones. Several of these are involved in regulation of food intake. For example, the 28-amino acid polypeptide ghrelin increases appetite. Antagonists of ghrelin have been proposed as antiobesity agents, and agonists at the ghrelin receptor are potential agents for treatment of anorexia. Peptide YY (PYY) reduces food intake, an effect opposite to that of ghrelin. Cholecystokinin, a peptide released in response to food intake, reduces food intake through action at CCK1 receptors on the vagal nerve. Amylin is a 37-amino acid polypeptide that can also reduce food intake. By and large, these and other gut peptide hormones act by signaling to the brain, notably to the hypothalamus and brainstem, centers associated with the control of body weight.

In sum, the control of food intake and body weight is extraordinarily complex and not fully understood. On top of a multigenic control background, we have layered on a family of adipose and gut hormones that act through the central nervous system in the context, in the United States at least, of an environment laden with an abundance of calorie-dense, nutrient-poor foods.

Encoded within our current environment is a genetic evolutionary relic. In an environment in which a regular supply of food is uncertain, think about the time of Christ or many places on the Earth now, it makes sense to develop a regulatory system that tends to store excess calories when they are available for possible use in times of food scarcity. So we have evolved a metabolism in which storage of calories is favored over their expenditure. Dieters will understand the result: the more weight you lose, the harder it is to continue to lose weight. As caloric intake decreases during a diet, the body senses the scarcity and responds by lowering the basal metabolic rate: that is, the body becomes more efficient in the use of its food fuel when that is scarce. This mechanism makes sense in terms of survival but it also makes it tough to lose weight. The body tends to try and maintain the maximal weight that it has achieved.

Cultural issues are important in weight control

Genetics and physiological controls are not the whole story for regulation of body weight and food intake. Cultural factors also play a major role. Culture influences levels of physical activity and, therefore, body weight. Culture also influences food portions. In the United States, there has been a marked tendency to increase food portion size. We have moved from simple hamburgers to double burgers, to quarter pounders, to third pounders, to half-pounders. Is the end in sight? French fries are now served in regular, large, and super sizes. It is common to see soft drinks served in 32-ounce (2 lb) or even in 44-ounce sizes. In movie theaters, popcorn is served in buckets. Finally, culture may value thinness, as it does in the United States for example, or obesity, as it does in Samoa.

There are a number of clear examples of the effect of culture, as opposed to genetics, on body weight. Japanese people gain weight on a Western diet. Since these are the same people, genetic differences cannot be a factor. It is the difference between the Japanese diet and the Western diet, a cultural matter, which makes the difference. Pima Indians are notable for their tendency toward obesity, doubtless a result in part at least of genetic factors. Nevertheless, Pima Indians in the United States weigh about 25 kilograms, (55 lb) more than do Pima Indians living in Mexico. Here, too, we have a clear case of the influence of cultural factors.

Energy expenditure is a key factor in weight control

Food intake is one side of the body weight equation; energy expenditure is the other. One aspect of energy expenditure is our basal metabolic rate. That is the rate of which we consume energy when at rest. Basal metabolic rates vary significantly from person to person. People with a high basal metabolic rate find it easier to maintain ideal body weight than do those with a low basal metabolic rate.

Basal metabolic rate is decreased under conditions of starvation (including dieting; as noted above one more reason why it is tough to lose weight) and increased under conditions of feeding. These responses tend to keep body weight constant. Thyroid hormone increases basal metabolic rate. Hyperthyroid people tend to be slender.

We are going to get back to matters of weight control and relate them to aspects of the metabolism of fats and oils. Before we can do that in a meaningful way, we need to say something about what fats and oils are. The key constituents of fats and oils are the fatty acids. So that is where we shall continue.

Fatty acids are key constituents of many structural classes of lipids

The fatty acids can be thought of as consisting of two rather different parts: a long hydrophobic, hydrocarbon chain and a hydrophilic carboxyl group: ($-COOH$). In water under physiological conditions, the carboxyl group loses a proton to form the negatively charged, highly hydrophilic carboxylate group ($-COO^-$):

$$R-COOH \rightarrow RCOO^- + H^+$$

So a simple fatty acid looks like:

$$CH_3(CH_2)_n-COOH$$

in which n varies from 4 to more than 20 or, under physiological conditions, like:

$$CH_3-(CH_2)_n-COO^-.$$

You will recognize that this is an amphipathic molecule, one with geographically distinct regions of hydrophilic and hydrophobic character, specifically a *soap*, with clearly separated hydrophobic, $CH_3-(CH_2)_n-$, and hydrophilic parts, $-COO^-$. As a consequence of their amphipathic character, these molecules may be soluble in water and can form interesting and unusual structures once dissolved.

The common fatty acids have a linear chain containing an even number of carbon atoms, which reflects that the fatty acid chain is built up two carbon atoms at a time during biosynthesis. The structures and common names for several common fatty acids are provided in table 18.1. Fatty acids such as palmitic and stearic acids contain only carbon-carbon single bonds and are termed *saturated*. Other fatty acids such as oleic acid contain a single carbon–carbon double bond and are termed *monounsaturated*. Note that the geometry around this bond is *cis,* not *trans*. Oleic acid is found in high concentration in olive oil, which is low in saturated fatty acids. In fact, about 83% of all fatty acids in olive oil is oleic acid. Another 7% is linoleic acid. The remainder, only 10%, is saturated fatty acids. Butter, in contrast, contains about 25% oleic acid and more than 35% saturated fatty acids.

Table 18.1 Names and structures for several common fatty acids.

Symbol	Common name	Systematic name	Structure
Saturated fatty acids			
12:0	Lauric acid	Dodecanoic acid	$CH_3(CH_2)_{10}COOH$
14:0	Myristic acid	Tetradecanoic acid	$CH_3(CH_2)_{12}COOH$
16:0	Palmitic acid	Hexadecanoic acid	$CH_3(CH_2)_{14}COOH$
18:0	Stearic acid	Octadecanoic acid	$CH_3(CH_2)_{16}COOH$
20:0	Arachidic acid	Eicosanoic acid	$CH_3(CH_2)_{18}COOH$
Unsaturated fatty acids (all double bonds are _cis_)			
16:1	Palmitoleic acid	9-Hexadecenoic acid	$CH_3(CH_2)_5CH{=}CH(CH_2)_7COOH$
18:1	Oleic acid	9-Octadecenoic acid	$CH_3(CH_2)_7CH{=}CH(CH_2)_7COOH$
18:2	Linoleic acid	9,12-Octadecadienoic acid	$CH_3(CH_2)_4(CH{=}CHCH_2)_2(CH_2)_6COOH$
18:3	α-Linolenic acid	9,12,15-Octadecatrienoic acid	$CH_3CH_2(CH{=}CHCH_2)_3(CH_2)_6COOH$
20:4	Arachidonic acid	5,8,11,14-Eicosatetraenoic acid	$CH_3(CH_2)_4(CH{=}CHCH_2)_4(CH_2)_2COOH$
20:5	EPA	5,8,11,14,17-Eicosapentaenoic acid	$CH_3CH_2(CH{=}CHCH_2)_5(CH_2)_2COOH$

Dietary saturated fatty acids have undesirable effects on plasma lipids

The effects of dietary lipids on human health are complex. However, in general dietary terms, monounsaturated fatty acids are considered to be more healthful than saturated fatty acids. A lot of people, concerned about their health, have decreased their dietary intake of butter in favor of olive oil or vegetable oils.

Let's pause for a moment and ask why health authorities recommend that we limit our intake of saturated fats. Note that "saturated fats" really means fats that contain saturated fatty acids, i.e., fatty acids that contain only carbon–carbon single bonds. It has been repeatedly demonstrated in clinical trials that dietary saturated fats increase the plasma level of low-density lipoproteins, LDLs, generally referred to as "bad cholesterol." They have no effect on the plasma level of high-density lipoproteins, HDLs, generally termed "good cholesterol." High levels of LDLs are an independent risk factor for the development of atherosclerosis and its sequelae: heart attacks, stroke, and peripheral artery disease. Drugs that reduce plasma LDL levels, such as the statins, have been shown to be effective at reducing the risks of these outcomes. It follows that dietary habits, including dietary saturated fats, that tend to raise the levels of LDLs are a threat to good health. Hence we have the sensible recommendation to limit their intake in our diet. Saturated fats are particularly abundant in meat and dairy products.

The adverse effects of saturated fats on the plasma lipid profile translate into adverse effects on human health. It has been demonstrated in clinical trials that there is a clear correlation between the percent of dietary energy as saturated fats and the incidence of coronary heart disease and mortality.[4] It follows that limiting your intake of saturated fats is a prudent course of action.

Some polyunsaturated fatty acids are essential in the human diet

Fatty acids, such as linoleic, linolenic, and arachidonic acids, contain two or more *cis* carbon–carbon double bonds and are referred to as polyunsaturated fatty acids. Several of these fatty acids, including linoleic and linolenic acids, are required nutrients for humans and must be part of a healthy diet. They are termed essential fatty acids, of which there are eight. These fatty acids cannot be synthesized by human beings but are essential to human health. Therefore, they must be consumed in adequate amounts in a healthy diet, specifically in the form of ingested plant-derived foods. A diet devoid of the essential fatty acids eventually results in a fatal condition characterized by inflammation of the skin (dermatitis), failure of wounds to heal, and poor growth. The essential fatty acids serve as precursors for complex molecules termed eicosanoids, to which we return below.

There are two general systems of nomenclature for fatty acids, as well as a few useful shorthand designations. Common names have their origins in history and are frequently used. Several examples are provided in table 18.1. Systematic names derive from the number of carbon atoms in the fatty acid: for example, lauric acid is systematically known as dodecanoic acid. The parent hydrocarbon dodecane has 12 carbon atoms as does lauric acid. Several other examples are provided in table 18.1.

For unsaturated fatty acids, we need to indicate where the double bonds are. We do this by numbering the carbon atoms, starting at the carboxyl end, and designating the position of the double bonds by the number of the carbon atom closest to the carboxyl end of the molecules. One example will suffice to illustrate the point. Properly numbered oleic acid (*cis*-9-octadecenoic acid) is:

$$^{18}CH_3{}^{17}CH_2{}^{16}CH_2{}^{15}CH_2{}^{14}CH_2{}^{13}CH^{12}CH_2{}^{11}CH_2{}^{10}CH{=}^9CH^8CH_2{}^7$$
$$CH_2{}^6CH_2{}^5CH_2{}^4CH_2{}^3CH_2{}^2CH_2{}^1COOH$$

Octadecane has 18 carbon atoms, as does oleic acid, and "enoic" implies a double bond (octadec*ane* is the saturated hydrocarbon while octadec*ene* has a double bond somewhere). The 9 indicates that the carbon atom closest to the carboxyl end that is involved in the double bond is number 9. Finally, *cis* establishes the stereochemistry at the double bond.

There is one nomenclature that is commonly used, particularly in news articles intended for the lay audience. It is useful to understand it. Here families of fatty acids are collected by designating where the first double bond occurs, starting at the methyl group (CH_3) end of the molecule, frequently known as the omega (ω) carbon atom. Thus, the omega-3 fatty acids look like:

$$CH_3 - CH - CH{=}CH - \ldots$$

And omega-6 fatty acids look like:

$$CH_3 - CH_2 - CH_2 - CH_2 - CH_2 - CH{=}CH - \ldots$$

Thus, α-linolenic acid and eicosapentaenoic acid are examples of omega-3 fatty acids, whereas γ-linoleic acid and arachidonic acid provide examples of omega-6 fatty acids (see table 18.1). As is developed below, these different families of fatty acids create families of more complex lipids that have distinct properties.

Above, I provided a short explanation for the recommendation to limit the dietary intake of saturated fats. That is not the only recommendation that we get from health authorities about dietary fats. Two more recommendations come to mind immediately: (a) limit your intake of *trans* fats and (b) consume substantial amounts of omega-3 fatty acids. Let's have a look at the basis for both recommendations.

There are good reasons for avoiding *trans* fats

Note that the term "*trans* fats" refers to fats containing *trans* fatty acids. At a couple of points earlier, I emphasized that all the common unsaturated fatty acids possess *cis* carbon–carbon double bonds. It is true that some common foods—beef and dairy products—contain very small amounts of *trans* fatty acids but it is also true that the vast preponderance of dietary *trans* fats come from processed foods. Here is one example of a normal *cis* fatty acid—oleic acid—and its *trans* isomer:

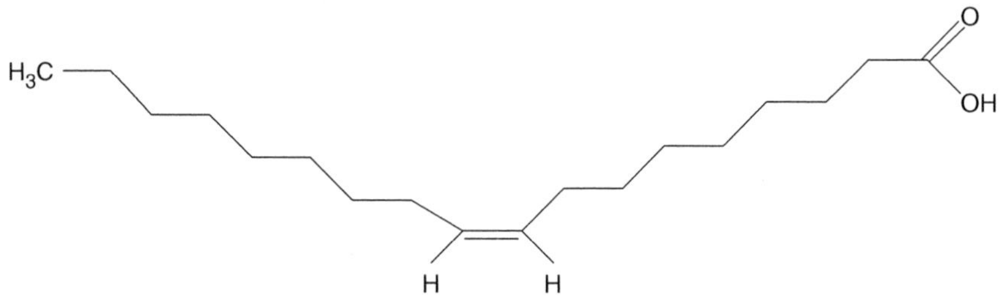

oleic acid

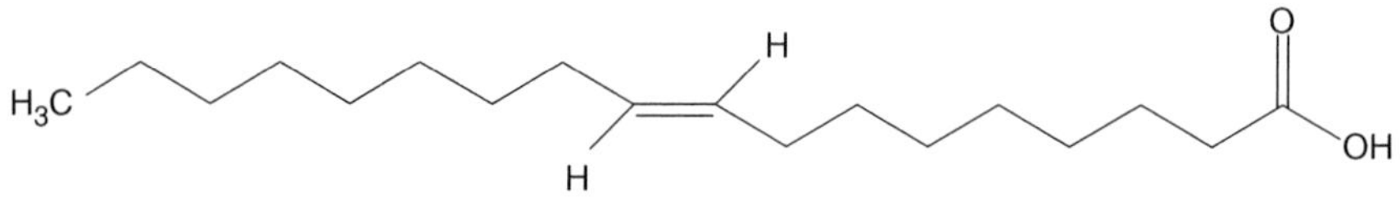

trans isomer of oleic acid

As you can see, the shapes of these molecules are different, as are their biological properties.

Trans fats are an undesirable by-product of the partial hydrogenation of liquid vegetable oils. This process converts the oils into solids. Hydrogenation converts unsaturated fatty acids into saturated ones that have higher melting points. Hydrogenation also destroys polyunsaturated fatty acids that have a tendency to oxidize and become rancid. This limits the shelf life of food products that contain them. To extend product shelf life and to provide more desirable taste, shape, and texture, food processors partially hydrogenate oils. *Trans* fats are an undesirable consequence. The bulk of *trans* fats that may find their way onto your dining table are found in margarine, shortenings, cookies, crackers, pastries, doughnuts, fried foods, and baked goods. Most commercial French fries are cooked in oils containing *trans* fatty acids and are a good source of a bad actor: *trans* fats.[5]

Effective January 1, 2006, the FDA in the United States required food companies to list *trans* fat content on the Nutrition Facts panel of all packaged foods. This should help you limit your intake of these fats. In addition, there are key words or phrases that you should be aware of: partially hydrogenated vegetable oils; shortening. These designations basically promise you a substantial dose of *trans* fats.

Metabolic studies have demonstrated that dietary *trans* fats increase the level of plasma LDLs, just as do saturated fats (as noted earlier). That is a negative for health concerns for the reasons mentioned above. But *trans* fats do more: they lower the level of plasma HDLs, good cholesterol. In contrast, saturated fats do not affect plasma HDLs. Since low levels of plasma LDLs and high levels of plasma HDLs are both good, the LDL/HDL ratio is a useful measure of the plasma cholesterol situation. Lower values of that ratio are favorable; higher values are not. Saturated fats in the diet tend to raise this ratio by increasing LDLs; *trans* fats in the diet tend to raise this ratio both by raising LDLs and lowering HDLs. Thus, the net effect of *trans* fats on this ratio may be greater than that for saturated fats.

This finding has been replicated several times in clinical studies. Let me cite one example. In a careful metabolic study carried out in 1990, Mensink and Katan determined the plasma LDL/HDL ratio when 10% of the energy from oleic acid was replaced in the diet by either the corresponding *trans* fat or the corresponding saturated fatty acid, stearic acid.[6] The resulting LDL/HDL ratios were 2.02 on the oleic acid diet, 2.34 on the stearic acid diet, and 2.58 on the *trans* fatty acid diet. This is one more example of the impact of small structural changes in molecules on their biological properties.

Omega-3 fatty acids are important for human health

Above, I have cited dietary saturated and *trans* fatty acids as negative influences on human health. Now it is time for a positive example: the omega-3 fatty acids. The initial clue that omega-3 fatty acids may have favorable effects came from the observation that Greenland Inuit, who have a diet very high in fat, have a very low incidence of coronary heart disease. Their dietary fat comes very largely from marine sources and these are high in omega-3 fatty acids. EPA, which has the unwieldy chemical name of 5,8,11,14,17-eicosapentaenoic (eye-cosa-penta-een-oh-ic) acid, is a prominent example:

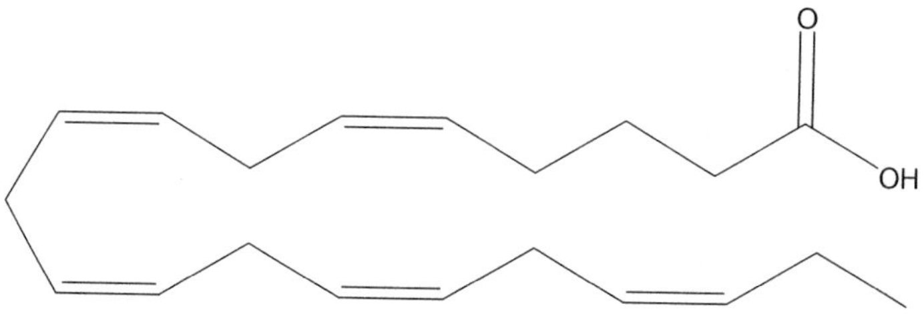

EPA

Note that all the double bonds are *cis* and that the double bond closest to the methyl group end of the molecule begins with carbon atom 3, numbering from the terminal methyl group. Hence EPA is an omega-3 fatty acid.

Since the initial observations on the Greenland Inuit, it has been established in clinical trials that consumption of omega-3 fatty acids, which occur in high levels in fish and other marine organisms, lowers the risk of atherosclerosis and sudden death from coronary artery disease. In addition, omega-3 fatty acids are known to inhibit the aggregation of platelets and thus impair formation of blood clots (thrombogenesis). They may also have antiarrhythmic properties. It is known that diets high in α-linolenic acid reduce the risk of coronary artery disease and stroke. Fish oils are prominent items in health food stores. To understand why, we need to talk about the eicosanoids. The name "eicosanoid" derives from the fact that all these molecules have 20 carbon atoms (Greek for twenty is *eikosi*).

Eicosanoids are a complex group of physiologically important lipids

Eicosanoids is a general term for five classes of lipids: prostaglandins, prostacyclins, thromboxanes, leukotrienes, and lipoxins. The first class to be discovered was the prostaglandins, detected in human semen by Ulf von Euler in 1930 based on their physiological effects. Von Euler believed that these compounds were synthesized in the prostate gland, hence the name. Actually, most of the prostaglandins in human semen are synthesized in the seminal vesicles, but the name has stuck. The influence of history frequently endures despite advances in scientific knowledge. Regardless of the name, the fact is that almost all mammalian cells make prostaglandins and other eicosanoids. They act locally at their receptors, near their site of synthesis, and are, therefore, paracrine hormones.

The eicosanoids have a broad spectrum of physiological activities and these are elicited at very low concentrations, 10^{-9} M or less. They mediate the inflammatory response, produce pain and fever, regulate blood pressure, initiate blood clotting, induce labor, and regulate the sleep/wake cycle.

The key substrate for synthesis of the eicosanoids is the omega-6, 20-carbon fatty acid arachidonic acid:

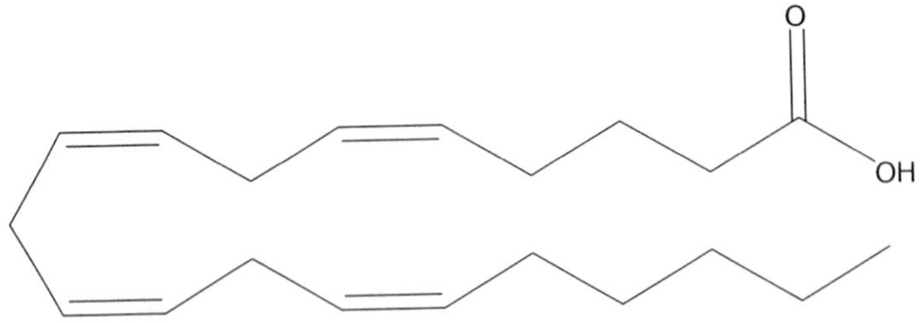

arachidonic acid

Note that arachidonic acid differs from EPA only in that the latter has one additional double bond near the methyl end of the molecule. Arachidonic acid is an omega-6 fatty acid.

Pathways for the synthesis of prostaglandins (PGs), prostacyclins (PGIs), and thromboxanes (TxBs) are provided in figure 18.1. Note that the numerical subscripts identify the number of carbon–carbon double bonds in the molecules: there are, for example, PGE_1, PGE_2, and PGE_3, containing one, two, and three double bonds respectively. PGE_2 derives, as shown, from arachidonic acid; the others come from related fatty acids. The key enzyme in this complex set of pathways is the very first one, PGH synthase, more commonly known as cyclooxygenase or simply COX.

The COX reaction is quite remarkable. Arachidonic acid has no chiral centers, but PGE_2 and PGI_2 have four and TxB_2 has five! The immediate product of the COX reaction, known as PGH_2, also has five chiral centers. This is an elegant example of the ability of enzymes to control the stereochemistry of reactions.

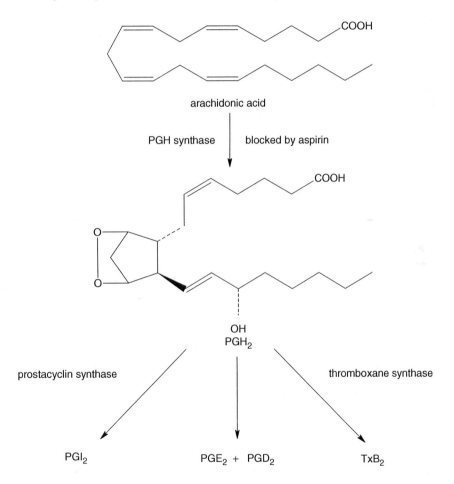

Figure 18.1 One pathway of arachidonic acid metabolism. The branches of this pathway lead to the prostaglandins (PGs), prostacyclins (PGIs), and thromboxanes (TxBs). The key reaction is the formation of PGH_2, creating five chiral centers from an achiral molecule. (Modified from D. Voet and J. G. Voet, *Biochemistry*, 3rd edn, 2004. Reprinted with permission John Wiley and Sons Inc.)

COX acts on substrates other than arachidonic acid to yield other families of eicosanoids. For example, the action of COX on α-linolenic acid yields eicosanoids containing one fewer double bond than those derived from arachidonic acid and on eicosapentaenoic acid to yield eicosanoids containing one more double bond (e.g., PGE$_3$). These differences are not trivial. The physiological effects of PGE$_1$, PGE$_2$, and PGE$_3$ are quite different. PGE$_2$ is an important mediator of pain and inflammation. PGE$_1$, in marked contrast, inhibits the release of mediators of pain and so has analgesic properties. PGE$_3$ also has several favorable physiological properties. Here again, we note the sensitive dependence of biological properties on chemical structure. We can begin to see why the nature of the polyunsaturated fats in our diet can have important health effects.

COX is an important target for drug discovery

Aspirin is a rather simple molecule that has important analgesic, anti-inflammatory, and antipyretic properties. It is known how aspirin works: it is an irreversible inhibitor of COX. In fact, there are three human COX enzymes; more about this follows below.

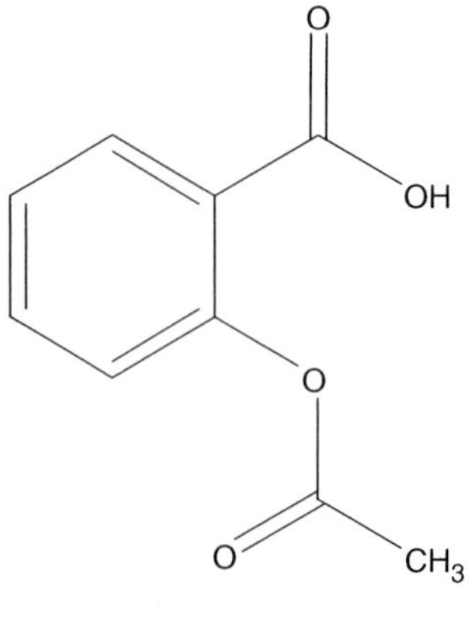

aspirin

An important analgesic, comparable to aspirin in structural complexity, is acetaminophen:

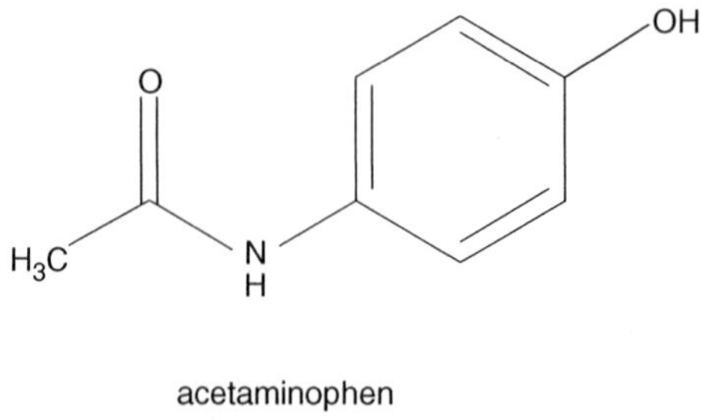

acetaminophen

Acetaminophen is marketed around the world under at least 60 trade names.[7] In the United States, it is most commonly marketed as Tylenol; in Europe, Panadol is a more common trade name. Acetaminophen is substantially distinct from aspirin in terms of its pharmacological properties. Both are effective in relieving pain and reducing fever but aspirin alone has effective anti-inflammatory properties. Acetaminophen is free from the potential of aspirin to cause gastrointestinal distress, though it is more toxic (though not very toxic) to the liver. These differences are pretty well understood: aspirin inhibits both COX-1 and COX-2.[8] Acetaminophen, in contrast, is rather specific for COX-3.

Other inhibitors of COX are collected under the general term nonsteroidal antiinflammatory drugs (NSAIDs). Several of these are available OTC, including ibuprofen (Advil, Motrin), naproxen (Aleve), and ketoprofen (Orudis). About 25 drugs in this class have been approved for use in clinical medicine in the United States, including the four just mentioned. Others are available by prescription only.

In the 1970s, John Vane discovered the mechanism of action of NSAIDs, inhibition of COX, an accomplishment rewarded with a share of the Nobel Prize in Physiology or Medicine for 1982.

Moderate use of NSAIDs is generally safe and effective for their intended use: control of mild to moderate pain and inflammation. At the same time, use of NSAIDs is associated with gastric discomfort (stomach ache) in some patients. People with more severe symptoms who ingest large quantities of NSAIDs may develop ulcers or experience life-threatening perforation of the gastrointestinal (GI) tract. Rarely, NSAIDs are associated with kidney damage. They also retard blood clotting, a result of their actions on platelets. Retardation of blood clotting can have positive effects. Specifically, it has been demonstrated that small doses of aspirin, 81 mg, taken daily lessen the probability of heart attack in men at high risk and lessen the probability of occlusive stroke in both sexes.

Before taking leave of the eicosanoids, I need to point out that a second enzyme, 5-lipoxygenase, also metabolizes arachidonic acid. 5-Lipoxygenase initiates the synthesis of the leukotrienes from arachidonic acid. There is a whole family of leukotrienes and these molecules have a spectrum of biological properties. I will focus on one important leukotriene, LTB_4.

LTB_4 is a potent bronchoconstrictor, as are several other leukotrienes. A 5-lipoxygenase inhibitor, Zileuton, is approved for therapy of asthma (though it is not much used for this purpose) as is a leukotriene blocker, montelukast, marketed as Singulair. Singulair is widely used by asthmatics as a preventive for asthma attacks. Certain corticosteroids are employed for the same purpose. Neither montelukast nor the steroids are effective in terminating an established asthmatic attack. Beta agonists are employed for that purpose (see chapter 17).

Up to this point, we have simply noted that fatty acids are key components of fats and oils. It is time to move on and consider just what fats and oils actually look like structurally. I begin that task in the next chapter.

Key Points

1. The body mass index, BMI, is the most reliable measure of overweight and obesity.

2. Obesity is a risk factor for diabetes, cardiovascular disease, and high blood pressure.

3. Energy intake is equal to the sum of energy expenditure and energy stored. Thus, if intake is greater than expenditure, energy will be stored, usually in the form of fat.

4. Adipose tissue, fat, is an endocrine organ. Through release of molecules such as leptin and adiponectin, adipose tissue influences body weight, the immune response, blood pressure, and bone mass, among other things.

5. The regulation of food intake in humans is complex. In addition to genetic and cultural influences, it involves the action of leptin, insulin, ghrelin, PYY, cholecystokinin, and amylin. Most of these molecules act by signaling to the brain.

6. There is a family of antiobesity drugs available but these are less effective than one would wish.

7. Energy expenditure comes in the form of basal metabolism, maintenance of body temperature, and physical exercise.

8. The fatty acid content of our diet has important consequences for health. Several fatty acid are essential dietary components. In general, saturated and *trans* fatty acids are inimical to good health. In contrast, omega-3 fatty acids are favorable to good health.

9. Eicosanoids include prostaglandins, prostacyclins, thromboxanes, and leukotrienes. Molecules in these classes have a spectrum of physiological activities.

10. Aspirin, acetaminophen, and many other nonsteroidal anti-inflammatory drugs (e.g., Advil) are inhibitors of cyclooxygenase, COX.

19

Lipids

The greasy stuff of life

Fats are a remarkably effective form of energy storage. Several classes of complex lipids play important roles in human physiology and are key constituents of biological membranes.

The fatty acids are important building blocks for more complex lipids. I turn to these in this chapter, including the role of fats and oils as energy storehouses and that for other complex lipids in the structure and properties of biological membranes.

Fats and oils are esters of glycerol and fatty acids

Fats and oils are hydrophobic, water-insoluble substances. The operational difference between them is simple: fats are solid at room temperature; oils are liquid at room temperature. Chemically, fats and oils are triglycerides, triesters formed from glycerol and three fatty acids. Let me remind you that an ester is a functional group derived by linking a carboxylic acid to an alcohol, with the elimination of a molecule of water:

$$R_1-COOH + R_2-OH \longrightarrow R_1-\overset{\overset{\textstyle O}{\|}}{C}-O-R_2 + H_2O$$

Glycerol is a molecule possessing three hydroxyl groups:

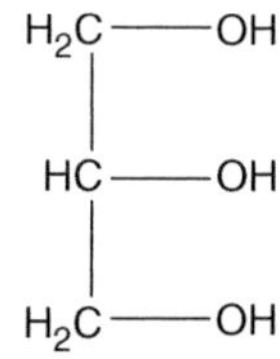

$$\text{H}_2\text{C} \longrightarrow \text{OH}$$
$$\text{HC} \longrightarrow \text{OH}$$
$$\text{H}_2\text{C} \longrightarrow \text{OH}$$

If we now link each of the hydroxyl groups to a carboxyl group of a fatty acid, we get a triglyceride that possesses three ester functional groups:

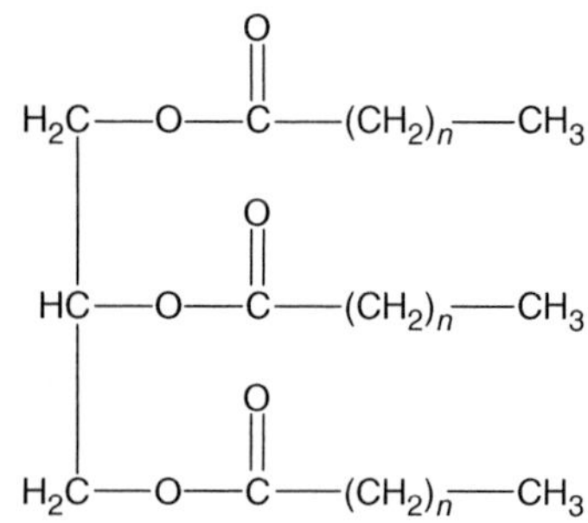

$$
\begin{array}{c}
\text{O} \\
\parallel \\
\text{H}_2\text{C}-\text{O}-\text{C}-(\text{CH}_2)_n-\text{CH}_3 \\[1em]
\text{O} \\
\parallel \\
\text{HC}-\text{O}-\text{C}-(\text{CH}_2)_n-\text{CH}_3 \\[1em]
\text{O} \\
\parallel \\
\text{H}_2\text{C}-\text{O}-\text{C}-(\text{CH}_2)_n-\text{CH}_3
\end{array}
$$

This structure shows a triglyceride with three identical saturated fatty acids. Tripalmitin, in which all fatty acids are palmitic acid ($n = 14$), provides one example of a fat. Triolein is an oil containing only oleic acid moieties esterified to glycerol. In contrast to these two examples, it is by no means necessary that the three fatty acid groups be derived from only one fatty acid. For example, we might have a triglyceride that contains one saturated fatty acid, say palmitic acid, one monounsaturated fatty acid, say oleic acid, and one polyunsaturated fatty acid, perhaps arachidonic acid.

Whether a triglyceride is a fat or an oil depends on two things: the length of the carbon chains and the degree of unsaturation of the constituent fatty acids. Specifically, the longer the fatty acids, the higher is the melting point and the more likely the triglyceride is to be a fat. The more unsaturation (carbon–carbon double bonds), the lower is the melting point and the more likely it is to be an oil.

For example, fats such as beef fat or butter have a high content of long-chain saturated fatty acids. Vegetable oils such as corn oil and sunflower oil, as well as olive oil, canola oil, and some fish oils, are highly unsaturated. Margarine is made by converting some of the carbon–carbon double bonds in vegetable oils into saturated (carbon–carbon single bonds) bonds, generating a fat from the oils and generating *trans* fatty acids in the fat.

In human physiology, fat plays several roles: energy reservoir, thermal insulation, and mechanical insulation. In addition, as noted in the last chapter, adipose tissue is an active endocrine organ.

Fat is the most important energy reserve in the human body. The energy is realized by conversion of the fats to fatty acids and glycerol and their oxidation to carbon dioxide and water. Let's see the consequences of these catabolic conversions.

Palmitic acid is a typical fatty acid. It contains a long straight chain of 16 carbon atoms:

$$CH_3CH_2CH_2CH_2CH_2CH_2CH_2CH_2CH_2CH_2CH_2CH_2CH_2CH_2CH_2COOH$$

or, more simply, $CH_3(CH_2)_{14}COOH$. Palmitic acid may be completely degraded to eight acetate units, CH_3COOH. Without worrying about the individual steps in this well-understood process, here is the net result:

$$CH_3(CH_2CH_2)_7COOH + ATP + 8CoASH + 7NAD^+ + 7FAD + 7H_2O \rightarrow$$
$$8CH_3CO\text{-}SCoA + AMP + PP_i + 7NADH + 7FADH_2 + 7H^+$$

Although there are numerous abbreviations in this equation, we have seen them all before. The equation as written consumes one molecule of ATP but yields 8 molecules of acetyl-SCoA and 7 each of NADH and $FADH_2$. The consequences of this for energy generation are pretty spectacular. Let's do the calculation.

From the electron transport chain, we generate $3 \times 7 = 21$ molecules of ATP from the reoxidation of NADH and $2 \times 7 = 14$ molecules of ATP from the reoxidation of $FADH_2$. Remembering that we consume one molecule of ATP, we get a net production of $21 + 14 - 1 = 34$ molecules of ATP per molecule of palmitic acid. If we consider that a single fat molecule contains three molecules of palmitic acid esterified to glycerol, complete oxidation will yield $3 \times 34 = 102$ molecules of ATP per molecule of fat. But that is not all. Back in chapter 17, we learned that the complete oxidation of acetyl-SCoA to carbon dioxide and water via the citric acid cycle generates an additional 12 molecules of ATP. So the total yield of ATP from oxidation of the acetyl-SCoA generated from one molecule of palmitic acid is $8 \times 12 = 96$ molecules of ATP. Since there are three molecules of palmitic acid in our triglyceride, we get a total of $96 \times 3 = 288$ molecules of ATP. Adding these to the 102 noted earlier gives us a grand total of $288 + 102 = 390$ molecules of ATP generated from the complete catabolism of one molecule of tripalmitin to carbon dioxide and water.[1] It should now be clear why fat is such a good storage form of chemical energy and why it is so difficult to lose weight stored in the form of fat.

Gram for gram, the energy yield from metabolism of fats is twice that from unhydrated glycogen. Beyond that, fat exists in an unhydrated state in adipose tissue, consistent with its hydrophobic nature. Adipose tissue contains very little water, in contrast to most other tissues of the human body. Glycogen, a hydrophilic carbohydrate, exists largely in a hydrated state (that is, associated with water) in liver and muscles. One gram of fat in adipose tissue provides as much energy as 6 grams of hydrated glycogen in liver or muscle. Having a substantial energy reservoir to draw on in times of caloric deprivation is basically a good thing. At the same time, many of us have a rather larger fat reservoir than we would like. Obesity is not simply a matter of weighing too much. All the excess weight is present as fat. The human body has fabulous capacity to store large amounts of energy in adipose tissue.

Fats and oils are among the structurally simpler lipids. Later we will have a brief look at some of the more complex classes. Before we get to that, let's consider some unusual structures formed by fatty acids and structurally related molecules. These structures provide an early look at some aspects of the structure of biological membranes.

Amphipathic molecules form bilayers

Molecules with sharply demarcated regions of hydrophilic and hydrophobic character are known as amphipathic molecules. Soaps provide an example. These form a variety of interesting structures. Such molecules may be thought of as schizophrenic, simultaneously struggling to satisfy two opposing natures. In water, amphipathic molecules will act so as to expose their hydrophilic structures to the aqueous environment while trying to find ways to hide their hydrophobic structures from it. One possibility among several is to form bimolecular layers or, more simply, bilayers.

The characteristic feature of a bilayer is that both faces are hydrophilic and are exposed to the aqueous environment. The hydrophobic hydrocarbon chains are hidden away inside the bilayer. Here is the idea:

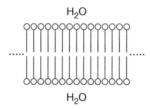

There are several classes of complex lipids

Complex lipids fall into several general structural classes: fats and oils; glycerophospholipids; sphingolipids; and steroids. We reserve consideration of the steroids for the following chapter. As usual, life is more complicated than this classification would suggest. For example, there are important molecules that are hybrids of carbohydrates and lipids, the glycolipids. One important class is glycosphingolipids, including the gangliosides, important constituents in nervous system tissue. In addition, there are complexes, covalent and not, of lipids and proteins.

In some aspects, the glycerophospholipids are related to the triglycerides: both use glycerol as a structural backbone and both have fatty acids esterified at two of the hydroxyl groups of glycerol. However, glycerophospholipids have a phosphoryl moiety attached to the third hydroxyl group, replacing a third fatty acid of triglycerides. This has a profound effect on the properties and functions of the glycerophospholipids. Specifically, a strongly hydrophilic group based on phosphate replaces a strongly hydrophobic fatty acid group. Thus, glycerophospholipids are, unlike triglycerides, amphipathic molecules, like soaps or detergents. This underlies their key role in the structure and function of biological membranes.

The simplest of the glycerophospholipids is phosphatidic acid, in which phosphate is linked to the third hydroxyl function, forming a phosphate ester. More complex glycerophospholipids are derivatives of phosphatidic acid in which one of several groups is attached: commonly choline, ethanolamine, serine, or *myo*-inositol. Structures are collected in table 19.1.

Although my central interest in these complex lipids is the role that they play in organizing the structure of biological membranes, these complex phospholipids also

Table 19.1 Structures for the major classes of glycerophospholipids

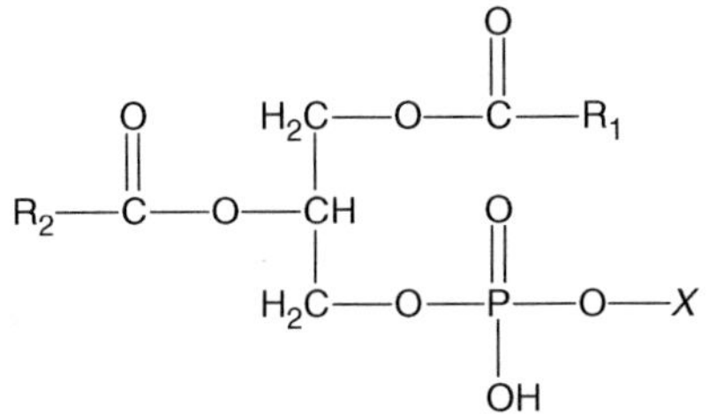

Name of X–OH	Structure of –X	Name of phospholipid
Water	–H	Phosphatidic acid
Ethanolamine	$-CH_2CH_2NH_2$	Phosphatidylethanolamine
Choline	$-CH_2CH_2N(CH_3)_3^+$	Phosphatidylcholine (lecithin)
Serine	$-CH_2CH(NH_2)COOH$	Phosphatidylserine
myo-Inositol	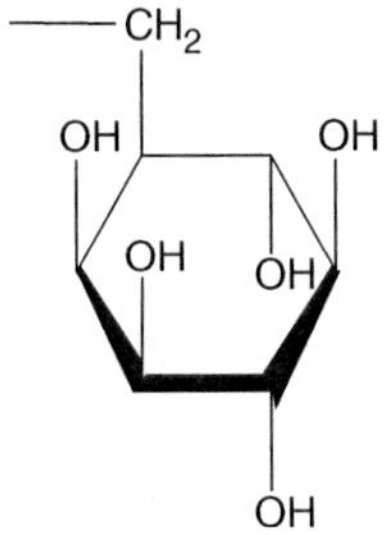	Phosphatidylinositol
Glycerol	$-CH_2CH(OH)CH_2OH$	Phosphatidylglycerol

have key metabolic roles to play. For example, in the phenomenon of programmed cell death a large number of cells undergo apoptosis and spill their contents into the milieu of healthy cells. This exposes phosphatidylserine, ordinarily hidden on the inside of biological membranes (see below), to the external environment. Receptors for phosphatidylserine on the surface of phagocytic cells interact with the exposed phosphatidylserine and trigger the phagocytes to clean up the cellular debris. Basically, the phosphatidylserine acts as an "eat me" signal to phagocytes. This keeps the area neat and clean and everyone happy.

Phosphatidylinositol is a key player in a signaling pathway known as the PI3K pathway. We will come across this pathway again in the chapter on cancer. It is one of the most important signaling pathways controlling cell proliferation and several proteins along it are targets for development of drugs against cancer. My general point is that the complex lipids collected in table 19.1 do more than act as structural elements for membranes.

Complex lipids are core components of biological membranes

We now know everything that we need to know to understand the structural organization of biological membranes and to begin to understand their functions.

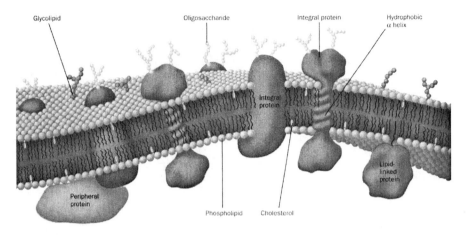

Figure 19.1 A schematic diagram of a plasma membrane. Integral proteins are embedded in a bilayer composed of phospholipids (shown, for clarity, in much greater proportion than they have in biological membranes) and cholesterol. The carbohydrate components of glycoproteins and glycolipids occur only on the external face of the membrane. (Reproduced from D. Voet and J. G. Voet, *Biochemistry*, 3rd edn, 2004. © 2004, Donald and Judith G. Voet. Reprinted with permission of John Wiley and Sons, Inc.)

At the outset, recognize that there are many types of biological membranes: some surround complex viruses; others occupy a space inside the bacterial cell wall; and still others isolate mammalian cells from their environment.

If we focus on the mammalian cell, we have the plasma membrane, which encompasses the cell and separates what is inside from what is outside. We also find the membranes that isolate the nucleus, the mitochondria, the lysosomes, and other intracellular organelles from the cytoplasm. All of these membranes have their own peculiarities and distinctions. However, it is not my purpose to make a catalog of known membrane types but to provide insights into the structural and functional features that are common to many membrane types. Since we need something specific to talk about, let's focus on the plasma membrane of mammalian cells. A schematic view of such a membrane is provided in figure 19.1.

We have encountered examples of simple lipid bilayers earlier. These bilayers are composed largely of amphipathic molecules. In water, they have their hydrophobic parts occupying the center of the bilayer and their hydrophilic parts occupying the bilayer surface. Such bilayers form a continuous and essential structural feature of virtually all biological membranes. We need to distinguish between that layer which faces out from the cell and is in contact with the external environment, the exoplasmic leaflet, and that which faces in and is in contact with the cellular contents, the cytoplasmic leaflet. As we shall see, these two aspects of the lipid bilayer are quite distinct.

Our first issue with respect to the lipid bilayer is its composition. This varies from membrane to membrane but generally includes several glycerophospholipids—phosphatidylcholine, phosphatidylethanolamine, phosphatidylserine—as well as cholesterol and glycolipids. The composition is complex. Beyond the basic complexity, the composition of the two bilayer leaflets is different. For example, all of the glycolipids are found in the exoplasmic leaflet, exposed to the external environment. In addition, most (but not all) of the phosphatidylcholine is also found in the exoplasmic leaflet. In contrast, most, but again not all, of the phosphatidylethanolamine and phosphatidylserine, which have neutral or anionic head groups, are found in the cytoplasmic leaflet. Cholesterol is found in both leaflets.

The bottom line here is that there is a basic asymmetry in the distribution of lipid between the two leaflets, being most dramatic in the case of the glycolipids, which appear exclusively in the exoplasmic one. As noted below, the asymmetry of the lipid distribution in the membrane is echoed in the asymmetric disposition of proteins that penetrate the membrane.

The lipid bilayer forms a barrier to transport of matter into and out of the cell. This barrier function is essential since cells need to be able to control their internal milieu, regardless of the external environment. (Some antibiotics work by disrupting the barrier function of bacterial membranes; see Chapter 23). At the same time, some communication of signals and materials across the bilayer must occur. Special mechanisms to do this are a key property of biological membranes. More specifically, these mechanisms are the province of proteins that one finds in these membranes.

As is depicted in figure 19.1, proteins may occupy the surface of the lipid bilayer or penetrate through it. Proteins in the latter class are termed integral proteins (some integral proteins penetrate only one of the two lipid leaflets). Integral proteins generally have three domains: an extracellular domain in contact with the outside world; a transmembrane domain that spans the membrane bilayer; and an intracellular domain that occupies a place within the cell. The actual situation may be considerably more complex than this simple picture suggests. For example, many biological receptors have seven transmembrane domains, G protein-coupled receptors (GPCRs), and some have even more. These weave back and forth between the outside world and the cellular interior seven times, exposing several polypeptide segments to both environments. There are about 1000 GPCRs coded for in the human genome.

This general organization makes basic sense. Think of a biological receptor as a device that communicates information between the outside world and the cellular interior. The insulin receptor, for example, provides this function. The extracellular domain includes a binding site for insulin. When an insulin molecule comes along and binds to this site, the structure of the extracellular domain is changed in some way, signaling that insulin binding has occurred. The transmembrane domain is a conduit for this signal that reaches the intracellular domain. This alters the structure of the intracellular domain in turn in a way that generates an enzymatic kinase activity. Specifically, the intracellular domain of the insulin receptor phosphorylates itself (using ATP) and a number of intracellular proteins, perhaps as many as 200. This turns on a sequence of intracellular events that culminate in the uptake of glucose molecules by fat cells and muscle and induction of glycogen storage in the liver. This is one important point of insulin action.

Now it would make no sense at all to have the insulin receptor turned end for end in a biological membrane. That would put the insulin recognition site on the inside on the cell, where it would be useless, and the intracellular domain with its kinase activity on the outside, where it would be equally useless. It follows that all insulin receptors have the same, specific orientation with respect to the bilayer leaflets, with the receptor site adjacent to the exoplasmic leaflet and the kinase site adjacent to the cytoplasmic leaflet.

The concept of a specific orientation for integral membrane proteins is quite general. All integral membrane proteins show such an asymmetry. This, coupled with the asymmetric distribution of lipids in the two leaflets, provides the two leaflets with different characteristics.

Integral proteins include receptors for hormones and other biologically active molecules, including a great many drugs employed in human medicine. Included are carriers for molecules that the cell needs but that would otherwise be excluded from it by the lipid bilayer and channels for ions that must get in or out of the cell. Finally, integral proteins act as signaling devices to other cells in the environment.

The French Paradox and resveratrol: curiosity or breakthrough?

To conclude this chapter, I would like to have a bit of fun with a new molecule— resveratrol.[2] There has been a substantial bit of news about this molecule and it is worth knowing something about it.

Nutritional orthodoxy in America tells us that high-fat foods, particularly those rich in saturated fats, are unhealthy. Americans are pretty much convinced intellectually if not in practice. A trip to any American supermarket will reveal a wealth of products labeled "light," "lite," "lo-fat," "reduced fat," "fat-free," and anything else that advertising folks can dream up. We buy tremendous quantities of these products and consume them. Many Americans obsess about food, seeing it not as a route to pleasure but as a collection of chemicals to be ingested at your own risk.

Now think about the French. By the standards of American nutritional orthodoxy the diet of the French leaves a good deal to be desired. The French consume butter, cream sauces, sausages, duck confit, foie gras, brie, camembert and 300-plus other cheeses, and rich pastries pretty much as they please. In fact, the French consume considerably more saturated animal fat than do Americans but they live just as long. Beyond that, they wash it down with wine. They seem to do so without worrying much about it at all: they view food as a genuine source of pleasure. The French obsess less about food than Americans; they enjoy it more.

Now think about the outcome. In America, about half of all adults are overweight and about one-quarter are frankly obese. In France, only 7% of adults are obese. The rate of heart disease in France is significantly less than that in America. So if we Americans are eating right and the French are eating wrong, how come they are slender and we are fat? How come we are in the coronary heart units of hospitals and the French are sitting in their restaurants enjoying their excellent cuisine? The French eat pretty much anything they want and live to tell about it. Americans obsess about the health aspects of their diet and waddle their way to the grave. This situation is widely known as the "French Paradox."

The French Paradox has generated a lot of interest. Clearly, despite the nature of their diet, the French are doing something right. It is of great interest to know what that is. Many people have suggested that no single factor is responsible for the French Paradox and that seems likely. The French eat smaller food portions than Americans and, unlike Americans, avoid eating between meals. Those factors may well be important in maintaining French health.

There is a more interesting, and possibly more important, explanation. A principal difference between American and French diets is the consumption of wine, specifically red wine. Could there be something in wine that accounts for the French Paradox? Attention has been focused on a class of compounds found in red wine called polyphenols (they are found in white wine as well but in much smaller amounts). These have potent antioxidant properties. Most of the attention has been focused on resveratrol:

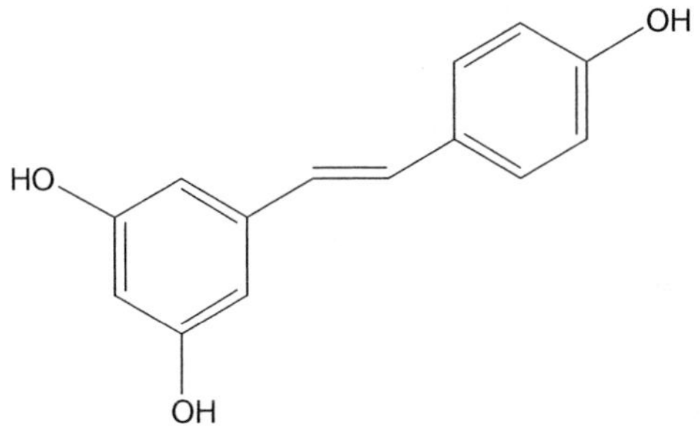

resveratrol

"OH" groups attached to benzene rings are termed phenols (Chapter 7). Since there are three such structures in this molecule, resveratrol is a polyphenol.

Resveratrol has a long history.[2] It was initially isolated from the roots of white hellebore in 1940. No one paid much attention. In 1963, it was isolated from a plant used for centuries in traditional Japanese and Chinese medicine. Again, this did not attract much attention. The story got a lot hotter in 1992 when the presence of resveratrol in red wine was suggested to be associated with the cardioprotective effects of red wine.

Subsequent studies in experimental animals have yielded provocative results. Resveratrol is known to extend the lifespan of a number of organisms from yeast to vertebrates. Resveratrol is also known to prevent or slow the progression of cancer, cardiovascular diseases, diabetes, inflammation, and ischemic injuries in experimental animals. In short, the suggestion that resveratrol in red wine may be responsible for favorable outcomes in human health is supported by a number of studies in experimental animals. However, the support is suggestive but certainly not definitive. Carefully controlled clinical trials in people will be required to establish the role, if any, of resveratrol or related small molecules in human health. Such clinical trials are currently underway.

How resveratrol works is not understood. There have been efforts to link resveratrol action to the effect of calorie restriction.[3,4] Here is why.

The only established way to prolong life is calorie restriction. Calorie restriction is known to prolong the life of yeast cells, human cells, the worm *C. elegans*, the fruit fly *D. melanogaster*, and rodents. It is not known if calorie restriction prolongs the life of humans. In addition, it has been established that calorie restriction also delays the development of age-related diseases, including cancer, atherosclerosis, and diabetes, in rodents. In mice at least, calorie restriction need not be a lifelong commitment. It is known that beginning calorie restriction in mouse late middle age (about 19 months) leads to a prolongation of life.

A central question is how calorie restriction works to prolong life and that is beginning to be understood. In yeast, calorie restriction leads to an increase in the activity of a protein known as Sir2, the product of the *SIR2* gene. Sir2 turns out to be an enzyme.[5] In this case, Sir2 catalyzes the chemical modification of proteins, the histones, that are complexed with DNA and regulate the expression of genes. It is quite plausible, though not certain, that the increase in the activity of the Sir2 enzyme induced by calorie restriction underlies the extension of life span in yeast.

It turns out that there is a mammalian analog of the *SIR2* gene; it is known as *SIRT1*. The enzyme products of *SIRT1* and related genes are collectively known as sirtuins and they all have activities similar to that of Sir2 in yeast. There are two questions: does upregulation of the activity of the sirtuins prolong life? Is there any way to upregulate the activity of the sirtuins other than calorie restriction? Perhaps, but the best evidence to date suggests that resveratrol does not activate or upregulate the activity of sirtuins.

Experimental studies have identified at least 26 potential molecular targets that are affected by resveratrol. Which of these, or what combination of these, is responsible for the beneficial actions of resveratrol in experimental animals is unclear. This will be a difficult puzzle to unravel. However, if ongoing clinical studies in people establish a clear role for resveratrol or a related molecule in human health, the mechanism of action will be of secondary importance.

There are a lot of stories of the resveratrol type: small, naturally occurring molecules for which there is evidence of important physiological properties. Resveratrol, or one of its chemical relatives, may prove to be a valuable compound for human health. It may also prove to be a curiosity. Time will tell: stay tuned. In the meantime, the resveratrol story is a boon for the makers and marketers of red wine. I do not know if we really need an excuse to drink a glass of red wine but it is convenient to have one.

Key Points

1. Fatty acids are key constituents of several structural classes of lipids: triglycerides, glycerophospholipids, and glycolipids.

2. Fats and oils are triglycerides: esters of glycerol and fatty acids.

3. Fat is an excellent form of energy storage. Complete oxidation of one molecule of tripalmitin to carbon dioxide and water generates at least 390 molecules of ATP.

4. A bilayer formed from complex lipids, largely glycerophospholipids, forms the core structure of biological membranes. This bilayer forms a barrier to penetration of exogenous molecules into the cellular interior. Proteins penetrate into or through this bilayer.

5. Resveratrol is known to extend the life span of a variety of living organisms. Its role in human health is not known.

20

Steroids

Sex and other good things

Steroids are derived biosynthetically from cholesterol. They play multiple roles in human physiology: sex characteristics, control of inflammation, embryo implantation, and control of salt and water balance. Cholesterol itself is an indispensable constituent of biological membranes.

The steroids are a remarkable family of molecules. They are essential for health, responsible for the characteristics of our genders, useful medicinal agents, and drugs of abuse. Modest structural variations on the central theme of steroid structure profoundly alter the physiological response to the action of these molecules. Small differences in structure can create big differences in function.

The steroids are unified structurally in having a basic molecular skeleton containing four fused rings, three having six members and the fourth having just five:

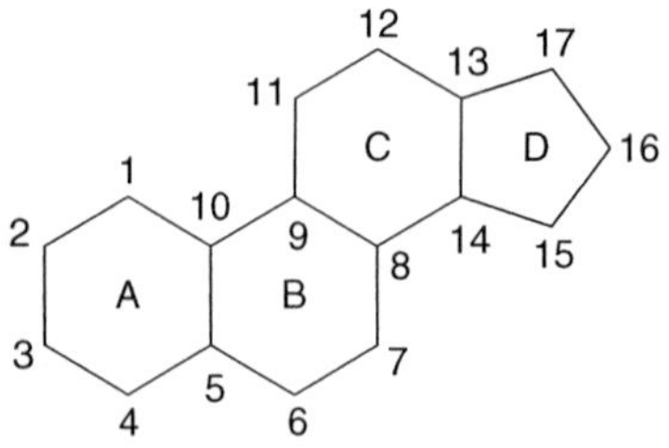

The steroid skeleton may be thought of as a kind of chemical Christmas tree upon which we can dangle ornaments pretty much as we please. In our case, dangling ornaments takes the form of hanging various functional groups on whichever atom or atoms that we choose in whatever stereochemistry suits us. For convenience, I have numbered the various carbon atoms of the steroid skeleton in the conventional way. This will permit me to designate precisely where chemical modifications to this skeleton occur. Many functional groups are available to us—varying from single atoms such as those of fluorine or chlorine, to alkyl groups such as methyl or ethyl, to more complex ones—and there are 17 carbon atoms available for decoration. Moreover, typical steroids have several chiral centers. Thus, there is enormous latitude available to us as we search for molecules having interesting and useful properties.

In addition to numbering the individual carbon atoms, I have labeled the four rings, A, B, C, and D, so that I can easily specify which ring I am talking about later. To anticipate, the structural difference between male and female sex hormones lies almost entirely in the A ring.

The steroid skeleton is decorated in two basic ways. The first is reflected in the variety of structures that are synthesized by living organisms. Humans and other animals synthesize quite a number of different steroids. Plants, insects, and other forms of life make many others. The second source of steroid skeletal decoration is that provided by chemists. Here we are not limited to what is coded in genomes but only by the knowledge and wit of scientists. Chemists have succeeded in making a huge number of different steroids, largely in the search for new or improved drugs for clinical medicine. In this search, novel compounds having fascinating and useful properties have been discovered.

Doubtless, the steroid most commonly referred to in daily life is cholesterol. It is the ideal jumping-off point for a discussion of steroids and their properties.

Cholesterol has two faces

Cholesterol is a famous and much maligned molecule:

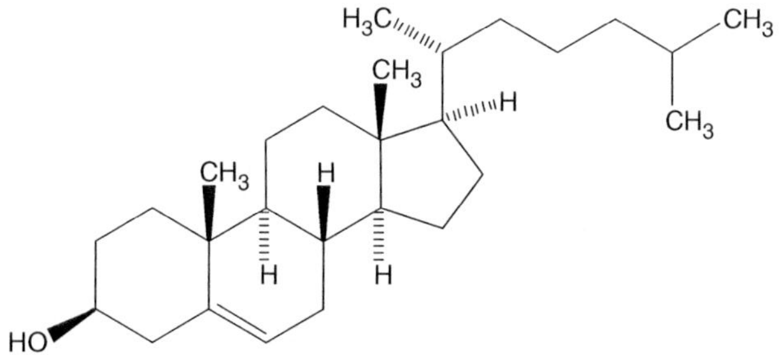

Cholesterol

Cholesterol is derived from the steroid skeleton by adding a hydroxyl group at C3, methyl groups at C10 and C13, an eight-carbon alkyl group at C17, and introducing a double bond between carbon atoms 5 and 6. I have taken some pains to indicate the stereochemistry at each of the eight chiral centers in this molecule; its shape matters.

Humans make cholesterol from a very simple precursor—acetate. Cholesterol synthesis occurs principally in the liver. We also consume varying amounts of cholesterol in our diet. The multistep biosynthetic pathway to cholesterol beginning with acetate gradually creates increasingly complex molecules, and, finally, yields cholesterol itself. The late Konrad Bloch worked out this biosynthetic pathway in some detail. Bloch shared the Nobel Prize in Medicine or Physiology in 1964 for this work. We now understand it in intimate and revealing detail. The pathway itself is of considerable interest, in part because it contains a number of branch points leading to other families of molecules of considerable interest. It is summarized in figure 20.1. I will have occasion to refer to this pathway in the context of drug discovery.

Cholesterol is the steroid from which all other steroids in human biochemistry are made. Thus, humans can synthesize this complex molecule from simple precursors and convert it into a large family of related complex molecules having critical roles in human physiology.

Most of the mention of cholesterol in the popular press positions this molecule as a threat to human health. Many foods are proudly labeled "cholesterol-free." People are properly warned to pay attention to their plasma cholesterol level,[1] particularly that carried in the low-density lipoproteins, LDLs, commonly known, with pretty good reason, as "bad" cholesterol. LDLs are lipoprotein particles containing a large protein known as B-100 associated with cholesterol, cholesteryl esters, phospholipids, and some triglycerides.

It is well-established that an elevated level of cholesterol, particularly that carried largely in the form of LDLs, is an independent risk factor for the development of atherosclerosis and its sequelae, including coronary artery disease (leading to heart attacks), strokes, and peripheral arterial disease.

Atherosclerosis is the result of formation of lipid-rich deposits on the intimal walls of medium and large arteries. Over time, these deposits grow and become fibrous and calcified. This leads to compromise in the free flow of blood along affected arteries. In addition to elevated plasma levels of LDL cholesterol, risk factors for development of atherosclerosis include a family history of atherosclerotic disease, hypertension, smoking, diabetes, obesity, and being male. Some of these risk factors are under our personal control to varying degrees. Others are not: no one gets to choose his or her parents and if you are male there is not a whole lot you can do about it. You can control your diet and body weight, eliminate smoking, and treat hypertension if necessary. Since cardiovascular diseases are among the most important causes of disability and death in humans, an enormous amount of attention is properly paid to their prevention and correction.

Cholesterol is absolutely required by human beings for two reasons. The first one we have already noted: cholesterol is the starting place for the synthesis of other steroids that are themselves required for normal human physiology. The second reason is that cholesterol is a key constituent of cell membranes in all cell types and tissues of the body.

Careful dietary restriction of cholesterol and saturated fat, as well as an increase in exercise, will result in diminution of plasma cholesterol, particularly that carried in LDLs. A prudent diet takes cognizance of this fact. The dietary guidelines of the American Heart Association (AHA) recommend a daily intake of cholesterol of no more than 300 mg.[2,3] At the same time, restriction of dietary cholesterol tends to

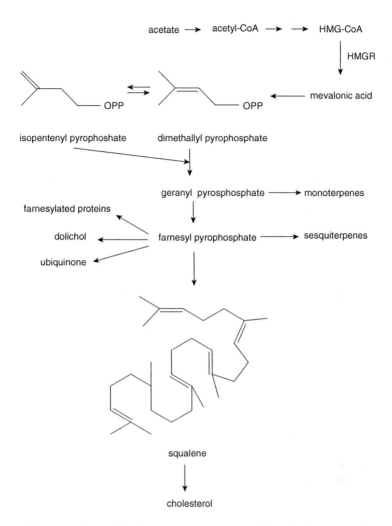

Figure 20.1 An outline of the biosynthetic route to cholesterol. Cholesterol biosynthesis begins with the two-carbon precursor acetate. In several steps, acetate is converted to hydroxymethylglutaryl-CoA, HMG-CoA, a five-carbon intermediate that is the substrate for hydroxymethylglutaryl-CoA reductase, HMGR, the molecular target for the statins. By way of mevalonic acid, HMG-CoA is converted into two equilibrating five-carbon precursors of cholesterol and many other molecules of animals and plants: isopentenyl and dimethallyl pyrophosphates. Note the carbon skeletons of these molecules, which are still recognizable in squalene. These pyrophosphates are condensed to generate the 10-carbon molecule geranyl pyrophosphate, the precursor in plants to a large number of molecules known as terpenes. A molecule of isopentenyl pyrophosphate is added to geranyl pyrophosphate to yield the 15-carbon compound farnesyl pyrophosphate, the precursor of sesquiterpenes in plants and dolichol, ubiquinone, and farnesylated proteins in humans. Dolichol is a carrier of sugar units in the synthesis of oligosaccharides and polysaccharides. Ubiquinone, also known as coenzyme Q, is an electron carrier in the electron transport chain. Two molecules of farnesyl pyrophosphate condense to yield the 30-carbon compound squalene, the terminal noncyclic precursor on the route to cholesterol. I have shown squalene as it is believed to exist prior to cyclization. I believe that it is easy to see how squalene can be "sewn up" to generate the steroid skeleton. It takes several steps to get from squalene to cholesterol.

increase the biosynthesis of cholesterol in the body, compensating in significant part for the reduced amount of cholesterol in the diet. This should not be surprising. Since cholesterol is absolutely required by people, the human body has developed more than one way to fulfill its need for this critical molecule. The bottom line for significant plasma cholesterol reduction through dietary control is that one must be very attentive to dietary cholesterol and saturated and *trans* fats. Increasing ones' exercise level also helps. Contrary claims for plasma cholesterol reduction by those marketing foods or dietary supplements should be viewed as being somewhere between exaggerations and outright fabrications.

For people who are at risk of cardiovascular disease due to high plasma LDL cholesterol levels, lifestyle changes to control plasma cholesterol levels are the first and best place to start. When efforts to control plasma cholesterol levels by diet and exercise fail, people frequently turn to drugs, some of which are effective in producing substantial lowering of cholesterol levels and realizing associated clinical benefits.

The most important class of cholesterol-lowering agents is the statins. These include lovastatin (Mevacor), simvastatin (Zocor), pravastatin (Pravachol), and atorvastatin (Lipitor), among others. These molecules work, in modest part, by inhibiting biosynthesis of cholesterol and, in larger part, by increasing the rate at which cholesterol is eliminated by the body. Let's have a look at this in more detail.

There are a lot of underlying causes of elevated plasma cholesterol levels. I am going to focus on one: a rather common human genetic disease termed type II hypercholesterolemia, also known as familial hypercholesterolemia. An important clue to the importance of LDL cholesterol levels in human health comes from this disease. These patients have defects in the liver (hepatic) receptor for LDL. Hepatic LDL receptors are the only important mechanism for removal of cholesterol from the human body. These receptors take up circulating LDL, incorporate it into liver cells where the LDL is broken down, and most of the cholesterol is converted to bile acids. These are then dumped into the small intestine, by way of the gall bladder, to aid in the digestion of lipids. Although most of these bile acids are reabsorbed and reused, a significant fraction is eliminated in the feces, which is how cholesterol gets out of the body. The defect in hepatic LDL receptors in type II hypercholesterolemia patients (there are many different defects at the molecular level) renders the LDL receptor ineffective. Michael Brown and Joseph Goldstein tracked down this story in molecular detail and were rewarded with the Nobel Prize in Medicine or Physiology in 1985 for their work.

Heterozygotes for type II hypercholesterolemia have one normal and one mutant LDL receptor gene. Thus, they have just half the normal number of functional hepatic LDL receptors. They are, therefore, less than fully effective at eliminating cholesterol from the body. These individuals have substantially elevated plasma LDL cholesterol levels and frequently experience heart attacks in their 40s and 50s.

A desirable total plasma cholesterol level is considered to be less than 200 mg/dl. This value corresponds to about 130 mg LDL cholesterol/dl. Values between 200 and 239 mg/dl are classified as borderline high; those above 239 mg/dl (190 mg LDL cholesterol/dl) are simply high. Type II heterozygotes frequently have plasma cholesterol in the range 400–600 mg/dl, very high indeed. Homozygotes, who have two mutant LDL receptor genes, have no functional hepatic LDL receptors. These people have no effective means of eliminating cholesterol from their bodies. They have immensely high plasma cholesterol levels, sometimes in the range 800–1000 mg/dl, and may have heart attacks at age 10–12 years old. Homozygotes for type II hypercholesterolemia are mercifully very rare. The basic point here is that elevated plasma cholesterol levels are correlated with elevated risk of coronary artery disease. Here are the essentials of the statin story.

The statins are an important class of cholesterol-lowering drugs

A key enzyme on the pathway to cholesterol is known by the obscure (to nonchemists) name of hydroxymethylglutaryl-CoA reductase, HMGR; figure 20.1. This enzyme catalyzes the first step in the pathway that is unique to cholesterol biosynthesis and, therefore, is a good target for inhibition of the pathway.[4] A Japanese pharmaceutical company, Sankyo, discovered the first potent and specific inhibitor for HMGR. It is known as mevastatin and was isolated from the broth in which a fungus had been grown. Mevastatin proved to be a potent cholesterol-reducing agent. However, mevastatin was later reported to cause tumors in experimental animals and Sankyo halted the clinical trials.[5]

Subsequently, a second, closely related molecule—lovastatin—was discovered by scientists at Merck in the United States in another fungal fermentation broth and by Sankyo in Japan. Lovastatin, marketed as Mevacor in the United States, proved both safe and efficacious for the intended use and was the first "statin" to be approved for human use. Several others, some mentioned above, followed. The history of discovery and development of HMGR inhibitors has been pulled together by Jonathan Tobert, who led the lovastatin and simvastatin clinical development effort at Merck.[6]

The statins have been demonstrated to markedly lower plasma LDL levels (and triglyceride levels to a lesser extent). In fact, statins were approved by the US FDA on the basis of a surrogate endpoint: reduction in plasma cholesterol levels. Since we know that increased plasma cholesterol levels are correlated with increased risk of coronary artery disease, it seems logical that reducing plasma cholesterol levels would lead to reduced risk. That turns out to be true in this case. However, see the case of hormone replacement therapy (HRT) for women for a more complex example, discussed below.

The important point for patients is not so much that plasma cholesterol decreases but that the clinical endpoints decrease: nonfatal heart attacks, coronary artery disease deaths, surgery for coronary artery disease, and total mortality. Subsequent to the approval of statins based on reduction of plasma cholesterol levels, several large, controlled clinical trials have been carried out to establish the effects of statin use on these clinical endpoints.[7]

These studies demonstrate that optimal doses of statins reduce the incidence of clinical events in patients with established coronary artery disease, in patients with elevated plasma LDL levels but without existing coronary artery disease, in individuals with normal plasma LDL levels without existing coronary artery disease, and in diabetics, a patient population at high risk of cardiovascular disease.[8]

This story of how the statins work to lower cholesterol is not quite complete. As is the case with many drugs, the human body reacts in a way that threatens to minimize the action of the drug: that is, to keep things the same. In the case of the statins, the liver senses that it is being deprived of cholesterol by the drug and reacts in two ways. The first thing that happens is the upregulation of HMGR and other enzymes on the pathway to cholesterol in an effort to negate the effect of the inhibitor. It does this quite well and the rate of hepatic cholesterol synthesis in people taking a statin is rather little changed. This is not a bad thing. The body requires a source of cholesterol: for cell membrane synthesis, as a precursor to all the steroid hormones, and as a precursor to the bile acids. Beyond that, several important human metabolites branch off the cholesterol pathway following the HMGR step (see figure 20.1).

The second thing that happens is a very good thing. The liver upregulates its LDL receptor level. In sum, the liver does the two things that it can do to maintain an adequate source of its cholesterol: pump up the synthesis (to normal or slightly subnormal rates only) and pump up the sequestration of cholesterol from the blood LDLs. The net effect is that the rate of excretion of cholesterol from the body is increased due to the increase in the number of LDL receptors expressed on hepatic cells without compromising the availability of cholesterol to meet cellular needs.

Now, returning to the case of type II hypercholesterolemics, this is exactly what we wish to have happen: these patients return to normal levels of LDL receptors. This is a beautiful outcome to a drug discovery story.

Sex hormones include the estrogens and androgens

In various tissues of the body, cholesterol is converted to several classes of steroid hormones. Let's begin with the sex hormones.

Androgens are male sex hormones; testosterone and dihydrotestosterone are good examples:

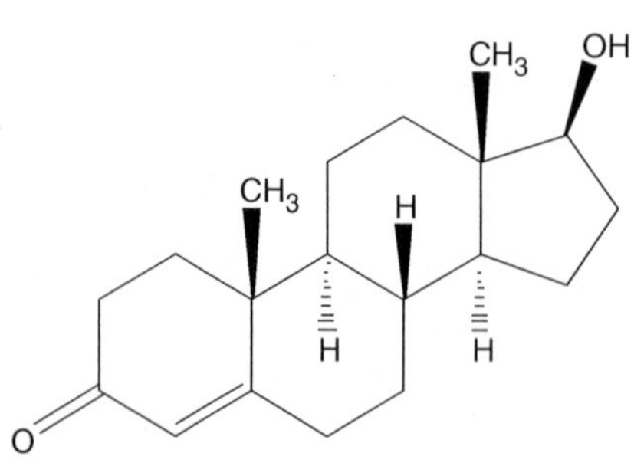

testosterone 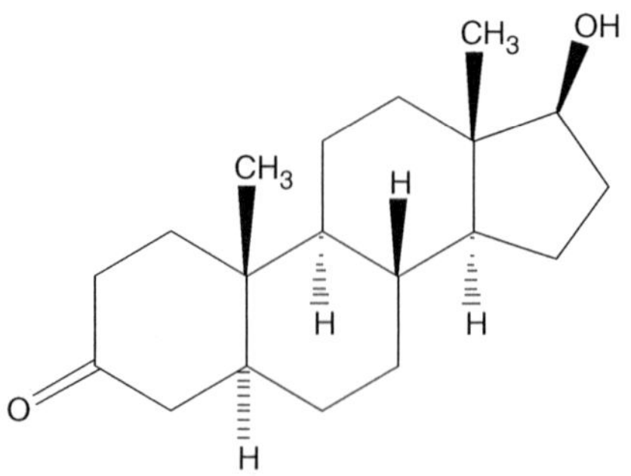   dihydrotestosterone

Key structural changes from cholesterol to testosterone include the replacement of the alkyl group at C17 with a hydroxyl group, conversion of the hydroxyl group at C3 to a keto group, and relocation of the double bond. Nothing else has changed. The double bond between C4 and C5 in testosterone is lost in the conversion to the more powerful androgen dihydrotestosterone.

The first androgen to be isolated in pure form was androsterone, in 1931, by the chemist Adolph Butenandt, the same guy who isolated the first pheromone (that for the female silkworm moth, bombykol), which we met back in chapter 6. Butenandt managed to isolate 50 mg (less than 0.002 ounces) of androsterone from 25,000 liters of men's urine! Some efforts are more heroic than others.

In men, testosterone is synthesized and released from the interstitial cells of the testis. In fetal development, androgens are responsible for degeneration of the female sexual organs and the development of the male ones (we all start life basically as females).[9] At puberty, androgens cause development of the secondary sex characteristics of males: facial hair, deepened voice, upper body musculature. In addition, testosterone is largely responsible for the sex drive in both men and women. The plasma testosterone level in women is normally about 10–12-fold less than in men but it nonetheless plays a significant role in female physiology. Quantitatively, the normal range for total testosterone in men is 270–1070 ng/dl. The corresponding value for women is 6–86 ng/dl. However, much of the plasma testosterone is carried bound to a steroid-hormone-binding globulin, SHBG. In this form it is unavailable to act at the androgen receptor. Normal ranges for unbound plasma testosterone are 15–40 pg/ml for men 20–40 years old, 13–35 pg/ml for men 41–60 years old, and 12–28 pg/ml for men 61–80 years old. The corresponding values for women are 0.6–3.1, 0.4–2.5, and 0.2–2.0 pg/ml, respectively.

Hormone replacement therapy for postmenopausal women has long been widely practiced for the relief of postmenopausal symptoms: more about this follows later. Newer is HRT for men.[10] As men age, their plasma testosterone level tends to fall over time, as noted above. For some men, the fall is far more precipitous than indicated by the range of normal values. The decrease in plasma testosterone over time may contribute to gradual diminution in sexual desire, loss of energy, and increase in body fat at the expense of body muscle (the extent of the contribution is controversial). In an effort to minimize these changes with aging, HRT is becoming increasingly popular with older men and, to a significant extent, with younger men. Some 30% of male testosterone users are between the ages of 18 and 45 years old. As of 2003, the best estimates suggest that about 2 million men in the United States were being prescribed testosterone. However, careful controlled clinical trials to establish the long-term effects of HRT in men have not been done. The health implications of long-term HRT in men are, therefore, not yet known. This is of particular concern for the younger men who may be users for decades.

As noted above, dihydrotestosterone, DHT, is a substantially stronger androgen than is testosterone, T, itself. T is converted to DHT by the action of an enzyme known as 5-alpha reductase. There is a rare genetic disease known as pseudohermaphroditism in which 5-alpha reductase activity is compromised, leading to abnormally low levels of plasma DHT.[11] Male babies afflicted with this genetic error are born phenotypically female, recognized as females by their parents, and raised as females. Stunningly, at the time of puberty they develop into phenotypical males: that is, a penis appears as does a scrotum into which testicles descend. These males develop the typical upper-body musculature of men and their voices deepen. Basically, a developmental process in utero that should have yielded a recognizable male baby was compromised by a deficiency of DHT. The developmental process was completed by the androgen drive at puberty. These men do not get acne, they do not experience male-pattern baldness, and they have a small prostate gland that never grows. These observations strongly suggest that DHT is the etiologic agent for acne, male-pattern baldness, and the benign growth of the prostate gland in aging men (benign prostatic hypertrophy or BPH). The prostate gland surrounds the urethra and when it grows it may compromise the ability of men to urinate, among other symptoms. This can be annoying, very annoying, or quite serious, as it can lead to complete urinary retention. There are several surgical approaches to the treatment of BPH, the most common of which is transurethral resection of the prostate, which I personally prefer not to think about. The observations recited above suggested that an inhibitor of 5-alpha reductase might be a useful pharmacological treatment for BPH, and perhaps male-pattern baldness and acne.

Such an inhibitor is finasteride, marketed as Proscar for treatment of BPH and Propecia for treatment of male-pattern baldness.

Androgen antagonists are useful in the treatment of prostate cancer

Most cancers of the prostate gland in men are androgen-dependent: that is, their growth depends on a supply of androgens. Androgen antagonists, antiandrogens, are frequently employed in the therapy of prostate cancer. Cyproterone (Androcur) is an example of an antiandrogen. Cyproterone and its ilk compete with physiological androgens, T and DHT, for the binding site on the androgen receptor. Fairly subtle structural changes between testosterone and cyproterone change an agonist into an antagonist. Testosterone binds to the androgen receptor and things happen; cyproterone binds to the androgen receptor and nothing happens. Beyond that, when cyproterone is present, nothing happens when testosterone is also present. Cyproterone sits on the androgen receptor where testosterone usually sits and blocks access. That is the difference between an agonist and an antagonist. Since molecules in this class block the action of all androgens, they abolish male sex hormone action. Of course, cessation of use relieves the consequences, including impotence, lack of sex drive, and gynecomastia (development of the breasts in the male).

Estrogens are female sex hormones

The flip side of androgens is estrogens: female sex hormones. Estradiol is a good example of a potent estrogen:

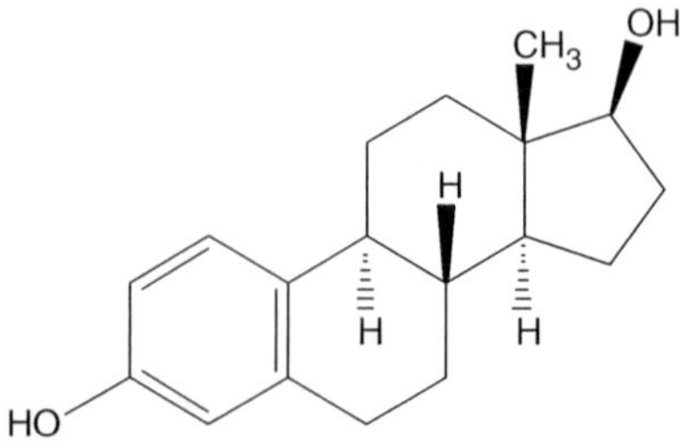

estradiol

Note the structural similarity of testosterone to estradiol. Except for changes in the A ring, the molecules are identical. In fact, estradiol is derived biosynthetically from testosterone through the action of an enzyme known as aromatase. The structural changes in the A ring elicited by aromatase have powerful biological consequences. Note particularly that the A ring in estradiol is a modified benzene ring and, therefore, planar. The A ring in testosterone, in contrast, is not. The critical consequence of these shape changes is that estrogen has affinity for the binding site of the estrogen receptor but not the androgen receptor and testosterone has just the opposite properties. Small structural changes can have large biological consequences.

Here too Adolph Butenandt played a key role. He isolated the first estrogen, estrone, in 1929, in minute quantities from the ovaries of 80,000 sows! Clearly, Butenandt was not deterred by major challenges. We have yet to hear the last from him.

Estradiol is synthesized in and released by the ovary, placenta, testis, and perhaps the adrenal cortex. Estrogens are responsible, among other things, for development and maintenance of the secondary sexual characteristics of females. Estrogens are very potent molecules. The circulating level of estrogens in premenopausal women varies a lot, depending on stage of the menstrual cycle: the normal range is 50–450 pg/ml. In postmenopausal women, the value drops to <59 pg/ml, comparable to the normal value for men, which is <50 pg/ml.

At menopause, the production of estrogens in females is markedly reduced. This leads to a number of unpleasant effects, including hot flashes and night sweats. More important for long-term health, the estrogen decrease is associated with an increased risk of cardiovascular disease and osteoporosis. The action of estrogens decreases plasma cholesterol and helps maintain good bone mineral density.

For decades, millions of postmenopausal women have been taking estrogen in one form or another (e.g., as Premarin) for relief of the symptoms of menopause. Stimulated by a publication of the Nurses' Health Study in 1985 reporting that women using estrogen had a major reduction in the frequency of heart attacks, a great many women elected to broaden their use of estrogens in the form of hormone replacement therapy, HRT. In HRT, the exogenous use of estrogen is designed to replace the endogenous estrogen that is lost at the time of menopause, not only to relieve menopause symptoms but also to promote heart and bone health. In fact, there are two basic forms of HRT for women: estrogen alone and a combination of estrogen and a progestin. The risks and benefits of these two forms of HRT are not the same, as developed below.

It has been established through clinical trials that HRT slows the development of osteoporosis (i.e. bone loss leading to fractures) in postmenopausal women, a notable benefit. The data from the Nurses' Health Study reinforced suggestions that HRT would have beneficial effects on cardiovascular health, since estrogens decrease blood levels of LDLs and increase those of high-density lipoproteins, HDLs. There were suggestions that women on HRT might be less susceptible to Alzheimer's disease and depression as well. For three decades there have been worries about estrogen use and cancer. It has been established, for example, that estrogen replacement increases the risk of development of endometrial cancer. On balance, however, the benefits of estrogen replacement therapy seemed to outweigh the risks, for most women at least.

In 1998, the medical world got a big surprise. The Heart and Estrogen-Progestin Replacement Study, HERS, was the first carefully controlled clinical study of HRT in women. It was found that estrogen–progestin replacement in women who had heart disease actually *increased* the rate of heart attacks (why it took so long to do a carefully controlled clinical study of HRT in women after decades of clinical use is another story).[12] In 2002, data from a controlled clinical study known as the Women's Health Initiative, WHI, revealed that HRT in postmenopausal women increased the risk of heart disease, stroke, breast cancer, blood clots, and, possibly, dementia.[13] Some scientists believed that these negative findings were the consequence of the progestin component of estrogen–progestin combinations. Subsequent clinical studies employing estrogen alone revealed that elimination of the progestin component did alleviate some of the negative findings but that some harmful effects did remain.[14] Finally, a 2007 report of a controlled clinical trial of HRT suggested that HRT might protect women from heart disease if started during menopause but is deleterious if started later. In sum, there remains considerable uncertainty concerning the risks and benefits of HRT, particularly for postmenopausal women. As a consequence, the long-term use of estrogen replacement therapy, particularly in combination with a progestin, in women has been reduced.

Let's ask why the 1985 results from the Nurses' Health Study have been called into serious question by later placebo-controlled clinical studies. The Nurses' Health Study is an observational study, a matter for epidemiologists, as opposed to a placebo-controlled clinical study. In an observational study, associations are sought between disease frequency and lifestyle factors, including matters such as diet, exercise, use of tobacco and alcohol, drugs, and the like. Usually a large cohort of people is followed for a period of years. The Nurses' Health Study, for example, was begun in 1976 and has enrolled a total of 122,000 nurses. The goal is to establish causation between one or more lifestyle factors and the frequency of one or more diseases. The underlying problem is that association does not establish causation.

It is useful to differentiate two classes of results. In a few cases the association of a lifestyle factor with disease frequency is overwhelming and causation is highly likely. Examples include the association of cigarette smoking with lung cancer and heart disease and the association of exposure to the sun with skin cancer. Both these cases have been followed up and there is no doubt about the validity of correlating association with causation. However, the more general case is the finding of a *weak* association of some lifestyle factor with disease frequency. In these cases, linking association with causation is highly suspect and sometimes wrong. In these cases, my strong inclination is not to believe that association implies causation until it has been confirmed by a controlled clinical trial.

There are several inherent weaknesses in observational trials. In controlled clinical trials, every effort is made to select patient populations in a way that excludes as many variables as possible between the placebo- and drug-treated groups, leaving drug treatment as the only variable. This ideal cannot be strictly realized in practice but experienced investigators come pretty close. Beyond that, controlled clinical trials state upfront what the primary and secondary endpoints of the study are. So you know what you are looking for. These conditions cannot be met in observational studies. There is no control group. Several bias factors may be important. One hunts for associations post hoc, not upfront. Observational studies frequently lead to subsequent controlled clinical trials. These may or may not confirm the original findings. The problem is that the initial observational study usually gets big play in the media. The subsequent controlled clinical trials get little play, whether or not they confirm the findings. This has major potential to lead the public astray. Now, it is time to get back to the main thread of our exploration of estrogens.

Estrogen antagonists are useful in the treatment of breast cancer

Some common tumors, including many breast cancers, are estrogen-dependent. Consequently, agents known as selective estrogen receptor modulators (SERMs) are frequently employed after surgery, chemotherapy, and/or radiation therapy to slow or prevent the recurrence of breast cancer. SERMs are also known to reduce the risk of developing breast cancer in women at high risk for breast cancer. The most widely used SERM in breast cancer or high-risk patients is tamoxifen, marketed under the name Nolvadex, among others. Tamoxifen and other SERMs have a complex pattern of tissue-specific estrogen antagonist and agonist activity.[15] In the breast, tamoxifen acts as an estrogen antagonist, as one would expect from its profile of utility in breast cancer. More about SERMs follows below.

There is an alternative to antagonizing the action of estrogens in breast cancer patients and that is to limit or prevent their synthesis. Preventing the synthesis of and antagonizing the action of estrogens are simply two ways of getting to the same endpoint: limiting estrogen action at specific sites in the human body. As noted above, estradiol (and other estrogens) is synthesized from testosterone under the action of the enzyme aromatase. Three inhibitors of this enzyme have been approved for use in breast cancer patients. Anastrozole (Arimidex) provides an important example.

Nuclear hormone receptors comprise a family of mechanistically related transcription factors

The superfamily of nuclear receptors includes about 150 proteins in the human genome. These are the targets of all principal classes of steroids as well as those of retinoids, vitamin D, and thyroid hormones.

Although the details of each receptor class in the superfamily vary somewhat, there is a central theme running through all of them. I shall use the estrogen receptor (in fact there are two) as an example. In contrast to many hormone receptors that are anchored in the plasma membrane of the cell, the nuclear receptors are found in the cell cytoplasm or occasionally in the cell nucleus. In the absence of ligand, the receptor proteins are present as large multiprotein complexes, including those proteins known as heat-shock proteins. Once the activating ligand is present, it binds to the receptor, which is then freed from the heat-shock proteins, undergoes a change in conformation, and generally dimerizes: that is, two of them get together and form a complex. It is this ligand-bound dimer that is the biologically active species.

As noted above, there are two estrogen receptors, unimaginatively named ERa and ERb. The structural organization of the two human ER receptors is similar. Starting at the N-terminal region, there is a domain that is responsible for activating transcriptional activity, followed by a domain that has DNA-binding properties, a hinge region, and finally a ligand-binding region to which estrogens and estrogen antagonists bind. The two receptors differ in their tissue distribution and patterns of transcriptional activation. ERb is less sensitive to the action of several antiestrogens than is ERa.

Once the ligand-bound estrogen dimer is formed, it migrates into the cell nucleus and the DNA-binding domain searches out the estrogen response element, ERE, on the nuclear DNA. The ERE consists basically of a set of six nucleotides followed by a spacer of zero to eight nucleotides followed by another six nucleotides. The two sets of six nucleotides serve as a recognition site for the estrogen receptor dimer.

Once the ERa dimer is bound to the ERE, a specific set of genes is turned on, the proper proteins are synthesized, and the consequences of estrogen action ensue. Remember that it is the proteins that, in the last analysis, determine the personality of the cell and, in the end, of the organism.

The activated ERb dimer complexed with estrogen when bound to the ERE has a very different effect: transcription is turned off rather than on. This accounts in part at least for the complex and perplexing actions of SERMs. Consider tamoxifen again: it binds to both ERa and ERb. In those tissues where ERa dominates, tamoxifen will act as an estrogen antagonist by blocking estrogen-activated transcription, as it does in breast tissue. However, in tissues where ERb activity dominates, tamoxifen will show estrogenic activity by blocking the actions of a transcription inhibitor: blocking a blocker is a lot like activating. Tamoxifen shows estrogenic activity in bone, the cardiovascular system, and the uterus. This activity in bone is good, as it tends to increase bone mineral density and ward off osteoporosis. The activity in the cardiovascular system is also good, as it reduces levels of LDLs and triglycerides, reducing the risk of atherosclerosis. The estrogenic activity in the uterus is unfavorable: tamoxifen increases the risk of uterine endometrial carcinoma in postmenopausal women.

The idea here is quite general: androgens find their androgen receptors, which have a similar functional organization. The dimerized receptor finds its androgen response element, ARE, to which it binds, and ultimately controls transcription. The same steps follow for glucocorticoids, mineralocorticoids, etc. Specificity is achieved at several levels: the structure of the steroid, the nature of the hormone-binding site in the receptor, the nature of the DNA-binding domain in the receptor, the structure of the response element, and, finally, the spectrum of proteins whose synthesis is activated. And that is pretty much how steroid hormones, among others, work.

Progestins are critical for a successful pregnancy

Progestins, of which progesterone is a prime example, induce endometrial maturation following ovulation, setting the stage for implantation of a fertilized ovum and pregnancy:

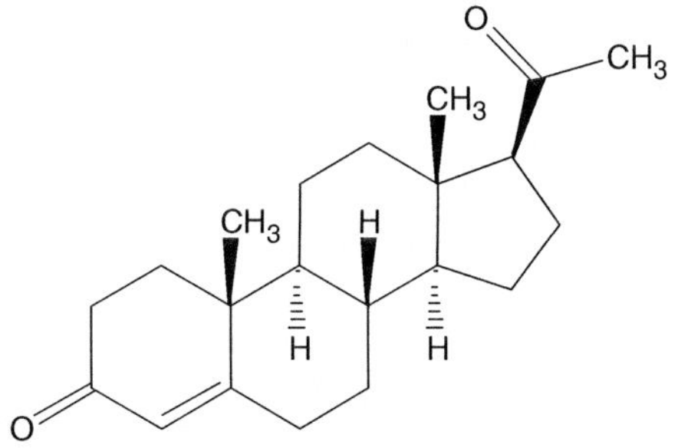

progesterone

Note that progesterone has a close structural relationship to testosterone. The single difference is the replacement of the C17 hydroxyl group of testosterone with the acetyl group of progesterone. Here is another striking example of the critical dependence of biological activity on nuances of chemical structure. Testosterone is, in fact, synthesized from progesterone.

Progesterone is synthesized in the corpus luteum in a cycle timed to the menstrual cycle. Synthesis continues should pregnancy develop. Progestins such as megestrol acetate and medroxyprogesterone acetate find use in therapy for metastatic breast cancer.

Here is one final triumph for Adolph Butenandt. He was the first to isolate progesterone; this in 1934. For his accomplishments, he shared the Nobel Prize for Chemistry in 1939. However, the Nazi regime prevented him from accepting the award. He was finally able to do so in 1949.

Cortisol and corticosterone are examples of glucocorticoids

Cortisol, synthesized in the adrenal cortex, is a glucocorticoid:

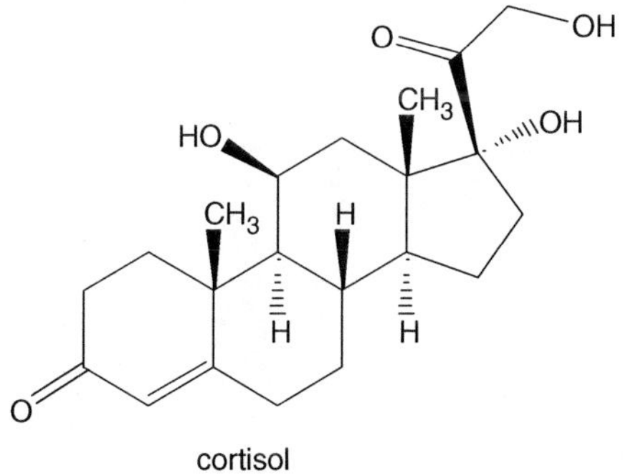

cortisol

A novel structural feature of cortisol not seen for other steroids thus far is the C11 hydroxyl group. Glucocorticoids have major influences on metabolism, including an increased rate of gluconeogenesis, glycogen deposition, and fat deposition. Noted at the beginning of this chapter is the potent anti-inflammatory activity of the glucocorticoids, including cortisol, corticosterone, and many of their derivatives. These drugs are employed in the treatment of diseases in which inflammation plays a major role. Philip Hench and Edward Kendall discovered the utility of cortisone for the treatment of rheumatoid arthritis. They shared the Nobel Prize in Physiology or Medicine in 1950 for their work. Cortisone is actually a prodrug and is reduced in vivo to cortisol, a far more potent anti-inflammatory agent. The availability of glucocorticoids for use in human medicine was greatly facilitated by the work of Lewis Sarrett, who carried out the first total synthesis of cortisone in the research laboratories of Merck. Sarrett subsequently won just about every relevant award in chemistry with the exception of the Nobel Prize. Glucocorticoids find a lot of use as topical anti-inflammatory agents.

Orally administered glucocorticoids have substantial adverse effects, including immunosuppression, suppression of the synthesis of endogenous glucocorticoids themselves (that can be a problem when glucocorticoid therapy is stopped), resorption of bone, and retention of water. Nonetheless, they are important drugs for control of inflammation. They also augment resistance to stress.

Mineralocorticoids regulate salt and water balance

Finally, we have the mineralocorticoids. Aldosterone is a good example:

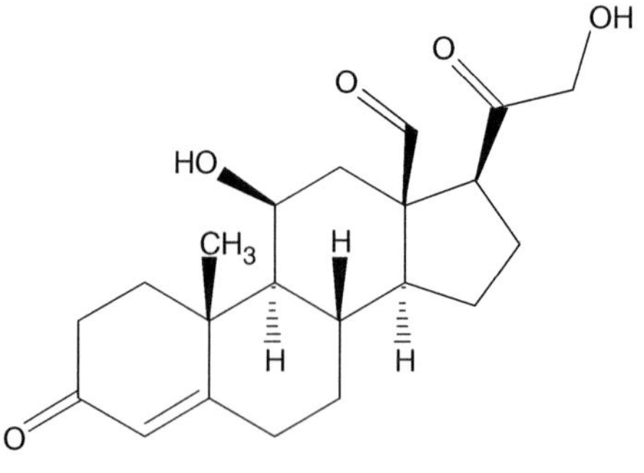

aldosterone

Aldosterone influences electrolyte balance in the body. Specifically, aldosterone increases the excretion of potassium by the kidney but decreases the excretion of sodium by this organ. One result is the net retention of water. The action of mineralocorticoids tends to increase blood pressure. Basically, the more sodium you retain, the more water you retain. Retaining water tends to increase the fluid level in the vascular system and that increases blood pressure. Think about the pressure changes that happen when you continue to fill a flexible container such as a balloon with water: the more water, the greater the pressure in the balloon. You might well imagine that an agent that antagonizes the action of aldosterone and other mineralocorticoids would be effective in hypertension. You would be correct. Aldosterone antagonists, such as spironolactone and epleronone, are employed for treatment of hypertension and edema.

This concludes what I have to say about the steroids. A constantly recurrent theme is the sensitive nature of biological activity to chemical structure. Details of chemical structure determine if and how small molecules such as steroids interact with the large molecules, say the steroid hormone receptors, to effect biological change. Such interactions are notably sensitive to molecular size and shape.

Having said that, let's change focus completely and have a look at the functions of the human brain and their control by small molecules.

Key Points

1. All steroids have a basic molecular skeleton containing four fused rings: three have six members and the fourth has five.

2. By attaching a variety of atoms or groups of atoms to the steroid skeleton stereospecifically, molecules are produced having an amazing spectrum of biological activity.

3. Cholesterol is the iconic steroid and is the steroid from which all others in human biochemistry are synthesized.

4. Cholesterol is absolutely required in human physiology: (a) as a constituent of biological membranes and (b) as the metabolic precursor to all other steroids.

5. An elevated level of plasma cholesterol, particularly that carried in LDLs, is an independent risk factor for developing cardiovascular disease. Cholesterol comes from dietary sources and is made in the body, largely in the liver.

6. Statins are the most important class of cholesterol-lowering drugs. They are inhibitors of an enzyme on the route to synthesis of cholesterol. As a result, there is an increase of LDL receptors in the liver, the only important mechanism for elimination of cholesterol from the body.

7. The statins have been demonstrated to reduce the incidence of clinical events (e.g., heart attacks) in patients known to have coronary artery disease as well as in patients with both elevated and normal levels of cholesterol without coronary artery disease.

8. Bile acids aid in the digestion of dietary lipids. They are made in the liver and secreted into the small intestine in the bile where they emulsify lipids.

9. Androgens and estrogens are, respectively, male and female sex hormones. They act through cytosolic receptors that, when activated, migrate to the nucleus and influence gene expression.

10. Hormone replacement therapy in aging women and men is common but the full range of effects, positive and negative, is complex and not fully appreciated.

11. Molecules that mimic or antagonize the action of sex steroids or inhibit their synthesis find substantial medical use: estrogen agonists for the treatment of the symptoms of menopause; aromatase inhibitors and estrogen antagonists for the prevention of breast cancer; androgen antagonists for the treatment of prostate cancer; 5-alpha reductase inhibitors for the treatment of benign prostatic hypertrophy and male-pattern baldness; and androgen agonists as anabolic steroids (legitimately for the treatment of cachexia and illegitimately to enhance athletic performance).

12. Nuclear hormone receptors, including those for the principal classes of steroids, retinoids, vitamin D, and thyroid hormones, are transcription factors that influence gene expression.

13. Progestins are critical for a successful pregnancy.

14. Glucocorticoids are potent anti-inflammatory agents that find multiple clinical uses, including treatment of arthritis and asthma.

15. Mineralocorticoids regulate salt and water balance in the human body.

16. LDL is a positive risk factor and HDL is a negative risk factor for development of cardiovascular disease. There are a number of additional, independent risk factors as well.

21

Your brain

What it does and how it does it

There are clear principles of the science of mind; these focus on the brain, neural circuits, nerve cells, signaling molecules in neural circuits, and conservation of these molecules throughout evolution. The nervous system receives signals, integrates them, and responds to them. The neuron is the unit of function in the nervous system.

The human nervous system is a marvel. Its functional beauty stirs the imagination. Its complexity both creates our intellect and challenges it. Our nervous system permits us to interpret our environment and respond to it. It is essential to the salient feature that makes human beings unique among living organisms: our amazing capacity for language. The human nervous system underlies our learning, memory, emotions, behavior, and consciousness. It is, in a certain sense, what it means to be human.

We understand many aspects of the anatomy, physiology, and biochemistry of the human nervous system.[1] The central points are the subjects of this chapter. As we come to understand them, much of great interest will be revealed to us. We will get important insights into how the nervous system functions and, in disease, malfunctions. We will also begin to understand why many molecules are effective in treatment of mental health disorders or induce abnormal states of consciousness in people.

That having been said, it is also true that there is much about the human nervous system, or that of simpler organisms, that we do not understand. A complete understanding of nervous system function is one of the most important scientific challenges that remain in the biomedical sciences. The nervous system is, and will continue to be, a focal point of intense research effort, leading to an increasingly complete understanding.

Two forces drive the study of the human nervous system. The first force is the human desire, encoded somehow in our nervous system, to accept the challenge to understand something as complex as the nervous system. Consider some of the underlying questions:

- How do we learn?
- How are memories stored?
- Why do we behave the way that we do?
- Why are we elated some days and sad others?
- What does it mean to be conscious?
- How is the state of consciousness created?

The second force is the desire to ameliorate, prevent, and cure mental illness. Mental illness is the cause of enormous human suffering, both to the victims and their families and friends. Beyond that, the social and economic costs are huge.

Here are the principles of the science of mind

Eric R. Kandel, a professor at Columbia University, is one of the world leaders in the science of the central nervous system, which he prefers to call the science of mind. A winner of the Nobel Prize in Medicine or Physiology in 2000, together with Arvid Carlsson and Paul Greengard, for his studies on the mechanism of learning, he has written the history of his life in science in an elegant book: *In Search of Memory*. In this book, Kandel defines five principles of the science of mind.[2] Here they are (and I quote directly):

1. Mind and brain are inseparable. Mind is a set of operations carried out by the brain.
2. Each mental function in the brain—from the simplest reflex to the most creative acts in language, music and art—is carried out by specialized neural circuits in different regions of the brain.
3. All of these circuits are made up of the same elementary signaling units, the nerve cells.
4. The neural circuits use specific molecules to generate signals within and between nerve cells.
5. These specific signaling molecules have been conserved—retained as it were—through millions of years of evolution.

The final principle, the conservation of signaling molecules over evolutionary time, has the greatest practical significance for understanding the science of mind. The structure of the human nervous system is a marvel of complexity. This is splendid in the sense that the complexity yields so much functionality. The complexity makes problems, however, for those who seek an understanding of the mind: learning, memory, consciousness.

The conservation of signaling molecules over time permits us to use organisms possessing far simpler nervous systems than our own as reliable models for the human nervous system. Here are three examples. The nervous system of the nematode *C. elegans* consists of just 302 nerve cells. These are individually identifiable. That permits identification and study of individual neural circuits within the nervous system of this worm. This worm uses some of the same neurotransmitters that we do. So does the sea slug *Aplysia*, which has proved to be an organism of great utility for the understanding of the mechanisms of learning and memory. In fact it is for work on *Aplysia* that Kandel received his Nobel Prize. We get back to this topic at the end of the next chapter.

For the present, remember the five principles enunciated by Kandel as we explore the chemistry of the nervous system. Since the focal point of this book is the molecules of life, I shall have more to say about the signaling molecules that act on and between nerve cells than about neural circuits. However, let's begin by identifying some of the mental functions of the brain.

The nervous system receives, integrates, and responds

Imagine that you are strolling on a country road in a wooded area of New England on a sunlit day early in October. The colors of the fall foliage are spectacular. Responding to a desire to capture this beauty for another day, you pull out your camera, focus on a stunningly red sugar maple backed by a multicolored hillside of trees, and snap the shutter. What has happened here?

The first thing is that you received sensory input. In this case, the sensory input is external, from the environment. The input comes in the form of light, received by the retina of the eye. External sensory input could have come in the form of sound waves received by the ear, music or noise, the kinetic energy of molecules impacting on the skin, heat or cold, activation of receptors in the nose, smell, or tongue, taste, or by pressure on receptors in the fingers, feel. Input can also come internally, from changes in the body itself. The body has receptors that respond to changes in body temperature, the pH of the blood,[3] the concentration of hormones, ... In sum, the first basic function of the nervous system is to receive input from external and internal environments.

The second thing that happened is that your nervous system integrated the sensory input from the environment. Through specific neural circuits in the brain, it created a visual image. The eyes receive the light but the brain forms the image of your environment. Beyond that, the brain created an emotion: appreciation for the beauty of the visual image and a desire to retain it. The brain also laid down a memory of the visual image. The second basic function of the nervous system is to integrate the sensory input into a useful form for storage and action.

Finally, you reacted to the nervous system integration of the sensory input by responding, taking action. Specifically, you recorded the scene for later enjoyment. This involved a number of complex physical motions as well as personal artistic judgment. Thus, the third and final function of the nervous system is to respond to the sensory input in a constructive manner.

These responses involve all organs and organ systems of the body. They are controlled by a complex system of neural circuits, molecular mechanisms, and feedback loops. Our task now is to develop an understanding of what mechanisms underlie the completion of these functions.

In passing, note that the way we experience the world is in a sense indirect. The signals that we receive from our external environment are detected by sensory cells and generate signals in our nervous system. However, we cannot and do not sense these signals. What we experience is generated for us by our minds in response to signals that we receive and this is subject to processing by genetic factors and our previous experiences. It is no wonder that different people respond differently to the same set of environmental signals.

Nervous system anatomy correlates with function

The central nervous system includes the brain and the spinal cord. We shall refer to it as the CNS. The peripheral nervous system, PNS, is composed of nerves, bundles of individual cells called neurons, which connect the CNS to the rest of the body. If we make a very rough analogy to a bicycle wheel, then the CNS is the hub and the PNS is the spokes. When the bike tire hits the road, force is generated at the hub and is transmitted by the spokes.

As we move forward, it will prove helpful to get some basic aspects of the human nervous system in place. An enormous amount of work has gone into making associations between brain anatomy and function. Starting with the three main parts of the brain, we know that the cerebrum is the seat of consciousness. It is divided into two hemispheres, which are linked by the corpus callosum. In a very general sense, the left hemisphere is associated with intellectual and the right hemisphere with emotional responses. Within the cerebrum, one can associate a number of brain areas (the prefrontal, frontal, temporal, parietal, and occipital lobes, for example) with functions including vision and hearing. One can make crude maps in which function is mapped onto brain structure.

There is nothing simple about such maps. For example, there are half a dozen regions in the brain, distant from each other, that are involved in converting signals from the retina of the eye into an image. These include the specialized neural circuits for vision. Nonetheless, these maps are useful. Within the cerebrum we know: certain long-term memories are stored in the cerebral cortex; the hypothalamus manages homeostasis, including regulation of thirst, hunger, and body temperature, and is essential for learning and memory; and the thalamus is a central relay point for incoming messages to the nervous system.

We associate the cerebellum with movement, balance, and muscle coordination, and the medulla oblongata with regulation of heartbeat, breathing, blood pressure, and reflex centers (coughing, sneezing, swallowing, ...). Although the associations of regions and structures within the brain with functions is useful, it is also true that most complex brain functions involve communication and coordination among several centers within the brain, comprising the specialized neural circuits, as in the case of vision.

An intact CNS is critical to the well-being of a person. Nature has gone to a good deal of trouble to protect the CNS from damage, both external and internal. The brain is protected from external trauma by the bony structure we call the skull. The spine is similarly protected by the bony spinal column. The CNS is also protected by the blood–brain barrier against penetration by molecules that might prove a problem. There are special facilitated transport mechanisms for specific molecules required by the CNS but which cannot penetrate the blood–brain barrier.

The PNS contains only nerves. These provide the connections between the CNS and, for example, the muscles, sensory organs (eyes, ears, nose, fingertips, ...), and internal body organs. Motor neurons activate muscles. Controlling neurons, interneurons, modulate the sensitivity of motor neurons to sensory neuron input.

There are two principal components to the PNS. First, there are afferent, or ingoing, pathways, also known as sensory pathways. Afferent pathways are those that transmit signals from the body to the CNS. For example, if you touch a hot stovetop with your hand, an afferent pathway will inform the CNS of that event. It carries the information from your hand to the CNS, the spinal cord, and the brain. The CNS will then receive this sensory input, integrate it, and respond.

The response to this painful stimulus comes by way of an efferent, or outgoing, pathway that leads from the CNS to the muscles and organs. In this case, the response will be to jerk your hand back off the hot stovetop and say "ouch." This will limit, but not eliminate, pain and tissue damage and may alert others within the range of your voice to the fact that you have done an unwise thing. The response to a sensory input may be generated in the spinal cord, jerking your hand back off the hot stovetop, or in the brain, saying "ouch" in this case, depending on its nature. Painful stimuli may be received, integrated, and responded to in the spinal cord. This limits the distance that the nerve impulse must travel and, therefore, minimizes the time delay between making a misstep and correcting it. This has an obvious advantage. In contrast, the nerve pathways linking the eyes to creation of a visual image are well known and involve the brain directly, not the spinal cord.

The neuron is the unit of function in the nervous system

Neurons and glial cells make up the bulk of the human nervous system. Glial cells come in several guises: astrocytes, oligodendrocytes, microglia, and others. Glial cells serve a number of important functions in the human nervous system.

The distinguishing features of a neuron are the two types of structures, known collectively as neurites, that extend out from its cell body: the axon and the dendrites. The function of the axon is to carry messages away from the cell body to recipient cells, including other neurons, an efferent function. The function of the dendrites is to receive messages from other cells, including neurons, and conduct them to the cell body, an afferent function.

The human brain contains about 100 billion (10^{11}) neurons. Although all neurons have a cell body, an axon, and dendrites, there is a lot of heterogeneity here. For example, the axons of some neurons are very short, reaching just to a few neighboring cells. Others are very long, as much as 4 ft in length, reaching from the brain to parts of the body that are remote from the brain. However, the heterogeneity of the neurons extends well beyond variation in the length of their axons. To appreciate the full scope of the heterogeneity, we need to take a closer look at both axons and dendrites.

At the ends of axons, the part farthest from the cell body, they branch into structures known as axon terminals or presynaptic terminals. The number of axon terminals varies from axon to axon. Some have thousands. The great majority of these axon terminals function to transmit signals to the dendrites, cell bodies, and, less commonly, to the axons of other neurons. A far smaller number transmit signals to muscle cells. We can begin to see the development of vast networks and circuits within the CNS.

Just as axons branch out into many axon terminals, dendrites branch out into dendritic spines. It is the dendritic spines that accept signals from other neurons and communicate them to the cell body. Individual neurons may have as many as 10,000 dendritic spines. The large number of dendritic spines per neuron, coupled with the large number of axon terminals per neuron, permits one cell to signal to perhaps a few thousand others. Given that there are some 100 billion neurons, the number of contacts among neurons through which messages may be exchanged is truly immense. It is the functioning of circuits formed within this vast network of cellular contacts that underlies the properties of the CNS.

The signals that pass from one neuron to another may be either stimulating or inhibitory. Within a given circuit, a stimulating signal from one neuron to another may cause an inhibitory signal to be sent back to the neuron that initiated the signal in the first place. This is an example of a negative feedback loop. This is a very common type of biological control mechanism. The point is that too much of a good thing can be a problem. (I note here that Mae West took the point of view that "too much of a good thing can be wonderful." Mae West had her charms but was not a prominent neuroscientist and we can continue with our argument). Basically, the human body strives toward homeostasis: keeping things pretty much the same under the same set of conditions. It follows that it makes sense to send a negative message back once a positive one has arrived. The negative message says "whatever you are sending has arrived and I have plenty. Stop sending."

The central message here is that the CNS is almost unbelievably complex. Billions and billions of neurons are sending both positive and negative messages to each other. Each neuron may send messages to thousands of other neurons at a time. The number of neuronal circuits is staggering. Yet, for the most part, all these circuits work together in harmony. This permits us to sense our environment and react to it in constructive ways. We think, learn, store memories, and generally act like reasonable human beings. There are exceptions, such as in mental illnesses, and we shall consider some of them later. So far, I have said nothing about the mechanism by which signals among neurons are sent and received. This follows.

Neuronal signaling occurs at synapses

A synapse is formed where the axon terminal makes an interface with a patch of cell membrane on a dendrite or cell body of another neuron. A schematic drawing of a

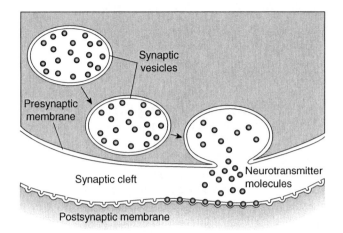

Figure 21.1 A schematic drawing of a synapse. The synaptic terminal is shown activated. Synaptic vesicles are fusing with the presynaptic membrane and releasing a neurotransmitter that diffuses across the synaptic cleft and binds to receptors on the postsynaptic membrane. This triggers a new nerve impulse. (Redrawn from D. Voet and J. G. Voet, *Biochemistry*, 3rd edn, 2004. © Donald and Judith G. Voet. Reprinted with permission of John Wiley and Sons, Inc.)

synapse is shown in figure 21.1. The axon terminal and the patch of cell membrane do not quite meet physically. The space between them is termed the synaptic cleft or gap junction. The axon terminal forms the presynaptic side of the synapse. The patch of cell membrane forms the postsynaptic side. So we have two cells separated by a small space; gap junctions are smaller that synaptic clefts. We need to signal across that space. That signaling is done either electrically across gap junctions or chemically across synaptic clefts using special molecules known as neurotransmitters. When neurons meet muscle cells, the analog of a synapse is termed the neuromuscular junction.

Neurotransmitters are stored in synaptic vesicles or secretory granules, depending on the chemical nature of the neurotransmitter. At the ends of the axon terminals, we find a collection of these small intracellular organelles. These organelles are surrounded by a membrane and contain the neurotransmitters that will signal across the synaptic cleft. When an electrical signal is transmitted from the cell body of a neuron down the axon, it causes the membrane of the synaptic vesicles or secretory granules to fuse with the neuronal plasma membrane at the synaptic cleft. When that happens, the contents of the synaptic vesicle or secretory granules, the neurotransmitters, are emptied into the synaptic cleft. They will quickly diffuse across the very small distance separating the presynaptic and postsynaptic sides and find their receptors on the postsynaptic neuron. When that happens, a message has been sent.

There are many neurotransmitters. The known neurotransmitters (and new ones are occasionally discovered) are a heterogeneous collection of molecules. Most fall into one of three structural categories: amino acids, amines, and peptides. Amino acid and amine neurotransmitters are stored in synaptic vesicles and are very quickly released upon arrival of an electrical signal. The larger peptide neurotransmitters are stored in the secretory granules. Their release is generally slower and requires the arrival of a chain of electrical signals.

Several of the neurotransmitters are small-molecule amines such as dopamine, serotonin, epinephrine, and norepinephrine. These neurotransmitters are synthesized in the cytoplasm of the axon terminal and subsequently transported into and stored within the synaptic vesicles. The amino acids glycine and glutamic acid are normal constituents of proteins and are present in abundance in the axons. These are also stored in synaptic vesicles. Each electrical impulse that arrives at the presynaptic side of a synapse will cause only a small minority of the synaptic vesicles to fuse with the plasma membrane and discharge their contents. The remaining synaptic vesicles remain, waiting for subsequent electrical impulses. At the same time, neurotransmitter synthesis continues, as does their storage in synaptic vesicles. This tends to restore the full complement of amine neurotransmitters at the axon terminal.

Other neurotransmitters are larger peptides, such as the enkephalins. These peptide neurotransmitters are synthesized in the cell body of the neuron, usually as part of larger polypeptides from which they are subsequently cleaved. Following synthesis, they are incorporated into secretory granules that bud off from the Golgi apparatus and are then transported down the axon to the terminals.

In the human CNS, glutamate is the most important excitatory neurotransmitter. Glycine is a major inhibitory neurotransmitter in the human CNS. Thus, these two amino acids, basic constituents of proteins, also function in other very important ways in behavior, emotion, learning, memory, and sensory perception. Nature uses its molecular constructs for more than one purpose. Among other neurotransmitters, dopamine, epinephrine, norepinephrine, and serotonin are derivatives of protein amino acids and are synthesized from them.

The movement of ions mediates electrical signaling

Above we noted that the fusion of synaptic vesicles or secretory granules with the plasma membrane of the synaptic terminal was caused by the arrival of an electrical signal flowing from the cell body down the axon. What is the nature of this electrical signal?

The basic answer to this question is that ions move across the plasma membrane of the neuron. Recall that ions are charged particles, frequently derived from single atoms by the gain or loss of electrons. The ions that are most important to us in understanding nervous system function are sodium ion, Na^+, potassium ion, K^+, calcium ion, Ca^{2+}, and chloride ion, Cl^-. If we compare the concentrations of these ions on the inside of the neuron and in the extracellular fluid that bathes the neuron, we find the neuron interior has a higher concentration of potassium ion than does the exterior fluid. In contrast, the exterior fluid has higher concentrations of sodium, calcium, and chloride ions than does the neuron interior. These concentration differences are referred to as concentration gradients.

Ions flow down their concentration gradients: that is, they will spontaneously flow from a region of high concentration to a region of lower concentration. The tendency is to equalize the ion concentrations everywhere. This is a reflection of the Second Law of Thermodynamics.[4]

There is one last, important point. The unequal distribution of ions across the plasma membrane of the neuron leads to a net negative charge on the inside of the neuron and a net positive charge on the outside. This separation of charge creates a potential difference, or voltage, across the neuronal plasma membrane. For a neuron at rest, this potential difference amounts to about 65 millivolts (mV), with the inside negative. By convention, we say that the resting potential of the neuron is −65 mV.

In sum, the natural tendency will be for sodium, calcium, and chloride ions to flow into the neuron and for potassium ions to flow out, and in so doing to reduce the membrane potential to zero. In reality, this is not so easy. The plasma membrane of the neuron is not very permeable to these ions. If it were, it would be impossible to sustain concentration gradients across it. The rate of passive diffusion of these ions across this membrane is very slow, though not zero, and different for each ion. So how do ions get across the neuronal plasma membrane rapidly? There are two ways: gated channels and active transport by pumps.

Ion channels are generally very specific. When open, the sodium channel will readily permit passage of sodium ions but excludes that for potassium ions. The converse is true for the potassium channel. This specificity is absolutely required for neuron function.

Ion channels can be open or closed (actually the situation is somewhat more complex than this but never mind). In the open state, ions can flow. In the closed state, they cannot. The resting state of an ion channel is closed, which helps to maintain the concentration gradients of the ions across the plasma membrane until there is reason to alter them. These are gated ion channels. There are two common ways to open the gates: ligands and voltage changes.[5]

Ligand-gated ion channels open following binding of the appropriate ligand to them. Ligands that open ligand-gated ion channels include a number of ions as well as the neurotransmitters. That is basically how neurotransmitters work, as I develop below. Voltage-gated ion channels open in response to a change in voltage, ordinarily depolarization of the neuronal plasma membrane. Depolarization means a change in membrane potential from the resting potential of −65mV to a less negative number. Erwin Neher and Bert Sakmann, both of Germany, discovered how gated ion channels regulate the flow of ions in and out of cells. They shared the 1991 Nobel Prize for Physiology or Medicine for this discovery.

One consequence of opening a ligand-gated or voltage-gated ion channel is that the initial ion concentration gradient will be partially or completely lost. As ions flow through the open channel, they diminish the concentration gradient that caused them to flow in the first place. Those concentration gradients need to be restored. Energy-dependent ion pumps do this. Energy is required because we are now working against the direction of spontaneous change. These pumps act to create or increase concentration gradients, which requires energy since we are moving in the direction of order creation. This energy is usually supplied in the form of ATP.

An action potential involves membrane depolarization followed by repolarization

A picture of an action potential is provided in figure 21.2: a plot of membrane potential as a function of time. The action potential involves a rising phase in which the resting membrane potential of −65mV rapidly rises and actually becomes positive, followed by a falling phase in which the membrane potential becomes slightly more negative than the resting potential, finally followed by a return to the resting potential. What is going on here?

Let us suppose that the neuron receives some sort of an electrical jolt, depolarizing the neuronal membrane, sufficient to open voltage-gated ion channels. The opening of sodium channels turns out to be faster than that for potassium channels. Thus, the rising phase of the action potential reflects an inflow of sodium ions down its concentration gradient. About the time that the sodium ion channels begin to close, the potassium ion channels open. Thus, the falling phase of the action potential reflects outflow of potassium ions, also down their concentration gradient. The falling phase undershoots the resting potential by a bit. Lastly, there is a restoration of the resting potential as ionic gradients and membrane permeabilities are restored to their original values. The whole affair takes about 2 milliseconds.

There are three additional points here. First, an electrical jolt is not the only way to get an action potential started. Alternatives include activation of ligand-gated ion channels or stress-gated ion channels. Stress-gating involves some type of physical deformation of the membrane: think about pressure waves hitting an auditory neuron, ultimately generating a sound; or physical trauma (e.g., sitting on a pointed object) stretching a sensory neuronal membrane, ultimately generating a pain.

Secondly, there is a threshold voltage change required to initiate an action potential. This voltage change is the difference between the resting potential of the neuron and a threshold potential required to get an action potential going. If the voltage change associated with membrane depolarization is smaller than the threshold, nothing happens in terms of an action potential. The normal resting potential is subsequently restored.

Finally, following an action potential, there is a refractory period in which the neuron cannot be induced to generate another action potential. The refractory period lasts about 1 millisecond (i.e., 10^{-3} seconds). Thus, there is a limit to the frequency with which a neuron can carry action potentials, about 1000 per second.

Membrane depolarization propagates the action potential down the neuron

Once you get an action potential generated, it will flow down the neuronal axon. Remember that an action potential is a sequence of membrane depolarization and repolarization events mediated by voltage-gated sodium ion and potassium ion channels. The basic idea is provided in figure 21.3. The initial depolarization will spread to adjacent voltage-gated ion channels, which will open, ions will flow, and so forth. Thus, the wave of depolarization moves down the axon. It is the means by

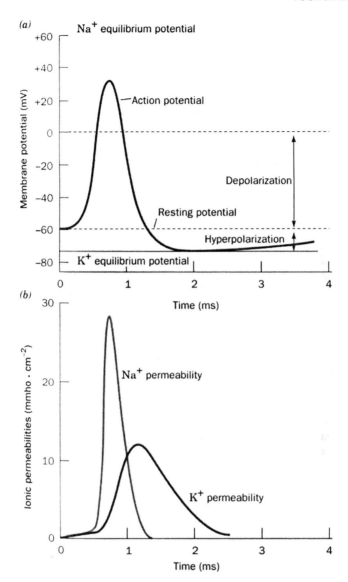

Figure 21.2 An action potential. The voltage across the plasma membrane of a neuron is plotted against time. The rising phase of the action potential is due to an inflow of sodium ions through an opened voltage-gated sodium ion channel. The resulting depolarization may spread to other voltage-gated ion channels in the membrane adjacent to this ion channel. This propagates the action potential down the axon of the neuron. Subsequently, voltage-gated potassium ion channels open and the resultant outflow of potassium ions repolarizes the membrane. A small undershoot occurs, followed by restoration of the resting potential. (Reproduced from D. Voet and J. G. Voet, *Biochemistry*, 3rd edn, 2004. © Donald and Judith G. Voet. Reprinted with permission of John Wiley and Sons, Inc.)

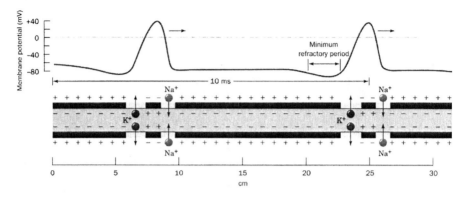

Figure 21.3 Action potential propagation along an axon. Membrane depolarization at the leading edge of an action potential triggers an action potential at the immediately downstream portion of the axon membrane by inducing the opening of its voltage-gated Na$^+$ channels. As the depolarizing wave moves further downstream, the Na$^+$ channels close and the K$^+$ channels open to hyperpolarize the membrane. After a brief refractory period, during which the K$^+$ channels close and the hyperpolarized membrane recovers its resting potential, a second impulse can follow. The indicated impulse propagation speed is that measured in the giant axon of the squid, which, because of its extraordinary width (∼1 mm), is a favorite experimental subject of neurophysiologists. (Reproduced from D. Voet and J. G. Voet, *Biochemistry*, 3rd edn, 2004. © 2004, Donald and Judith G. Voet. Reprinted with permission of John Wiley and Sons, Inc.)

which a signal originating in the neuron cell body is sent to other neurons or muscle cells through the neuronal axon. The action potential can move along the axon at a rate up to 100 meters per second. This is about 10 times as fast as the fastest human can run over short distances.

So far, we have described the sequence of events leading from an action potential traveling down an axon to the release of a neurotransmitter into the synapse. Now what happens?

The receptors for neurotransmitters on the postsynaptic neuron or muscle cell are ligand-gated ion channels

Suppose that a specific neurotransmitter arrives at its ligand-gated ion channel, say a sodium ion channel. It will open and sodium ions will flow into the postsynaptic neuron, depolarizing its membrane. If this depolarization exceeds the threshold level, this will open voltage-gated sodium and potassium ion channels, generating an action potential that will flow down the dendrite to the cell body, and so on.

One last point: neurotransmitters can be inhibitory as well as activating. For example, γ-aminobutyric acid (GABA) and glycine are inhibitory neurotransmitters. Their receptors on the postsynaptic side are ligand-gated chloride channels. Opening this channel will permit chloride ions to move down their concentration gradient from the outside into the cell. Since the chloride ion bears a negative charge, this will increase the negative charge inside the cell and decrease it outside the cell. The net result is to increase the potential difference across the cell membrane (hyperpolarization), making it more difficult to open voltage-gated ion channels to initiate an action potential.

Acetylcholine is an important neurotransmitter

Let's see how neural transmission works in terms of a specific example—acetylcholine:

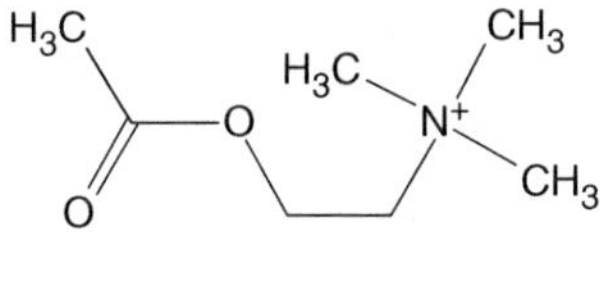

acetylcholine

The concept of a neurotransmitter originated in the 1920s with the acetylcholine molecule. Henry Dale and Otto Loewi originated the concept of chemical transmission of nerve impulses. These scientists shared the 1936 Nobel Prize in Physiology or Medicine for this work. Acetylcholine was also the first neurotransmitter for which the structure was determined. Otto Loewi accomplished that task, also in 1936.

Acetylcholine is synthesized from choline and acetyl-SCoA in a reaction catalyzed by choline acetyltransferase:

$$choline + acetyl\text{-}SCoA \rightarrow acetylcholine + HSCoA$$

Once synthesized, acetylcholine is stored in synaptic vesicles until time for its use. Once liberated into the synapse, acetylcholine diffuses across the synaptic cleft in about 100 microseconds (10^{-4} seconds; one ten-thousandth of a second), where it interacts with its receptor, and then dissociates from it in the next 1 or 2 milliseconds. Once liberated, acetylcholine is degraded by a second enzyme, acetylcholinesterase, a target for drug discovery (as I develop a bit later).

The amount of acetylcholine present in the synapse and the amount of time that it remains there are critical. For example, the venom of the black widow spider is highly neurotoxic. It contains a protein known as α-latrotoxin that elicits the release of massive amounts of acetylcholine at the neuromuscular junction. Too much of a good thing can be a serious problem.

Just as too much acetylcholine activity can be a problem, so can too little. Botulinum toxin is a mixture of eight proteins that act to inhibit the release of acetylcholine. This toxin is the product of the anaerobic bacterium *Clostridium botulinum*. Ingestion of this toxin causes the life-threatening food poisoning known as botulism.

We know that there are two general classes of acetylcholine receptors, called nicotinic and muscarinic receptors. This insight came from the study of drugs that influence the functioning of the human nervous system. These molecules fall into two classes: agonists and antagonists. Agonists mimic the action of neurotransmitters; antagonists, in contrast, block the action of neurotransmitters.

Muscarine is an acetylcholine agonist:

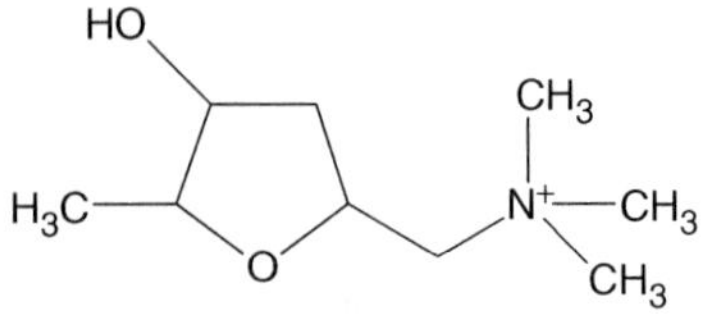

muscarine

Note that it shares some structural features with the acetylcholine molecule. Specifically, both molecules possess an oxygen atom separated by two methylene (–CH$_2$–) groups from a positively charged nitrogen atom. Muscarine was isolated from *Amanita muscaria*, a poisonous mushroom, more than a century ago. Muscarine causes sweating and pupillary constriction. These are some of the same effects that are caused by acetylcholine itself, confirming that muscarine is an acetylcholine agonist.

In contrast to muscarine, atropine blocks these actions of acetylcholine and muscarine. Atropine is, therefore, an acetylcholine antagonist. It binds where acetylcholine binds and therefore prevents the binding of the latter but does not activate it. Two molecules cannot occupy the same binding site at the same time. Atropine is isolated from the plant *Atropa belladonna*. Extracts of this plant have been used for millennia for a variety of purposes. Although large doses are poisonous (Atropos is the name of the Fate who cuts the thread of life), small doses causes dilation of the pupils, a consequence of its action as an acetylcholine antagonist, and has been used for cosmetic purposes by women. In Italian, *belladonna* means beautiful woman.

Things turn out to be a bit more complex than I have suggested thus far. Acetylcholine is also known to cause contraction of skeletal muscle and to slow the rate of heartbeat. However, muscarine does neither of these things nor are these actions of acetylcholine blocked by atropine. Another plant-derived molecule, nicotine from tobacco, proved to be an acetylcholine agonist at skeletal muscle and heart.

In sum, there are two kinds of acetylcholine receptors: they are termed the muscarinic cholinergic and the nicotinic cholinergic receptors. As the designations suggest, muscarine is an agonist at the former and nicotine is an agonist at the latter. The nicotinic cholinergic receptor is a ligand-gated ion channel. Acetylcholine acts as a classical neurotransmitter at this receptor. The muscarinic cholinergic receptor is a G protein-coupled receptor (GPCR). Acetylcholine acts as a neuromodulator at this receptor.

There are three faces of acetylcholinesterase inhibitors

We noted above that too much acetylcholine in the synapse or at a neuromuscular junction can be a problem: black widow spider venom works that way by causing massive release of this neurotransmitter. There is another way to accomplish the same thing: inhibit the normal route by which acetylcholine once released is subsequently removed. That route is degradation by acetylcholinesterase, an enzyme that catalyzes the hydrolysis of this substance to acetate and choline. In fact, small molecular inhibitors of acetylcholinesterase come in three flavors.

Highly potent, irreversible inhibitors of acetylcholinesterase are weapons of chemical warfare: nerve agents. These were first synthesized by the Germans during World War II but never used in that war. They fall into a structural class known as organophosphates. They are dangerous to make, dangerous to handle, and dangerous to use.

A more benign use of acetylcholinesterase inhibitors is as pesticides, used in agriculture. These also are generally organophosphates but are far less potent than chemical warfare agents. Nonetheless, they are nerve agents—the reason for their effectiveness as pesticides—and are potentially harmful to humans. They should be handled with care. Trade names for pesticide organophosphates include Orthene, Guthion, Dursban, Lorsban, Spectracide, Niran, Abate, and Counter, among several others.

Acetylcholinesterase inhibitors are used to treat Alzheimer's disease

The brightest side of acetylcholinesterase inhibitors is their use in the treatment of Alzheimer's disease, AD, a progressive brain disorder, named for Alois Alzheimer, who first described it early in the twentieth century.

Although AD sometimes afflicts younger people, it is usually a fatal degenerative disease of old age. There are far better ways to depart this world. Although the progress of the disease manifests itself somewhat differently in different patients, there is progressive dementia leading to death. The disease slowly destroys memory, learning, communication, and judgment. Patients may become anxious, agitated, suspicious, and delusional. They gradually lose the ability to carry out normal daily activities and care for themselves. At some point, total care 24/7 becomes required. Being the caregiver of an advanced AD patient is a heavy load to bear. AD is no respecter of intellect or prominence. One of my college biology professors was George Beadle, Chairman of the Division of Biology at Caltech, Nobel Laureate, and later President of the University of Chicago. This eminent and humane man died of AD. So did my mentor in graduate school, William P. Jencks, a brilliant enzymologist. Ronald Reagen, the thirty-ninth President of the United States, met the same fate. There is significant evidence that remaining intellectually active will delay the onset of AD and decrease the probability of becoming its victim. So do your crossword puzzles, read books, or perhaps write one.

There are two typical structural changes seen in the brains of AD victims. First, one finds beta-amyloid plaques, structures formed from insoluble aggregates of protein. The plaques of AD are largely composed of a fragment of a normal protein that has been abnormally processed: that is, two enzymes inappropriately chop out a fragment of the precursor protein. That fragment then aggregates around the neurons in the form of plaques. The enzymes that do the chopping have been characterized and scientists are hard at work to find useful inhibitors of them as possible drugs for treatment of AD. Secondly, neurofibrillary tangles are seen within neurons. It is not quite clear whether the plaques and tangles are causes of AD or merely manifestations of it.

Many scientists are seeking cures, as noted above, on the assumption that these are causes of neuronal damage.

At the biochemical level, there are numerous changes in the brains of AD patients. Prominent among them is degeneration of neural pathways containing acetylcholine neurons. Thus, it makes good sense to seek ways to augment the activity of acetylcholine in these neural pathways. The simplest way to do that is to find inhibitors of acetylcholinesterase, the enzyme that degrades acetylcholine, safe enough for human use. In fact there are three such inhibitors approved for treatment of AD: donepezil (Aricept), rivastigmine (Exelon), and galantamine (Reminyl). Unlike the acetylcholinesterase inhibitors used as chemical weapons or pesticides, these are reversible inhibitors, not irreversible ones. Clinical trials have shown that these drugs retard the progression of mild, moderate, and severe AD. Rivastigmine is also approved for treatment of mild to moderate dementia associated with Parkinson's disease. They are emphatically not cures for AD and have only modest clinical utility. The sad fact is that they are among the best treatments currently available.

There is one additional drug approved for use in moderate to severe cases of AD: memantine, marketed as Namenda. Memantine works by a distinct mechanism: it has moderate affinity for N-methyl-D-aspartate (NMDA) receptors and is the only approved molecule for moderate to severe AD in this structural class. We should be grateful for the AD drugs that we do have but the fact is that there is an enormous need for far better ones.

The nervous system controls muscle action

It is time to build a little more on the structure that we have established. To provide a concrete example, we shall consider one aspect of the peripheral nervous system, PNS. As discussed above, the efferent PNS carries information from the CNS to the muscles: movement results.

There are two classes of movements in the human body: voluntary and involuntary. Voluntary movements are pretty clear: they are the movements that we can control. Reaching for the French fries, swinging a baseball bat, turning on the TV, and typing at a computer keyboard provide obvious examples. Involuntary movements include those movements that we cannot readily control such as heart beats, vascular contraction, and movement of the gut muscles, and they basically control the internal environment of the body. Voluntary movements are controlled by the somatic nervous system. Involuntary movements are controlled by the autonomic nervous system, to which we now turn.

The autonomic nervous system is itself divided into two parts: the sympathetic and parasympathetic nervous systems. The sympathetic nervous system serves several glands and involuntary muscles. The primary neurotransmitter of the sympathetic nervous system is norepinephrine, which acts through α and β adrenergic receptors.

The primary neurotransmitter for the parasympathetic nervous system is acetyl-choline, a molecule that we have just met. Now let's compare the pharmacology of these two neurotransmitters.

Norepinephrine opens the pupil of the eye; acetylcholine narrows it. Note that an antagonist of acetylcholine at the muscarinic receptor, atropine, has the same outcome as norepinephrine (see above). Norepinephrine is a bronchodilator; in contrast, acetylcholine is a bronchoconstrictor. Norepinephrine increases the heart rate, chronotropy, whereas acetylcholine slows the heart rate, bradycardy. Norepinephrine decreases the rate of intestinal movements, whereas acetylcholine increases them. In all these cases, the effects of these neurotransmitters are opposed.

But not everything that norepinephrine and acetylcholine do is opposed. In one case, they cooperate. Acetylcholine stimulates penile erection by increasing blood flow to that organ. Norepinephrine controls ejaculation. In still other cases, the pharmacology of these neurotransmitters is unique. For example, acetylcholine induces urination by causing the bladder to constrict. Acetylcholine also induces secretion of saliva and tears. Norepinephrine has nothing to do with these functions.

The point here is that individual neurotransmitters have multiple physiological functions. These are frequently opposed to each other. This provides for exquisite control of physiological function through balancing the effects of the two. Subtle changes in relative importance of two or more neurotransmitters can elicit physiological changes that permit the human body to adjust to environmental or internal changes.

This pretty much sums up the key features of the nervous system of humans and identifies some of the key molecules involved one way or another with its functions. In the next chapter, I turn attention largely to diseases of the nervous system but conclude with a vignette about the mechanisms for learning and memory.

Key Points

1. There are clear principles of the science of mind. These focus on the brain, neural circuits, nerve cells, signaling molecules in neural circuits, and conservation of these molecules throughout evolution.

2. The nervous system receives signals, integrates them, and responds to them.

3. Nervous system anatomy correlates with brain function. At the same time, neural circuits associated with some functions extend to several brain regions.

4. The brain is protected from trauma by the skull and spinal column and from potentially toxic molecules by the blood–brain barrier.

5. The central nervous system, CNS, includes both neurons and glial cells. The peripheral nervous system, PNS, has no glial cells.

6. The PNS contains afferent neural pathways that bring signals to the CNS and efferent neural pathways that conduct signals from the brain to the periphery.

7. The neuron is the unit of function in the nervous system. It possesses a cell body, an axon, and dendrites.

8. The CNS is amazingly complex. Each neuron may send messages to thousands of other neurons at a time and these may be either positive or negative.

9. Neuronal signaling occurs at synapses. These are small gaps between neurons or between neurons and muscle cells.

10. Signaling across synapses may be electrical or chemical. The chemical signal carriers are known as neurotransmitters. They are released from the presynaptic side and recognized by their receptors on the postsynaptic side.

11. Neurotransmitters include molecules such as glutamate, GABA, dopamine, serotonin, acetylcholine, and norepinephrine as well as a family of peptides.

12. The receptors of neurotransmitters are ligand-gated ion channels.

13. Electrical signaling is mediated by the movement of ions: Na^+, K^+, Ca^{2+}, and Cl^-.

14. Activation of a neural membrane generates an action potential, a depolarization of the membrane followed by repolarization.

15. Membrane depolarization propagates the action potential down the neuron.

16. Acetylcholine is an important neurotransmitter. It acts at nicotinic and muscarinic receptors. Antagonists at these receptors find use in clinical medicine.

17. Inhibitors of acetylcholinesterase are useful as pesticides and for treatment of Alzheimer's disease. In addition, they are chemical warfare agents.

18. The nervous system controls muscle action. Norepinephrine and acetylcholine are important neurotransmitters at the neuromuscular junction.

22

Your brain

Good things and not-good things

Diseases of the central nervous system remain a challenge for medicine, both in the sense of understanding the molecular basis of the pathology and having tools in hand to deal effectively with them. We have a useful understanding of learning and memory. The content of this chapter is elegantly summarized in the following poem:

The problem for the chemists
Was explicity defined.
Just synthesize a lot of drugs
And hope that you will find
A patentable compound (that
Will benefit mankind).

So they made some novel structures
In some novel permutations
And multiple derivatives
In strange configurations,
Which gave them many chemicals
That no one else had seen
Including something special called
A phenothiazine

Which, much to their amazement,
(And their company's great gain)
Was accidentally found to cause
Marked changes in the brain,

Deleting the delusions
Of the chronically insane.

Oh, what a stir these chemists caused
In finding this new potion
Their seminal discovery
Set others into motion
(Perhaps there were some chemicals
That, in the proper doses
Might prove to be superior
In clearing up psychosis).

And so they mixed more mixtures
Manufactured variations,
And conjured up imipramine
By these improvisations;
Which didn't help pyschosis
But gave other welcome news:
That a bottle of imipramine
Obliterates the blues!

Then other types of drugs were found
With other applications,
Like lithium's preventing both
Depressions and elations;
But great confusion followed from
This flood of information
(Since cures caused by these compounds
Came without an explanation).

Till they found a simple answer
(But an incomplete solution):
Most drugs must bind receptors,
Making simple substitution
For a natural transmitter
With such specificity
The receptor grants admittance
(As a lock admits a key).

And so, for schizophrenia,
(Whose key is dopamine)
You give a strong antagonist
(The phenothiazines)

That binds a D receptor, thereby
Getting in between
A site on the receptor
And the catecholamine

It's different with depression—
There you bind amine transporters
That translocate transmitters
(Which makes their actions shorter);
And by yet another tactic
You may switch from flight to fight
By increasing GABA's binding
To a GABA binding site.

So praise the clever chemists
For the drugs they have confected
(Since changing minds with chemicals
Was really unexpected).
Now, with their new technology
That's sure to expedite
Additional discoveries—
All minds may soon work right.

Samuel H. Barondes in Molecules and Mental Illness

For certain, one thing is clear: life is not fair. If you have parents who love and care for you, you have an immeasurable advantage in life over those who lack such nuturing. I was fortunate to have such parents but, beyond that, I had two extraordinary grandparents, my paternal grandfather (I get around to him in a later chapter) and my maternal grandmother.

My grandmother Minnie was a woman of exceptional patience and goodwill. I never heard her say a negative word about anyone. She spent a good many hours reading to me before I figured out how to read by myself. Being a widow, she lived with my family for many years, helping where help was needed, doing what needed doing. She was a figure of kindness and stability.

One of the sad passages in my life was watching her decline mentally. Over time, she became increasingly forgetful. Happily, her forgetfulness never progressed to become a really major problem: she passed away in her eighties.

Her case is one of a great many in which some decline in optimal mental function compromises life. These compromises fall into two general categories, whose distinction may be less real than apparent. On the one hand, there are neurological diseases in which there is some detectable, definable lesion in the nervous system. In the last chapter, we met one notable example: Alzheimer's disease. As mentioned then, Alzheimer's disease is characterized by amyloid plaques external to neurons and fibrillary tangles within them—clear lesions. On the other hand, there are psychiatric diseases in which the pathology is clear but for which we can detect no lesion in the nervous system. Depression and schizophrenia provide two important examples.

Whether psychiatric diseases in fact involve no nervous system lesions or whether we have simply not been clever enough to detect and define them remains an open question.

In this chapter, I turn first to a few additional examples of central nervous system diseases. I begin with two examples that have a clear genetic basis and move on to others for which there is clear genetic disposition but an unclear genetic basis. Later, I will provide some examples—good and bad—of the actions of small molecules on the nervous system. Finally, I turn attention to the issues of learning and memory.

Huntington's disease and the Fragile X syndrome are single-gene genetic diseases involving triplet repeats

Back in chapter 11, we examined the molecular basis for a specific genetic disease: sickle cell anemia. That disease is caused by a single base pair change in the gene coding for the β subunit of hemoglobin, leading to a single amino acid substitution: a valine replaces a glutamic acid. Single base pair changes in protein-coding regions are by no means the sole molecular basis for genetic diseases in human beings. Here are two alternative molecular mechanisms associated with a small family of diseases.

Many genes contain triplet repeats in either coding or noncoding regions: that is, a specific triplet, say CGG, is repeated sequentially ... CGGCGGCGGCGGCGG .... This is normal and usually causes no problem. However, in some cases, the triplet repeat segment of the gene is expanded. This growth may increase in successive generations of families. If the triplet repeat number becomes too great, disease can result. Here are two specific examples, Huntington's disease and Fragile X syndrome.

The defect in Huntington's disease is in the *HD* gene and the affected protein is known as huntingtin, a protein that contains 3145 amino acid residues. The *HD* gene normally carries 11–34 repeats of a CAG triplet. That triplet codes for glutamine and so huntingtin normally has a substantial run of glutamine residues in its amino acid sequence. In victims of Huntington's disease, the number of triplet repeats is increased, reaching into the 80s in some cases. The critical number of repeats is 41 in the sense that everyone with 41 or more repeats becomes a disease victim. Some victims have a bit fewer than 41 repeats.

Huntington's disease is quite dreadful. It is a neurological disease that leads to progressively worse movement disorders, cognitive decline, and emotional disturbances. The course of the disease process is long, 15–20 years typically, and uniformly fatal. There is therapy of modest efficacy designed to provide some symptomatic relief. However, nothing is known that stops the progression of the disease.

Fragile X syndrome has some things in common with Huntington's disease. The genetic defect is in the *FMR1* (Fragile X mental retardation 1) gene. The affected protein is FMRP (Fragile X mental retardation protein).

The *FMR1* gene is located near the end of the long arm of the X chromosome. The entire gene spans 38,000 base pairs, of which only about 4000 actually code for amino acids in FMRP. These are collected in 17 separate exons. The bulk of the gene is composed of noncoding sequences, introns, that are edited out in the process of mRNA maturation. The repeated sequence in this case is CGG and the triplet repeats occur in one of the introns, a noncoding region, near the beginning of the gene. This is in contrast to the Huntington's disease case where the triplet expansion occurs in the coding region. Normally, there are about 30 repeats, 5–44 is the typical range, of this sequence in the *FMR1* gene intron. Patients with Fragile X syndrome have more than 200 repeats of this triplet and may have several hundred. When this occurs, a methyl group is added to the C in each triplet and synthesis of the FMRP is turned off. The excessive number of triplet repeats is also reflected in the X chromosome structure: it appears to have a pinched region near the end of the long arm of the chromosome. The chromosome can actually break at this point: hence, the name of the disease.

Fragile X syndrome is the leading inherited cause of mental retardation in humans. It affects about 1 out of 2000 males and about 1 out of 4000 females. The degree of mental retardation is quite variable, from a mild mental impairment to severe mental retardation. The disease is frequently more serious in males than females; males have only the defective X chromosome, whereas females have one normal X chromosome to go along with the defective one. There is no specific treatment for Fragile X syndrome.

Huntington's disease and Fragile X syndrome are classic examples of a single gene mutation that leads to a neurological disease in humans. Those diseases considered later in this chapter have in common a genetic disposition, generally multigenic, to fall victim to disease. In these cases, the connection between genetics and disease is far less clear.

Depression is a major mental health issue

Some days are better than others. On our good days, we feel happy, even elated. On the not-so-good days, we feel sad, blue, depressed. All this is perfectly normal. Mood swings can be influenced by any number of events and are part of life. However, extreme mood swings are not normal. Sadness may be bitter, prolonged, black; so profound that it prevents normal functioning. Sylvia Plath was a noted poet who was subject to episodes of profound depression. She eventually committed suicide at age 31 years old, putting her head in the oven and turning on the gas. Here are her words in *The Bell Jar*:

> I hadn't slept for seven nights...
>
> The reason I hadn't washed my clothes or my hair was because it seemed so silly.
>
> I saw the days of the year stretching ahead like a series of bright, white boxes, and separating one box from another was sleep, like a black shade. Only for me, the long perspective of shades that set off one box against the next had suddenly snapped up, and I could see day after day glaring ahead of me like a white, broad, infinitely desolate avenue.
>
> It seemed silly to wash one day when I would only have to wash again the next.

It made me tired just to think of it.

I wanted to do everything once and for all and be through with it.

These are not the thoughts of a person having a bad day. They are the thoughts of a person in the grip of a major depressive episode. Several people have put down in book form their feelings and reactions to major depressive episodes.[1] Their stories are enlightening.

About one person in 20 will suffer one or more episodes of major depression at some time during their life. Women are afflicted about twice as frequently as men. Major depressive episodes are life threatening. About 20% of victims end their lives by suicide. Susceptibility to major depression is family related. Although the genes that sensitize a person for major depressive illness have not been identified, it is clear that there is a genetic component to this disorder.

So what has gone wrong in the nervous system biochemistry of victims of depression? The answer to that question is not completely clear but we have substantial insights.

The mechanism of action of useful drugs for major depression provided key clues to an understanding of the biochemical basis of depression. Two key drugs are iproniazid and imipramine. The neurotransmitters dopamine, norepinephrine, and serotonin all possess a single amino group. They are collectively known as monoamines. Enzymes known as monoamine oxidases, or MOAs, degrade monoamines. There are two of them: MAO-A and MAO-B. Iproniazid is an inhibitor of the MAOs: that is, it reduces the rate at which MAO degrades these monoamines into nonfunctional products. It follows that iproniazid should augment the activity of the monoamines by slowing their degradation. This finding suggests that a deficit in the action of one or more of these monoamines may be responsible, in part at least, for major depression.

This conclusion is supported by the mechanism of action of imipramine. Once a neurotransmitter has been released into the synapse, there are two ways to terminate its action. The first is to degrade it to inactive products, by MAO for example. The second is to remove the neurotransmitter through reuptake into the presynaptic neuron. This mechanism is the predominant one for clearing the synapse of serotonin, norepinephrine, and dopamine. Specific proteins embedded in the neuronal plasma membrane mediate the reuptake of these monoamine neurotransmitters. Imipramine is a nonspecific monoamine reuptake inhibitor: that is, it slows the reuptake of all three of these monoamines, which enhances the activity of these neurotransmitters. This also suggests that a deficit in the activity of one or more of the monoamines underlies the problem of depression.

A breakthrough in the treatment of major depression was the discovery of fluoxetine, marketed as Prozac. Fluoxetine has a mechanism of action similar to that of imipramine with an important exception. It is a selective serotonin reuptake inhibitor, an SSRI. This strongly suggests that, in some sense, the symptoms of major depression result from a deficit in serotonin specifically. By inhibiting its reuptake from the synapse, the activity of serotonin is enhanced. Two other important drugs for major depression, sertraline (Zoloft) and paroxetine (Paxil), among several others, are also selective serotonin reuptake inhibitors. The SSRIs are the preferred treatment for depression. These drugs do not work for everyone, which suggests that there is more than one underlying biochemical abnormality leading to depression.[2]

Schizophrenia is arguably the worst disease of mankind

Schizophrenia is a chronic and debilitating mental disorder.[3] It is common, disabling, strikes early, and usually lasts a lifetime. The costs in human suffering are enormous as are the economic costs. We have a collection of drugs that help the schizophrenia patient, none of which is truly satisfactory in relieving symptoms and all of which have potentially serious adverse effects. We need to do better.

One of the striking stories of schizophrenia is that of John Nash Jr, a brilliant mathematician and economist who won the 1994 Nobel Prize for Economics for work done decades earlier. In the interim, he suffered severely from schizophrenia, a condition from which he miraculously recovered. Sylvia Nasar has described his life in *A Beautiful Mind*.[4] His story has been made into the hit movie of the same name.

There are two general classes of clinical characteristics of schizophrenia. First, there are the positive symptoms that include auditory hallucinations (voices) and delusions, often paranoid. Second, there are the negative symptoms; these include disorganization, loss of will, inability to pay attention, social withdrawal, and flattening of affect. The relative roles of positive and negative symptoms for a particular victim vary over time. The positive symptoms may predominate for a period to be followed by one in which the negative symptoms are more prominent. About 10% of people with schizophrenia commit suicide.

Like major depression, there is a genetic component to susceptibility to schizophrenia. It tends to run in families. Among identical twins, there is a 48% concordance for schizophrenia: that is, if one twin is schizophrenic, there is about a 50/50 chance that the other twin will also be schizophrenic. That statistic tells us two things very clearly. The first is that there is a genetic component to schizophrenia and, second, that genetics is not the whole story. A number of genes that code for specific proteins have been identified that are, in one way or another, linked to schizophrenia. However, the story of the genetic basis of this disease is incomplete. Better insights will lead to better therapy.

It is not clear that the different variants of schizophrenia have a common etiology. It may well be that there are important differences in the underlying causes. However, one thing seems clear: schizophrenia is associated with an excess of dopamine activity. Here is the supporting evidence.

The two pioneer drugs for schizophrenia are chlorpromazine and reserpine. Reserpine is known to reduce the brain levels of norepinephrine, dopamine, and serotonin. Since reserpine is also effective in coping with some of the symptoms of schizophrenia, perhaps an abnormally high concentration of one or more of these monoamines is a contributing factor to this disorder.

Chlorpromazine (Thorazine) was among the first drugs employed to treat schizophrenia: it is a dopamine antagonist. This finding focused attention on dopamine specifically. Indeed, all drugs in current use for schizophrenia are dopamine antagonists, though several have other biochemical activities as well. In some sense, excess dopamine activity contributes to the symptoms of schizophrenia. Chlorpromazine is an example of a phenothiazine antipsychotic, or neuroleptic, agent. There are several antipsychotics in this structural class that have found use in treatment of schizophrenia: perphenazine (Trilafon), trifluoperazine (Stelazine), and thiothixene (Navane), among others. Another prominent first-generation antipsychotic agent is haloperidol (Haldol), in a distinct structural class known as butyrophenones.

The major adverse effects of first-generation drugs for schizophrenia are involuntary movement disorders. Symptoms include tremor, rigidity, restlessness, and slowness of movement, strongly reminiscent of the movement disorders in parkinsonism, about which more follows later. The worst of these movement disorders is tardive dyskinesia, an irreversible movement disorder.

Some first-generation agents, such as haloperidol, are rather specific for one subtype of dopamine receptor, D_2. This suggests that some degree of both efficacy and side effects are associated with dopamine antagonism at this receptor. However, the situation is complex, as usual. There are five classes of dopamine receptors known: D_1 through D_5. To complicate matters further, several of these classes have subclasses. In total, there are at least 15 dopamine receptors. Which of these is important for relief of the symptoms of schizophrenia? Which is responsible for movement disorders? The answers to these questions are incomplete. We do have a few hints.

The first of the second-generation, or atypical, antipsychotics was clozapine. Clozapine (Clozaril) is relatively free of the movement disorders that characterize the first-generation drugs. This is true of, and defines, second-generation, atypical antipsychotics. It was a significant breakthrough for schizophrenia patients.

Clozapine has its own problem. In 1–2% of schizophrenic patients treated with clozapine, there is a severe reduction of white blood cells. This can be a life-threatening condition, owing to the potential for infection. Indeed, several patients have died of massive infectious disease while on clozapine. Patients on clozapine require routine monitoring of their white cell count. Clozapine has been followed by several other second-generation antipsychotics: risperidone (Risperdal), olanzapine (Zyprexa), quetiapine (Seroquel), ziprasidone (Geodon), aripiprazole (Abilify), and amisulpride (Solian). All but the last two of these antipsychotics have biochemical activity at both D_2 and 5-HT_{2A} receptors. This strongly suggests that antipsychotic activity is associated with muting both excess dopamine and serotonin activity. However, amisulpride is an antagonist at D_2 and D_3 receptors but without antiserotonin activity.

It is not quite clear just how much better these second-generation agents are in terms of efficacy, though their advantages in terms of patient safety seem clear. In a careful study of perphenazine against several second-generation drugs, only olanzapine showed a clear efficacy advantage.[5] Olanzapine appears to have a unique disadvantage as well: many patients on this drug experience highly significant weight gain during the first year of treatment. In extreme cases, this can amount to 100 lb or more.

There are a few things here that are clear: first, some progress has been made in the treatment of schizophrenia; second, the optimal mechanism of action for drugs for schizophrenia has yet to be clearly defined; and third, there remains great progress to be made.

Parkinsonism is a disease of diminished dopamine activity

Parkinsonism, or Parkinson's disease, is named for James Parkinson who first described the disease back in 1817. It is usually a disease of the elderly characterized by a spectrum of movement disorders: involuntary movements, rigidity, slowness, and loss of balance. It may progress to mental impairment, including depression. This is basically the same spectrum of movement disorders sometimes seen in schizophrenia patients taking dopamine antagonists. This suggests that parkinsonism may reflect, in some manner, a deficit in dopamine activity.

That conclusion is correct and has been known for quite some time. Back in 1960, examination of the brains of parkinsonism victims revealed that a pathway in the brain termed the nigrostriatal pathway had degenerated. This pathway proceeds from a group of cells called the *substantia nigra* to the striatum and is known to be important for the control of movement. The neurotransmitter employed in this pathway is dopamine.

Treatment of Parkinson's disease patients involves some form of dopamine agonism: the goal is to replace what has been lost through the degeneration of dopaminergic neurons. It would seem that the simplest thing to do would be to employ dopamine itself as the agonist. That, unfortunately, does not work. To be effective in parkinsonism, a dopamine agonist must get to the damaged part of the brain. Dopamine does not cross that blood–brain barrier. So it was necessary to come up with an alternative, indirect way to get dopamine where it needs to be. One solution comes in the form of L-dihydroxyphenylalanine, usually known more simply as L-dopa. L-dopa is a precursor of dopamine: loss of a carboxyl group in the form of carbon dioxide produces dopamine:

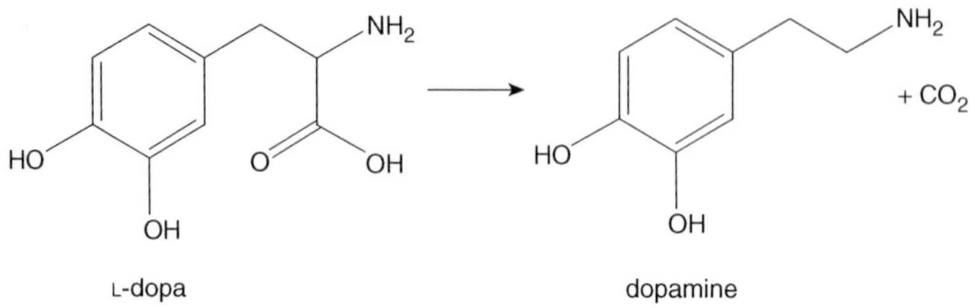

L-dopa   dopamine

Now the great thing about L-dopa is that it does get across the blood–brain barrier. Once in the brain, an enzyme known as dopa decarboxylase converts it to dopamine. The results for patients can be dramatic: have a look at stories in Oliver Sacks' book *Awakenings*.[6]

There is one further elaboration on this therapeutic scheme well worth knowing about. Dopa decarboxylase occurs in peripheral tissues and blood as well as in the brain. Close to 90% of administered L-dopa can be converted to dopamine by this enzyme before it gets into the brain. That is not good since a huge fraction of the L-dopa is lost. This problem was overcome in a very neat way. Carbidopa is an inhibitor of dopa decarboxylase that does not penetrate the blood–brain barrier:

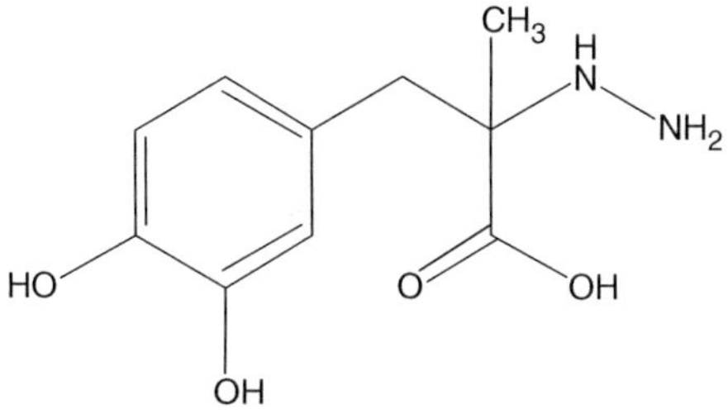

carbidopa

Carbidopa therefore inhibits the conversion of L-dopa to dopamine in the peripheral tissues and the blood while leaving it unchanged in the brain. The combination of L-dopa and carbidopa is marketed as Sinemet and was the treatment of choice for parkinsonism for many years. Note the structural similarity between carbidopa and L-dopa. Carbidopa looks enough like L-dopa to occupy the L-dopa site on the decarboxylase. At the same time, the subtle structural differences render carbidopa indifferent to the enzyme. It is an inhibitor, not a substrate.

L-dopa is not a panacea. After extended periods of use, parkinsonism patients may find that they experience periods in which the drug works well and periods in which it does not. This on-and-off phenomenon can be a serious problem for patients, requiring careful observation and care.

Here is one last word about L-dopa. It delivers dopamine to the entire brain, not just the nigrostriatal pathway. We know from our previous discussion that excess dopamine activity is associated with schizophrenia. It should not surprise you then that a rather common side effect of L-dopa is some of the symptoms of schizophrenia, including hallucinations.

An alternative to finding some means to slip dopamine into the brain is to discover dopamine agonists that pass the blood–brain barrier. There are a family of these, including bromocriptine (Parlodel), pramipexole (Mirapex), ropinirole (Requip), and pergolide (Permax). These are particularly useful in the early stages in parkinsonism and may delay the time that therapy with L-dopa is initiated. All these dopamine agonists are complex natural products in the alkaloid category.[7]

Morphine is a complex natural product with potent analgesic properties

Morphine is a natural product. Morphine was isolated as a pure compound from crude opium, an exudate of the poppy, in 1804. Morphine is quite a complicated molecule.

It took 119 years from the time of its isolation in pure form, until 1923, for its structure to be elucidated.

Here are the positive talking points for morphine. First, morphine is a highly effective agent for the treatment of severe pain. It is generally given by injection and relief is very fast. Second, morphine is widely available. Finally, morphine is cheap. For the latter two reasons, morphine is used worldwide, not just in developed countries that can afford more expensive (and sometimes extremely expensive) medications.

In addition, morphine has served as a point of departure for the discovery of many medically useful derivatives. These include codeine, a pain reliever and cough suppressant, levophanol, an orally active analgesic (morphine is not active when given orally and is usually given by injection), and many other modern and highly potent opiate analgesics.

There are downsides to morphine, as there are to basically all medications, those for pain and those for other uses. Here they are. First, morphine is addictive. Morphine is an opioid and all opioids currently in medical use are addictive, though there is hope for the discovery of opioids that relieve pain without addiction potential. Morphine is addictive in two senses. In the first place, users get a high and there is strong motivation to repeat the pleasant experience. In the second place, cessation of morphine use by those who are addicted is accompanied by a highly unpleasant, painful withdrawal experience. As noted earlier, most people tend to avoid pain and unpleasantness and so the temptation to continue use is strong. Hence, morphine is a drug of abuse as well as a clinically useful molecule. There are several opioids that are actually abused far more than is morphine (e.g., heroin and oxycodone (OxyContin)). Second, morphine is a respiratory depressant: that is, morphine depresses those regions in the central nervous system responsible for maintaining breathing. Take too much morphine, an overdose, and the consequences may be fatal.

Morphine is the chemical basis for heroin. A single, simple chemical reaction converts morphine into heroin. In chemical terms, heroin is simply diacetylmorphine. Here is how it works. Opioids work in the brain and they have to get into the brain in order to relieve pain or create a high. The brain is protected against the entry of many molecules by the blood–brain barrier, very tight junctions between cells that line the vasculature of the brain. As it turns out, heroin is very much better than morphine at penetrating the blood–brain barrier. Once heroin enters the brain, enzymes remove the acetyl groups, regenerating morphine. Thus, heroin is just a device for getting a lot of morphine into the brain very quickly. For that reason, heroin is a much more important drug of abuse than is morphine.

Morphine and other opioids work by activating a family of opioid receptors in the brain. These fall into three classes and morphine activates all of them. It has proved possible to design and synthesize opioids that are more-or-less specific for subsets of the opioid receptors. However, none of these has proved to have the right set of activities to retain the potent analgesic power of morphine without the addiction potential and respiratory depression potential. Finding such a molecule remains on ongoing challenge.

We have important insights into the molecular basis of learning and memory

At the beginning of the previous chapter, I took note of Eric Kandel's book: *In Search of Memory*. In it, Kandel describes his life's work on helping to unravel the molecular basis of learning and memory. Kandel is a leading figure in this field. It is a great story and I am going to try to relate the essential parts of it here. Before we get to the molecular part, there are a few things that we need to know about memory.

There are two kinds of memory: explicit and implicit. Explicit memory includes concrete things such as people and their physical characteristics, places, facts, events, odors, and the like. When you recognize that it is your child sitting in your car parked in your driveway by the side of your home, you are using explicit memory. When you fail to recall the name of an acquaintance or coworker, it is your explicit memory that has failed. Explicit short-term memories are laid down in the prefrontal cortex, converted into long-term memories in the hippocampus, and these are stored in that part of the cerebral cortex where the relevant information that led to the memory was initially processed. Thus, if the environmental information came to you through your eye, the long-term memory will be stored in the visual cortex. If the information came to you through your ears, say a particular melody, then the long-term memory will be stored in the auditory cortex.

The flip side of explicit memory is implicit memory. Implicit memory refers to things like skills and habits. When you drive from your office or school to your home every day, you are using your implicit memory to guide you. When you shoot free throws on the basketball court, it is your implicit memory that guides your actions. Long-term implicit memories are stored in various places in the brain: the amygdala, cerebellum, or striatum.

How about learning? Here there are three learning paradigms of direct interest: habituation, sensitization, and classical conditioning. Habituation is a type of learning involving recognition of a harmless stimulus so that one no longer reacts to it. If you are at a New Year's Eve party and a balloon pops, you may well react to this unexpected noise with a little jump. But if you then hear 10 more balloons pop, you will quickly recognize the harmless nature of the sound and stop reacting. That is habituation.

Sensitization is basically the opposite of habituation: it is learned fear. If you are walking alone in the dark and hear a gun go off, you are likely to react in an exaggerated manner to other stimuli, say a stranger tapping you on the shoulder. That is sensitization.

Finally, classical conditioning is pretty well known from Pavlov's famous studies of dogs many years ago. Pavlov demonstrated that when dogs were conditioned to associate an unpleasant stimulus with one that would usually elicit no response, they learn to react to the latter response. Thus, if the sound of a bell is quickly followed by an electric shock, and this is repeated a number of times, the dogs will learn to react (yelp, say) to the sound of the bell alone. The dogs have been conditioned to associate the sound of the bell with an unpleasant experience.

So we have three kinds of learning, leading to two kinds of memories that may be stored: short-term and, perhaps, long-term memories. Note that most short-term memories are not converted to long-term memories. If they were, your brain would shortly be crammed with totally useless, unimportant memories.

The next issue is how to go about the study of learning and memory. Kandel made a brilliant decision that would have eluded most of us: he chose the giant sea snail, *Aplysia*, as his experimental system. You may well wonder why Kandel chose such an exotic organism, but there are solid reasons for his choice. First, the brain of *Aplysia* contains 20,000 cells, a manageable number. The brain of a human contains, as noted earlier, about 100 billion cells, an unmanageable number. The 20,000 brain cells of *Aplysia* are organized into nine ganglia. It follows that, on average, each ganglion contains about 2200 cells.[8] This modest number makes mapping of neural circuits a whole lot easier than in the case of a more complex system. Secondly, the brain cells in *Aplysia* are large; those of the mammalian brain are small. The size of the *Aplysia* brain cells makes the insertion of microelectrodes for recording nerve cell functioning and for mapping out neural circuits far easier and with far less potential to damage the cells under study. Third, and this may seem surprising at first reading, *Aplysia* can learn in two of the three modes mentioned above. *Aplysia* proved to be an inspired choice. Note that success in science frequently depends on the proper choice of a model system for study. The system of real interest here is the nervous system of human beings. The point of choosing a model system is to find one that retains key features of the more complex system of interest while being simple enough to permit you to make headway in understanding it.

The next task for Kandel was to identify an *Aplysia* behavior that could be modified by learning. There were several possibilities. Upon reflection, he selected the gill-withdrawal reflex in this organism. *Aplysia* uses its gills, sensitive and important organs worthy of protection, to breathe. It employs its siphon to expel seawater and waste from a structure known as the mantel cavity, and apparently as an antenna for detecting potential danger. If the siphon is touched, *Aplysia* reacts by withdrawing the gills and the siphon into the mantel cavity. This reflex protects the gills from potential damage. A single ganglion, the abdominal one, controls this reflex.

Aplysia exhibits habituation. If the siphon is repeatedly touched lightly, the reflex withdrawal of gills and siphon gradually weakens. The snail has become habituated to the touch and its defensive reaction weakens as it no longer perceives a threat.

Aplysia also exhibits sensitization. The paradigm is to give the snail a strong shock, either to the head or tail. This is perceived as noxious by the animal, which then reacts to a subsequent light touch on the siphon with an exaggerated reflex withdrawal. In both cases of learning, both short-term and long-term memory can be achieved. The latter requires repeated experiences interrupted with periods of rest.

To summarize: there are two learning paradigms in the context of a reflex action controlled by a single ganglion in an experimental animal in which electrical measurements of nerve action are relatively easy. This experimental system permitted Kandel to map the neural circuit responsible for the gill-withdrawal reflex. The neural circuit mediating this reflex consists of just 12 cells: six sensory neurons and six motor neurons. Here is the neural circuit:[9]

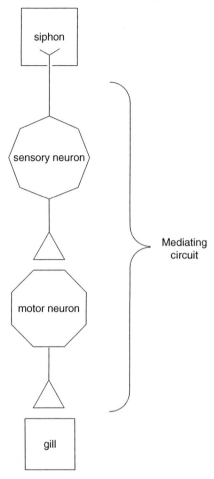

Each sensory neuron is attached to the siphon and synapses onto a motor neuron, which is linked to the gill. A touch to the siphon is detected by the sensory neurons. These transmit the message to the motor neurons, which in turn, cause the gill to withdraw. This is a beautifully simple system. Employing this system enabled Kandel to demonstrate that learning led to a change in strength of the synaptic connections. Specifically, habituation led to a weakening of the synaptic potential between the sensory and motor neurons; sensitization led to a strengthening of this potential.

The neurotransmitter at the sensory nerve–motor nerve synapse proved to be glutamate (incidentally, also the major excitatory neurotransmitter in the human brain). Further research established a basic molecular event associated with short-term learning: habituation caused the sensory neuron to release less glutamate into the synapse; sensitization caused the sensory neuron to release more glutamate into the synapse. Thus, the amount of neurotransmitter released into the synapse correlates with the strength of the motor response. The release of glutamate induces an action potential in the motor neuron. When this arrives at the neuromuscular junction, a synapse between the motor neuron and a muscle cell, acetylcholine is released and the muscle contracts. The more glutamate released into the synapse between sensory and motor neurons, the stronger the signal sent down the motor neuron, and the stronger the muscle contraction.

The neuronal circuitry proved to be a bit more complicated than I have let on so far. Shocking the tail of the sea snail proved to modulate the response of the organism to touches at the siphon. Thus, there appears to be a modulatory neural circuit between the tail and the sensory neuron. The complete neural circuit is:

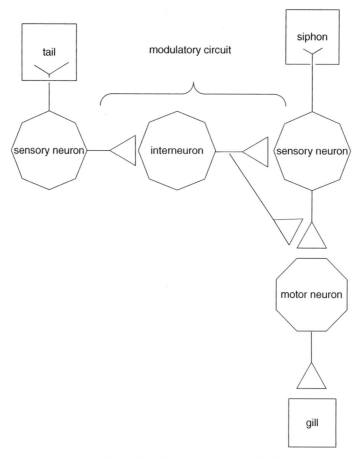

The sensory neuron from the tail makes a synapse with the modulating neuron, known as an interneuron. Its job is to fine-tune the response of the sensory neuron to stimulation. Note that the interneuron synapses with both the cell body and the presynaptic terminal of the sensory neuron. The neurotransmitter released by the interneuron into these synapses is serotonin (also a key neurotransmitter in the human nervous system). The net effect of the serotonin release is to strengthen the connection between the sensory neuron and the motor neuron. It remains to explain how this happens: that, too, has been tracked down in molecular detail.

Serotonin released by the interneuron is sensed by its receptor on the sensory neuron. The serotonin•receptor complex activates adenylate cyclase, the enzyme that catalyzes the synthesis of cyclic AMP (cAMP) from ATP. We encountered cAMP back in chapter 17 as a second messenger important in hormonal signaling; here, nature finds a different role for it to play. An increase in cAMP levels activates protein kinase A, PKA. PKA is an enzyme that adds a phosphate group from ATP to proteins, in this case, to a potassium ion channel, closing it. This potassium channel is involved in maintaining the resting membrane potential when it is open. Potassium ions flow out of the neuron through the channel during an action potential. When this channel is closed, the outward flow of potassium is slowed and the duration of the action potential is increased. This provides an opportunity for calcium ions to flow into the presynaptic terminals, which potentiates the release of glutamate into the synapse, strengthening the response of the motor neuron. Finally, PKA acts directly on the synaptic vesicles, also increasing the release of glutamate. Altering the strength of key synapses is how short-term memories are created in the course of learning. This applies with equal strength to the case of the human nervous system.

It remains to say something about laying down long-term memories. It has been known for many years that laying down long-term memories requires protein synthesis. There is no requirement for protein synthesis for making short-term memories. Kandel attacked the issue of the mechanism of long-term memory formation by looking at preparations of nerve cells from *Aplysia* in culture. He was able to study the simplest possible fully functional learning circuit: one sensory cell, one motor cell, and one modulatory interneuron.

When the creation of long-term memories—repeated stimulations interrupted by rest periods—was simulated in this preparation by repeated pulses of serotonin, anatomical changes occurred. Specifically, new synaptic connections were created. It is likely that there are two underlying components to formation of new synaptic connections. One is local protein synthesis in the nerve terminal and the other is CREB (cAMP response element binding) dependent transcription in the neuronal nucleus. Of course, serotonin pulses also stimulated the release of glutamate. So now the question is how repeated pulses of serotonin are related to protein synthesis and formation of new synapses.

We already know that serotonin increases the synthesis of cAMP, which activates PKA. Here is what happens next. PKA and a second kinase, known as MAP kinase (mitogen activated protein kinase) migrate to the cell nucleus. There, PKA activates (by phosphorylation) a gene regulatory protein known as CREB-1 (cAMP response element binding protein 1). Activated CREB-1 has a binding site on promoter regions of a number of genes. Once bound, CREB-1 turns on the synthesis of the mRNAs coded for by these genes. So one consequence is that a number of genes are turned on as a result of learning (in the context of laying down long-term memories).

The flip side of that coin is that MAP kinase inactivates, by phosphorylation, a second gene regulatory protein known as CREB-2. Once inactivated, the genes turned on by CREB-2 binding to promoter sites are turned off. Thus, learning turns off a number of genes as well. The laying down of long-term memories requires both the synthesis of new proteins and the inhibition of synthesis of other proteins. The net result is the creation of new synapses, strengthening communication, and creating long-term memory. Here too what has been learned through study of *Aplysia* applies fully to the case of the creation and storage of human long-term memories. The fact that studies on relatively simple organisms such as the sea snail can yield important insights into the physiology of human beings is one marvelous demonstration of the unity of life at the molecular level.

And that is all I have to say about the molecular events underlying learning and memory.

Key Points

1. Depression is a major mental health issue. Serotonin reuptake inhibitors are widely used for treatment of depression.

2. Schizophrenia may be the worst disease of mankind. All drugs employed to treat schizophrenia have dopamine antagonist activity, though many have additional actions as well.

3. Dopamine agonists are employed to treat parkinsonism, a disease in which there is a deficit of dopamine activity in the nigrostriatal pathway.

4. Many simple molecules have profound effects on the central nervous system. Several of these are drugs of abuse: methamphetamine, cocaine, and heroin, among others.

5. Morphine and other opiates are powerful analgesics but have dependency and abuse potential.

6. We have important insights into the molecular basis of learning and memory.

23

Antibiotics

The never-ending war against infectious disease

The fight against infectious pathogenic bacteria is ongoing. The principal weapons in this battle include a variety of β-lactam antibiotics as well as quinolines, tetracyclines, macrolides, and aminoglycosides.

Cemeteries can be useful sources of insight and information. The next time that you have the occasion to explore a cemetery have a look at the tombstones for people born around 1900. A lot of those tombstones will reveal that the date of death is within 10 years of the date of birth. Put succinctly, a lot of people born around 1900 did not live to be 10 years old. Now have a look at the tombstones for people born in 1940 or later. A much smaller fraction of people born in that era died before the age of 10 years old. A goal of this chapter is to explore and explain the reasons for this striking difference.

Great progress has been made against infectious diseases

Life expectancy in the United States increased markedly during the twentieth century, from about 46 years old in 1900 to about 76 years old in 2000. There are many reasons for the addition of three decades to the average duration of life. Among them are better public health measures, healthier lifestyles (better diets, more exercise, avoidance of tobacco, moderation in the intake of alcohol), availability and use of vaccines, improved surgical procedures, novel technologies for diagnosis of disease (magnetic resonance imaging, ultrasound imaging, computed tomography, for example), and powerful new drugs, specifically including several classes of antibiotics. These and other innovations have not only markedly increased the duration of life but also have contributed to a better quality of life. Here is one way to think about what we are trying to achieve: to remain healthy until the day you die and to postpone that day as long as reasonably possible.

Not only has life expectancy in the United States and elsewhere in the developed world increased markedly over the past century but the nature of life-terminating events has changed as well. In 1900 major causes of death included infectious diseases: smallpox, diphtheria, scarlet fever, pneumonia, tuberculosis, and others. These are the diseases that, by and large, are responsible for the frequency of early deaths in people born around 1900, noted above. For the most part, people born in 1900 in the United States did not live long enough to die of heart disease or cancer, now our major killers. We have made notable progress on one front, infectious diseases, and changed our focus to new disease targets: prevention and treatment for the chronic degenerative diseases of old age. The continuing fight against infectious disease has also found new targets: many of the major infectious diseases of 1900 are not those of the twenty-first century.

The focal point of this chapter is antibiotics, specifically antibacterials. There are two principal ways that science attempts to deal with infectious diseases: vaccines that prevent or reduce the frequency and/or severity of infectious diseases; and antibiotics that are usually employed as an aid in achieving a cure once an infectious disease has been acquired.[1] In addition, public health measures such as clean water, clean air, and pathogen-free food are critically important in limiting infectious diseases.[2]

Infectious pathogens come in many categories: bacteria, viruses, fungi, and parasites. Thus, we have several general classes of antibiotics: antibacterials, antivirals, antifungals, and parasiticides. Together with a host of vaccines and better public health measures, fabulous progress has been made against the scourge of infectious diseases during the past several decades.

Our job in this chapter is to understand what the important antibacterials are and how they work. If you want the whole story in all its elegant scientific detail, Christopher Walsh has brought it together in his book: *Antibiotics: Actions, Origins, Resistance.*[3]

The fight against infectious diseases is long term, maybe forever

Prior to the advent of frequent international travel and trade, many infectious diseases were largely confined to their sources. With a great many people and many items of trade now traveling from country to country each day, geographical limitations on spread of infectious diseases have ceased to be so important.[4] Consider human immunodeficiency virus 1 (HIV-1) infection and its end result—acquired immunodeficiency syndrome (AIDS). It had its origins in Africa but has spread to be a plague in most parts of the world. Drug-resistant tuberculosis is spreading globally.

We experience pandemics of influenza; these originate at some specific location but spread worldwide. In the United States, there are frequent reports of cases of exotic viral diseases (e.g., Ebola virus, West Nile virus, and Hantavirus infections) that had their origins far from our shores. We are all in this together.

The European bubonic plague of 1347 killed one-third of the population of Europe. It is the largest single plague ever recorded. The disappearance of the Aztec civilization was spurred by smallpox and measles introduced by Hernando Cortés and his band of Spanish invaders. The same diseases also decimated Native Americans in what is now the United States.[5] Much more recently, the influenza epidemic of 1918 killed an estimated 40 million people worldwide. Malaria continues to be a major problem for people and their countries today in areas in which it is endemic. AIDS, tuberculosis, influenza, hepatitis, pneumonia, and a lengthy list of parasitic infections continue as important constraints on the welfare of people throughout the world.

This is not to argue that infectious disease cannot be effectively fought. Smallpox has been wiped out on Earth; the last case was recorded in Ethiopia in 1977. The incidence of poliomyelitis on Earth has been reduced by 99.9% over the past few decades and there are reasonable, short-term prospects for its total elimination. Many of the important infectious diseases of the early twentieth century in the United States have been brought under effective control: smallpox, scarlet fever, whooping cough, and others. The discovery and use of novel vaccines and antibiotics over the past few decades have been enormous boons in the fight against infectious diseases. We now live longer and live better because of them.

However, the war against infectious disease continues and will continue for the foreseeable future. Here are three reasons why. First, drug-resistant organisms continue to develop. Methicillin-resistant *Staphylococcus aureus*, MRSA, provides one important example. MRSA is resistant to a host of potent antibiotics. For several years, the drug of last resort against MRSA and other life-threatening microbes was vancomycin. Vancomycin-resistant MRSA have been discovered in the last few years so that the last barrier has been breached. It seems reasonable to conclude that resistance will develop to any antibiotic that we come up with in the future. A final solution to the infectious disease problem is going to require something quite new indeed, a technology well beyond what we can envision currently.

Second we have the issue of bioterrorism. There are many other potential threats: smallpox, anthrax, tularemia, plague, a family of viral hemorrhagic fevers, among others. In addition, there are a limitless number of virulent bacteria that might be constructed employing the modern techniques of genetic engineering.

Finally, new infectious organisms arise continually. The fact is that we do not understand the microbial diversity on Earth. Here is an amazing, but not reassuring, example. The shotgun DNA sequencing of filtered water sampled from the Sargasso Sea revealed 1.2 million new genes and 1800 new bacterial species.[6] Most of these are doubtless harmless to man but some have the potential to develop or acquire virulence factors and become a threat.

Truly novel antibiotics are tough to find: only two antibacterials based on a novel chemical skeleton have been brought into clinical practice in the last 20 years. The discovery and development of vaccines is less profitable than that for drugs and adverse reactions to vaccines bring a lot of lawsuits down on the head of the manufacturers. Work continues and progress is made but on a scale smaller than the nature of the problem demands.[7]

The discovery of novel antibiotics that are effective against resistant microorganisms is not a trivial task. Let me provide one example why. A group of German scientists evaluated 700 enzymes, each a potential target for antibiotic development, in the pathogen *Salmonella enterica*.[8] More than 400 of these enzymes proved to be nonessential for virulence in this organism, narrowing the number of potential targets. Of the fewer than 300 enzyme targets that are essential for *Salmonella* virulence, 64 were identified as being conserved in other important human pathogens. These 64 enzymes are, therefore, potential targets for drugs that would have a useful spectrum of activity across pathogenic bacteria. Discouragingly, almost all of these were found to belong to metabolic pathways that are already targeted by antibiotics now in routine clinical use. It follows that the scope of the field of activity in the search for novel antibiotics has been substantially narrowed by earlier discoveries.

Getting around these issues, scientific and economic, is going to require the cooperation of pharmaceutical companies, biotech companies, governments, universities, and public health organizations.

There are historical antecedents to modern antibiotics

Of critical importance was the work of Louis Pasteur in France, Robert Koch in Germany, and Joseph Lister in England in the nineteenth century that established the germ theory of disease, a demonstration that took an amazing amount of time to catch on. Too many people, including scientists, do not understand that some fraction of what they know is just not so. People hold onto their misinformation with remarkable tenacity, retarding the acceptance of new knowledge.

The little black bag of physicians did not have a whole lot of useful medicines in 1900. The role of the physician at that time was diagnosis and prognosis far more than therapy. Nonetheless, some progress in chemistry in the service of human health had been made. Quinine, morphine, salicylic acid, digitalis, antipyrine, and ephedrine were known in 1900, though their utility, and their liabilities, in treatment of human disease was not fully appreciated.[9]

Natural products provided some of the earliest effective drugs

In the same period, vendors of proprietary medicines took advantage of our love of the natural to peddle a variety of plant-derived products. These were advertised as safer and more pleasant than those typically employed by physicians. For example, Mrs. Winslow's Soothing Syrup and Kopp's Baby Friend had morphine sulfate as their basic ingredient. Hostetter's Bitters was a 78 proof (39% ethanol) cocktail. This theme is being replayed currently as alternative medicine. This time around, controlled clinical trials are being carried out, slowly, to establish what works and what does not and the associated safety profiles. For example, a placebo-controlled, double-blind clinical trial has established beyond reasonable doubt that *Echinacea* is of very limited or no utility for prevention or treatment of the common cold.[10] Likewise, saw palmetto has been found to be ineffective for prostatism.[11] This is not to argue that alternative medicines have no role in human health. It does establish the need for careful evaluation of such products, which fall outside the purview of the Food and Drug Administration (FDA) in the United States.

During the nineteenth century, chemists had a good deal of success in isolating and purifying natural products from plant sources. Morphine was isolated as a pure compound from crude opium in 1804. Quinine was isolated from the bark of the cinchona tree in 1820 and was initially employed as a fever reducer. However, its effectiveness against malaria was soon discovered and it found an alternative highly important medical use. Sodium salicylate was isolated from the bark of the willow tree in 1821 and was also shown to have analgesic, antipyretic, and anti-inflammatory properties. It took an additional 76 years, until 1897, to synthesize the acetyl derivative, acetylsalicyclic acid, commonly known as aspirin.

These and other discoveries of biologically active substances in plant material are one source of modern medicinal agents. Natural products continue to provide us with important medical advances.

The dye industry of Europe provided another source of drugs

The other source of modern drugs was the European dyestuff industry of the nineteenth and early twentieth centuries. The goal of this industry was to make useful dyes, principally for fabrics. In the course of handling novel molecules, scientists occasionally make unexpected observations (serendipity) that suggest novel utilities. For example, the commonly used sweetener aspartame was discovered by accident when a scientist licked a finger containing a bit of this substance. It turned out to be surprisingly sweet.

Under the intellectual leadership of Paul Ehrlich, the dye industry provided arsphenamine, the first effective agent against syphilis. Syphilis is an infectious disease spread by sexual contact. It is caused by the spirochete *Treponema pallidum*. Syphilis runs various courses over many years and can result in death if not treated. Penicillins are now the drugs of choice for syphilis. Chemists discovered two other medicinal agents in the early years of the twentieth century: tryparsamide for trypanosomiasis, a parasitic disease, and oxophenarsine, also for syphilis.

Subsequently, other drugs have been developed for the same uses that provide advances in efficacy and safety. Nonetheless, these molecules were important at the time of their discovery and were key steps down the path to improved agents.

Sulfa drugs were the first important antibacterials

The first really useful antibacterials were the sulfonamides or sulfa drugs. A collection of useful sulfonamides is shown in figure 23.1.

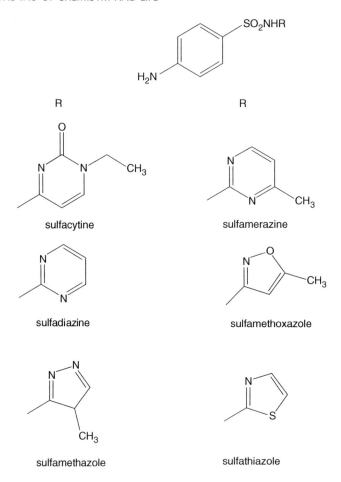

Figure 23.1   Structures for several clinically useful sulfonamides.

These agents were discovered in the 1930s as an outgrowth of a search for therapeutically useful dyes. Specifically, Gerhard Domagk found that a red dye known as prontosil rubrum was an active antibacterial agent in mice and rabbits, protecting both species from otherwise lethal doses of staphylococci and hemolytic streptococci. For this key observation, Domagk was awarded the Nobel Prize in Physiology or Medicine for 1939.

Domagk did not stop with his findings in mice and rabbits. The big question was whether or not prontosil rubrum would be an effective antibacterial agent in human beings. Unexpectedly, his daughter contracted a severe streptococcal infection. As a desperate measure, Domagk gave her a dose of prontosil rubrum and she made a complete recovery. Domagk did not report this finding until some years later when others confirmed efficacy in clinical studies.

Domagk also observed that prontosil rubrum was ineffective against bacteria in test tube studies. This was an unusual finding. In current practice, activity at the molecular or cell level is generally a required prerequisite to a search for activity in an animal model of disease. Current drug discovery protocols would not have found the antibacterial activity of prontosil rubrum. Research revealed that prontosil rubrum is metabolized in the mouse by liver enzymes to a simpler molecule—sulfanilamide:

prontosil rubrum   sulfanilamide

Sulfanilamide proved to be a safe and efficacious antibacterial agent. Prontosil rubrum is a prodrug: that is, a molecule itself devoid of the desired biological activity but which is metabolized to one or more compounds that do have this activity. The age of wonder drugs had dawned. Gerhard Domagk was the father of the sulfa drug revolution. A huge family of derivatives of sulfanilamide was subsequently synthesized and several of these came into medical practice (figure 23.1).

Have a look at the structures in figure 23.1 to get an insight into drug discovery. The parent compound, sulfanilamide, is the simplest one. All the other structures were derived from sulfanilamide by replacing one hydrogen atom on the amino group by a variety of suitably decorated heterocycles. Although modern science and experience with what works and what does not has given those involved in drug discovery a leg up, the process of searching around for new molecules by replacing one group for another continues. It will continue for the foreseeable future. This is not a random process. A wealth of experience gained over the years has encouraged medicinal chemists to favor some groups over others, for reasons of efficacy or safety or both. The availability of high-resolution structures, largely from X-ray diffraction studies, of the molecular targets of drug discovery frequently provide useful insights into which molecular replacements will prove effective for increasing potency.

Let's think of the molecular target as a lock and the small molecule that turns it on or off as a key. So one part of the drug discovery problem is to design a key that fits the lock, an exercise in molecular recognition. It is easier to design a key to fit a lock when you know the structure of the lock. X-ray diffraction studies provides that structure. For example, refer back to the structure of an enzyme complexed with an inhibitor provided in figure 9.1. Despite all our experience and the structural information, there does remain a good bit of chemical flailing about in search for the optimal molecules for clinical trials. In the usual case, chemists in the pharmaceutical industry will make about 2000 molecules prior to finding one suitable for clinical study or simply giving up the search and finding something more rewarding to do.

Given a molecule such as sulfanilamide that is an active antibacterial, why go to all the trouble of modifying the structure to find other active molecules in the same structural class? There are numerous answers: to find safer molecules, more potent molecules, molecules active against a broader spectrum of bacteria, molecules active longer so that doses may be taken less frequently, molecules that may pass the blood–brain barrier and be effective against central nervous systems infections, molecules that concentrate in the urine and are effective against kidney infections, and so on. Of course, there is also the goal of having something to sell rather than conceding the market to the pioneer discoverer.

Sulfa drugs continue to find modest utility in current medical practice. The discovery of novel agents and the development of allergic reactions to sulfa drugs have dimmed their glow.

Sulfa drugs inhibit bacterial folic acid synthesis

Now we come to the question: what is the basis of the antibacterial action of the sulfa drugs? Many useful drugs in human medicine are enzyme inhibitors, small molecules that frequently bear a structural resemblance to the substrates or products of enzyme-catalyzed reactions. So it is with the sulfonamides.

Folic acid is a vitamin, as we developed in chapter 15. It is a complex molecule that serves as an essential precursor for coenzymes involved in the metabolism of one-carbon units. For example, folic acid-derived coenzymes are critically involved in the biosynthesis of thymidine for nucleic acid synthesis and methionine for protein biosynthesis. The synthesis of both demands donation of a methyl group and they come from folic acid-derived coenzymes.

In contrast to humans, bacteria have the biochemical ability to synthesize folic acid from simpler molecules. Here we have a clear biochemical difference between human beings and infectious organisms that we can exploit to our benefit. The reaction catalyzed by an enzyme known as dihydropteroate synthetase, in which a complex heterocycle is linked to p-aminobenzoic acid, is key. Now recognize the structural similarity between sulfanilamide, or other sulfonamides, and p-aminobenzoic acid:

sulfonamides   p-aminobenzoic acid

Given this structural similarity, it should not be surprising to learn that sulfanilamide competes with p-aminobenzoic acid for a binding site on the surface of dihydropteroate synthetase. Put another way, sulfanilamide binds to the enzyme where p-aminobenzoic acid should bind but no reaction occurs. The consequence is that a step in folic acid biosynthesis is disrupted and the bacterial cell is deprived of adequate folic acid. Nucleic acid synthesis, among other things, is disrupted, leading to a cessation of cell growth and division. The human immune system can mop up what remains. No similar consequences befall the human host since it cannot make folic acid in the first place and must get an adequate supply of this vitamin in the diet.

The underlying idea of the mechanism of action of the sulfonamides is quite general in the search for agents effective against pathogens but safe for humans: find a clear biochemical difference between the host and its infectious agent and seek ways to exploit that difference to the benefit of the host and the detriment of the infectious agent. This is exactly what happens with the β-lactam antibiotics, to which we now turn.

The β-lactam antibiotics are a second great class of antibacterials

Following the sulfa drugs, a second discovery of the greatest importance for infectious bacterial disease was made: penicillin. Penicillin is the first of the β-lactam antibiotics, so named because each of these molecules contains a four-membered lactam ring:

A β-lactam

The story of the β-lactam antibiotics began in 1928 with the chance observation by Sir Alexander Fleming that accidental contamination of a bacterial culture plate by the mold *Penicillium notatum* created a clear area around the mold, the consequence of killing of the bacterial cells.[12] It was clear that the mold created and secreted into the culture medium some agent toxic to bacteria. As that agent diffused through the culture medium, it killed bacterial cells in its path, creating the clear, bacteria-free, area. The responsible molecule proved to be penicillin, as demonstrated by Howard Florey and Ernst Chain 11 years later. Isolating and characterizing the active principle made and secreted by *Penicillium* was no easy task for a number of reasons. Penicillin is a potent molecule and not much is required to kill bacteria. It is by no means the only molecule made by this mold and secreted into the culture medium. Thus, it was necessary to separate and purify penicillin away from a family of other molecules. Finally, penicillin is a somewhat fragile molecule, not the most stable molecular construct of nature. Given that the highly sophisticated technologies for isolation, purification, and characterization of molecules available to chemists today were unavailable at the time, their success was enormously notable. Wit, energy, and determination substituted for technology. These remain highly effective human traits today in drug discovery. For their pioneering discovery of the parent molecule of a major class of antibacterials, Fleming, Chain, and Florey shared the 1945 Nobel Prize in Physiology or Medicine.

Structurally, penicillin is a penam, see figure 23.2. A great many penams have subsequently been synthesized and tested and many have found clinical use: penicillins G and V, amoxicillin (Trimox), ampicillin (Pen A), methicillin (Azapen), oxacillin (Bactocill), cloxacillin (Cloxapen), carbenicillin (Carbapen), and piperacillin (Zosyn), among others. Like the sulfonamides, these molecules are a structural variation on a common theme. Novel penicillins had improved breadth of antibacterial action, improved tissue distribution, efficacy against organisms resistant to earlier agents, or improved safety profiles, including freedom from allergic reactions to earlier penicillins.

Of particular note is the combination of amoxicillin and clavulanic acid. The latter is a potent inhibitor of enzymes that degrade amoxicillin and many other β-lactam antibiotics. This combination, marketed as Augmentin, increases the efficacy of amoxicillin against organisms that would be otherwise resistant to it. It is among the most widely used antibiotics in the United States.

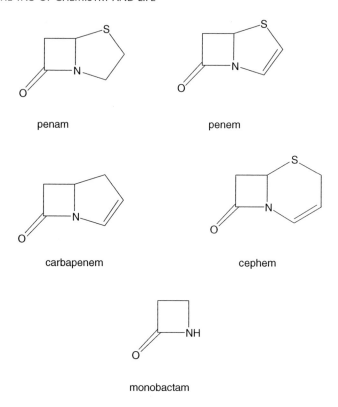

penam   penem

carbapenem   cephem

monobactam

Figure 23.2 Structural skeletons for the principal classes of β-lactam antibiotics. For example the penicillins are all penams and the cephalosporins are all cephems.

Antibiotics based on a related ring system, termed the cephems (figure 23.2), are highly useful. Here the five-membered ring has been expanded to six and the double bond of the penems retained. These are the cephalosporins and include such important antibiotics as cephalothin (Keflin), cefazolin (Ancef), cephalexin (Keflex), cefamandole (Cefam), cefuroxime (Zinacef), cefeclor (Ceclor) and cefoxitin (Mefoxin).

There are two additional important variations on the theme of β-lactam antibiotics. These are the carbapenems, of which imipenem (Primaxin) is the outstanding example, and the monobactams, of which aztreonam (Azactam) is the outstanding example (see figure 23.2).

Both of these antibiotics are notable. Imipenem is derived from a natural product, thienamycin.

The discovery of thienamycin created great excitement: it is a structurally novel β-lactam antibiotic of outstanding potency and has a remarkable spectrum of activity. It was the broadest spectrum antibiotic of its day. There was, however, a major problem: thienamycin is not a stable molecule. Merck scientists were faced with the touchy problem of modifying thienamycin chemically to create a stable molecule while maintaining all its remarkable properties. Following considerable effort, they solved that problem by adding a four-atom fragment to thienamycin. The marketed product, Primaxim, is actually a combination of imipenem and a second molecule that protects it from degradation in the human kidney.

Like cefoxitin and imipenem, aztreonem is a chemically modified product of a natural product. Aztreonem is highly resistant to a family of antibiotic-degrading enzymes (the β-lactamases) and is highly effective against Gram-negative aerobic bacteria.

β-Lactam antibiotics inhibit bacterial cell wall synthesis

The penicillins, cephalosporins, carbapenems, and monobactams all work in basically the same way. However, they have different therapeutic uses which derive from several factors: which organisms are susceptible to which agent, which may be taken orally and which must be injected, which do and do not penetrate into the central nervous system, and so forth. The target of these antibiotics is the bacterial cell wall.

Human cells have cell membranes but no cell walls. Bacteria, in contrast, have both: a cell membrane that is surrounded by a strong cell wall. The details of the structure of the bacterial cell wall differ somewhat from organism to organism, and in a rather major way between the Gram-positive and Gram-negative organisms. At the same time, there is an underlying, unifying structure: a complex polysaccharide cross-linked by chains of amino acids, the peptidoglycans. No related structures occur in mammalian cells. It follows that the bacterial genome codes for a series of enzymes that catalyze the assembly of the bacterial cell wall. These enzymes are not present in mammalian cells. Enzymes that are unique to the bacteria and required for bacterial cell division are targets for inhibitor discovery.

The pathway from simple molecules to the peptidoglycan of the bacterial cell wall is lengthy and complex. Many of the details are well known but need not concern us here. Suffice it to say that long carbohydrate chains are synthesized, subsequently decorated with shorter amino acid chains, and these are finally cross-linked to provide a strong structure. It is this final cross-linking step that is inhibited by the β-lactam antibiotics. The consequence is that cell wall biosynthesis cannot be completed and cell death ensues. Again, the mammalian host carries out no similar reactions so that similar consequences do not ensue for the host organism.

Quinoline antibiotics inhibit bacterial DNA gyrase

The quinoline antibiotics provide a third major class of useful antibacterials. These include norfloxacin (Floxacin), ofloxacin (Floxin), ciprofloxacin (Cipro), levofloxacin (Levaquin), and moxifloxacin (Avelox). These antibiotics are named for the core heterocyclic ring structure, quinoline:

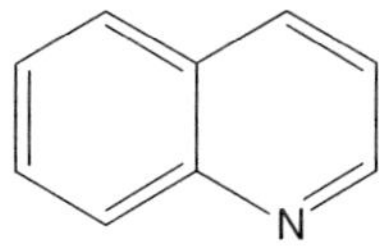

quinoline

With the quinoline antibiotics, we have another variation on the theme already developed: exploitation of a biochemical difference between an infectious organism and its host. In this case, the difference is in the nature of an enzyme involved in nucleic acid replication, as developed below.

The quinoline antibiotics are relatively late arrivals on the antibiotic scene. The parent compound of this class of drugs is nalidixic acid. The downside of this molecule is the rapid development of resistance by pathogenic bacteria. A major step forward was the introduction of a single fluorine atom at a key position, to yield new molecules such as ciprofloxacin, ofloxacin, levofloxacin, and moxifloxacin. Of these, levofloxacin and moxifloxacin have the best combinations of spectrum of action, potency, and pharmacokinetic properties. In time, they are likely to largely replace the current quinoline antibiotic of choice, ciprofloxacin. Since these molecules work by a different mechanism than do the β-lactams, organisms that become resistant to the latter are generally susceptible to the former.

The molecular target of the quinoline antibiotics is an enzyme, DNA gyrase, involved in the supercoiling of DNA. The biochemical distinction between host and bacterium is less clear here than in the two cases considered above, sulfa drugs and β-lactam antibiotics. Both mammalian and bacterial cells have enzymes involved in DNA supercoiling, generally termed topoisomerases II. However, the bacterial enzyme that performs this function is distinct from the human enzyme performing the same function. Thus, it is entirely possible to discover and employ enzyme inhibitors that are specific for the bacterial enzyme. That is what the quinoline antibiotics do. They are potent inhibitors of bacterial DNA gyrase but have little or no effect on the corresponding human enzyme topoisomerase II (which is itself a target for a class of antitumor agents; see chapter 24).

Tetracyclines, macrolides, and aminoglycosides are important classes of antibacterials

In the area of antibacterials, there are three important classes not yet mentioned. First, there are tetracyclines, the name reflecting their structural skeleton. Tetracycline itself provides one example (note the four rings):

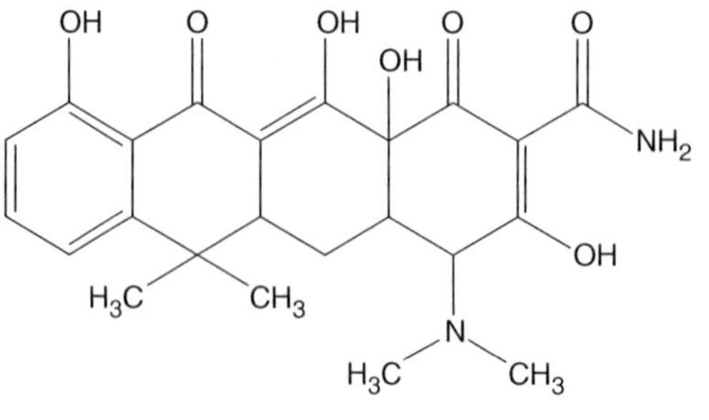

tetracycline

Important tetracyclines include chlortetracycline (Aureomycin) and oxytetracycline (Terramycin). Tetracyclines are broad-spectrum antibiotics, which means that they are effective against a wide variety of microorganisms. The tetracyclines are protein synthesis inhibitors. They act by binding specifically to the bacterial ribosome, the site of protein synthesis, and inhibiting its function. Tetracyclines have little or no affinity for the mammalian ribosome and specifically affect protein synthesis in the infecting agent.

The macrolide antibiotics provide a second distinct class. The first among them was erythromycin, a natural product isolated from *Streptomyces erythreus*. The macrolide antibiotics are named for the multi-membered ring that provides their core structure. That ring is amply decorated with a family of groups. The structural complexity is striking. Here again, we have an example of the complexity of a structure synthesized by a microorganism: multiple functional groups and many chiral centers, 17 in this case.

Newer and more generally useful macrolide antibiotics include azithromycin (Zithromax) and clarithromycin (Biaxin). These too are wide-spectrum antibiotics and both are semisynthetic derivatives of erythromycin. Like the tetracyclines, the macrolide antibiotics act as protein synthesis inhibitors and also do so by binding specifically to the bacterial ribosome, though at a site distinct from that of the tetracyclines.

Finally, the aminoglycosides form an important group of antibiotics. Here too we have a class of antibiotics that are protein synthesis inhibitors. However, the aminoglycosides are distinct from tetracyclines and macrolides in an important respect. The aminoglycosides are rapidly bactericidal: that is, they kill bacteria. The tetracyclines and macrolides, in contrast, are bacteriostatic: that is, they prevent the growth of bacteria. Once growth is prevented, the normal, uncompromised human immune system mops up the bacteria and terminates the infection. However, bacteriostatic agents do not work well in seriously ill, immunocompromised (e.g., AIDS or cancer) patients. In these cases, aminoglycosides are often used owing to their potency, spectrum and rapid bactericidal activity.

First among the aminoglycosides was streptomycin, one of several antibiotics isolated from *Streptomyces* species by Selman Waksman—this from *S. griseus* in 1944. Waksman proved to be an enormously effective seeker of antibiotics in natural products. In addition to streptomycin, he discovered neomycin, another widely used antibiotic. Less important discoveries include actinomycin, clavacin, streptothricin, grisein, fradicin, and candidin. Waksman received the Nobel Prize in Physiology or Medicine in 1952.

Newer examples of aminoglycoside antibiotics include amikacin, neomycin (Neosporin, Cortisporin), and tobramycin (TOBI, TobraDex). Injectable tobramycin is used in the treatment of serious infections at many body sites. It has also been formulated in an inhalable dosage form that has a very specific use: to treat cystic fibrosis patients having *Pseudomonas aeruginosa* lung infections. In the form suitable for inhalation by the patient, it delivers the antibiotic directly to the site of infection.

For each structural class of antibacterial discussed, there are many members in clinical use. I have selected a few as representative examples of each class. There are several additional classes of antibacterials as well. Notable is vancomycin (Vancocin), like the β-lactam antibacterials an inhibitor of cell wall biosynthesis although the site of action is different. In addition, there are a number of peptide antibiotics typically found in preparations for topical use.

Linezolid and daptomycin are recent, novel antibacterials

The discovery of a novel structural class of antibacterials is notable, as these are few and far between. The sulfa drugs, β-lactams, quinolines, tetracyclines, macrolides, and aminoglycosides have been around for decades. Multiple improvements have been made over time in each of these classes but without breaking out into new structural classes. There are two notable, recent examples of new structural classes of antibacterials and these are worth knowing about.

Linezolid (Zyvox) is an oxazolidinone, a five-membered heterocyclic ring that forms the core of the linezolid structure. The approval of linezolid by the FDA in 2000 marked the first new structural class of antibacterial introduced into medical practice in the United States in 40 years. It is notable for its activity against methicillin-resistant *Staph aureus*, MRSA, and vancomycin-resistant *Enterococcus faecium*, VRE. It is bacteriostatic rather than bactericidal but finds significant use in patients with an intact immune system. Like several other classes of antibacterials, linezolid is an inhibitor of protein synthesis. It interacts specifically with the RNA component of a bacterial ribosome subunit to prevent initiation of protein synthesis.

Even newer is the natural product daptomycin (Cubicin), a complex cyclic lipopeptide structure, approved for use in the United States in 2003. Daptomycin has a spectrum similar to that of linezolid and specifically includes MRSA and VRE. In contrast to linezolid, daptomycin is bactericidal for these Gram-positive organisms. It is, like vancomycin, a parenteral antibiotic and is given intravenously. It is indicated for treatment of complicated skin and skin structure infections and for some cases of bacteremia, including endocarditis. Daptomycin may be thought of as an alternative to vancomycin.

Daptomycin has a unique mechanism of action: it disrupts the bacterial membrane of Gram-positive organisms by forming channels across it. These channels permit the leakage of intracellular ions, eventually leading to cell death. Resistance to daptomycin is rare, at least so far.

It is clear that we need more novel antibacterials such as linezolid and daptomycin to counter resistance development and appearance of new pathogenic bacteria. These will not be easy to come by but continued control of infectious diseases requires them. It is also clear that we need a continuous flow of effective vaccines as a means to protect us from infections in the first place.

Key Points

1. The fight against infectious disease is never-ending: new pathogens; old pathogens in new places; emergence of drug-resistant pathogens; failure of public health measures; the threat of bioterrorism; and so on.

2. The discovery of novel agents against infectious diseases has become increasingly difficult: many of the most attractive molecular targets have already been exploited.

3. Effective antibiotics have come from isolation of natural products (e.g., penicillins and cephalosporins) as well as the work of medicinal chemists (e.g., fluoroquinolines).

4. Sulfa drugs were the first important antibacterials. Sulfa drugs act by inhibiting bacterial folic acid synthesis.

5. Beta-lactam antibiotics are a second great class of antibacterials: penicillins, cephalosporins, carbapenems, and monobactams. They act by inhibiting bacterial cell wall synthesis.

6. The quinoline antibiotics act by inhibiting bacterial DNA gyrase. These include Cipro, Levaquin, Floxacin, and Floxin.

7. Other key classes of antibacterials include the tetracyclines (Aureomycin, Terramycin), macrolides (erythromycin, Zithromax, Biaxin), and aminoglycosides (streptomycin, amikacin, neomycin). These antibacterials are protein synthesis inhibitors.

8. The newest antibacterials include Zyvox and Cubicin.

24

Cancer

*What it is and what we can do
about it*

---

*A cancer is a clone of cells arising from a single cell that has escaped
controls on growth and division, the consequence of genetic errors.
Environmental agents, such as tobacco, or oncogenic viruses may cause
these genetic errors. Drugs and technologies to deal effectively with
cancer are improving but leave a lot to be desired.*

Cancer! It is a most disagreeable word. Cancer is one of the most feared of all medical
diagnoses. It rightly conjures up images of radical, and sometimes mutilating, surgery,
lengthy and debilitating courses of drug treatment, incapacitation, the end of normal
activities of life, and death. Dying of cancer is often a painful and protracted process.
Cancer is feared for good reason. About 1 million people die of cancer each year in
the United States.

Robert Weinberg, a leading cancer scientist, has put it eloquently in his elegant
book *One Renegade Cell*:[1]

> Cancer wreaks havoc in almost all parts of the human body. Tumors strike the brain
> and the gut, muscles and bones. Some grow slowly; others are more aggressive and
> expand quickly. Their presence in human tissues signals chaos and breakdown of normal
> function. Cancer brings unwelcome change to a biological machine that is perfect,
> marvelously beautiful, and complex beyond measure. Wherever tumors appear, they
> take on the appearance of alien life forms, invaders that enter the body through stealth
> and begin their program of destruction from within. But appearances deceive. The truth
> is much more subtle and endlessly interesting.

My Grandpa Hank was a splendid grandfather: he was wise, funny, generous, and kind. Hank Cordes was a great outdoorsman: fishing and hunting were a big part of his life. He lived a very active life and was in excellent health until he reached the age of 89 years old: colorectal cancer struck. He survived the surgery but lost much of his pleasure of life although he eventually did regain a substantial measure of independence. The cancer recurred a few years later and he died a slow, unpleasant death at age 93 years old. Although he lived a long and happy life, it was nonetheless enormously saddening to see cancer put a bitter end to it. None us are going to get out of this alive but there are far better ways for life to come to an end. Sadly, almost everyone has a related tale or two to tell, some far more tragic than mine when cancer strikes down the young.

Cancer is not a uniformly fatal disease. Slow progress has and is being made against the family of malignancies that we call cancer. We have better methods for early diagnosis, advanced tools for staging of the disease, improved surgical techniques for elimination or reduction in size of tumors, better means of delivering therapeutic radiation to the sites of tumors, and improved drugs for chemotherapy. Childhood leukemia, once a death sentence, is now cured in the great majority of cases. Testicular carcinoma, also once invariably fatal, is now cured in 95% of men with the disease. Perhaps the most notable case is that of the bicycle racer Lance Armstrong. Armstrong was a victim of testicular cancer that had spread to his lungs and brain. Aggressive chemotherapy eliminated the disease and Armstrong went on to win the most demanding bicycle race in the world, and perhaps the single most challenging athletic event of all, the Tour de France, seven consecutive years 1999–2005, prior to retiring from competitive cycling. This is a most dramatic case of recovery from a life-threatening cancer. In some cases, there is not only life after cancer but strength, endurance, accomplishment, and fame.

For several other cancers, the cure rate is up, the disease-free interval is lengthened, life is prolonged, or the quality of life is improved. Colorectal cancer falls into that category, though it remains a life-threatening disease. In 1994, the only Food and Drug Administration (FDA) approved chemotherapy for colorectal cancer in the United States was 5-fluorouracil. Ten years later, we had six agents approved by the FDA for this disease. In addition to 5-fluorouracil, there are capecitabine (Xeloda), irinotecan (Camptosar), oxaliplatin (Eloxatin), cetuximab (Erbitux), and bevacizumab (Avastin).[2] Patients have benefited from the discovery and development of these new agents.

Despite the progress, the stark fact remains that the major solid tumors of people—lung, breast, colorectal, ovarian, and prostate—claim a great many lives and degrade the quality of life in the process. The drugs that we employ to battle these malignancies leave a lot to be desired.

Most drugs employed in the chemotherapy of cancer are pretty crude weapons. As a class, they are known as cytotoxics: molecules that kill cells. Clearly, the goal is to kill the cancer cells, leaving the normal, healthy cells of the human body untouched. Unhappily, we do not know how to do that very well. The cytotoxics employed in cancer chemotherapy have varying degrees of selectivity for killing cancer cells rather than normal ones but the selectivity is less than one would like. None of them correct the underlying error or errors that led to the tumor in the first place. Many cytotoxics are themselves mutagenic or carcinogenic. There are a few examples of targeted, as opposed to cytotoxic, drugs for cancer and I will get around to these important advances later.

Here is an introduction to the language of cancer

We need to understand a bit of the language of cancer. Cancers attack basically all the tissues of the body: lungs, liver, brain, gonads, stomach, pancreas, colon, breasts, kidney, prostate, and so on.

To begin with, a cancer is a clone of cells arising from a single cell.[3] From time to time a single cell will escape from the set of mechanisms that control cell division and evade the immune response. That cell can then divide quite independently of the needs and welfare of the organism that harbors it. The progeny of that cell can proliferate willy-nilly as well, creating a clone of cells. Each member of the clone is identical to the parent cell. This process happens in a stepwise manner: that is, some genetic error permits limited expansion of a clone of cells. A second error in one of these cells then permits a second expansion. This is repeated until enough errors are accumulated to permit uncontrolled cell division: cancer.

Most tumors are solid. That is, they form a discrete, stable, identifiable structure within some tissue or organ. Tumors of the lung, liver, brain, breast, and so on, are all solid. In contrast, leukemias and lymphomas, cancers of the blood cells, are liquid: that is, the tumor is a collection of cancer cells that circulate but do not form a discrete, identifiable structure.

Tumors fall into two general classes: benign and malignant. Benign tumors do not invade their surrounding tissues and do not form distant colonies, termed metastases. Benign tumors are generally small and localized. Benign tumors, polyps, of the intestine remain in the intestine, for example. Benign tumors do not generally pose a significant threat to their host. In some cases, they may grow large enough to interfere with normal organ function and must be removed surgically. In other cases, they may have the capacity to turn malignant. The polyps of the large intestine fall into this category and are removed surgically when detected. The cells of benign tumors generally resemble quite closely those of the tissue from which they derive.

Malignant tumors are another matter entirely. In contrast to the cells of benign tumors, those of malignant tumors frequently bear only a passing resemblance to their progenitors. In general, they have lost many of the aspects of their progenitor cells and have gained some capacities not possessed by them. Malignant tumors have the capacity to invade tissue and to form metastases, which are new tumor sites distant from the site of origin of the tumor (the primary tumor). Malignant tumors are generally life-threatening. In fact, most people that succumb to malignant tumors do so as a result of the metastases, not as a result of the primary tumor. The primary tumor is localized and can frequently be effectively removed by surgery. Disseminated tumors are, in general, not removable surgically and are tough to defeat with radiation or chemotherapeutic agents. This emphasizes the need to detect tumors early in the their natural history when they can be effectively handled, before they have spread to secondary sites.

The division between benign and malignant is useful but too simple. To be a little more precise, we can add the classification of dysplastic tissue. Dysplastic cells are not yet fully malignant but are on their way to getting there. Finally, the term neoplastic is frequently used to describe malignant cells or tumors. So we have the sequence: normal, benign, dysplastic, malignant (neoplastic).

That is enough for a few of the generalities of cancer. Now we need to begin to understand the origins of cancer. This understanding derives from two very different lines of investigation: effects of environmental agents and effects of viruses. I begin with the former.

Environmental agents may be carcinogenic

One of the most dramatic and important accomplishments of biomedical sciences in the last half of the twentieth and early years of the twenty-first century is the development of an understanding of the molecular processes that lead to cancer.[4] This understanding, not yet complete but enormously exciting, has come from many lines of research. It promises many new insights into the prevention, diagnosis, staging, and treatment of cancer.

We all know that things that we encounter in the environment can elicit cancer. The use of tobacco is responsible for about one-third of all tumors of people. These include tumors of the lung, bladder, esophagus, head, and neck. It follows that tobacco is a cancer-causing substance, a carcinogen. This has been known since the 1950s. It took the tobacco industry in the United States half a century to own up to this simple fact. Half a century that is replete with denials known to be false, fraudulent research, intimidation of health proponents, advertising of cigarettes to children, and the continued vigorous promotion of tobacco use outside the United States. It is not a happy story.

The history of the relationship between environmental agents and cancer is a long one. To begin with, the British surgeon Percival Pott noted in 1775 that young boys employed in London as chimney sweeps suffered an abnormally high frequency of scrotal cancer. Clearly, there was something in the coal tar to which they were exposed in their miserably filthy work that elicited this cancer.

A bit later in eighteenth century London, snuff users were observed to have an abnormally high frequency of nasal cancer; thus, the earliest evidence that tobacco is carcinogenic is more than 200 years old!

Many correlations of environmental exposure with certain cancers followed in the nineteenth and twentieth centuries. These include German pitchblende miners with lung cancer; X-ray workers with leukemia and skin cancer; painters of radium onto watch dials with tongue cancer (they licked their paintbrushes); smokers with lung and other cancers; tobacco chewers with cancers of the lips and tongue. We know that asbestos and nickel elicit lung cancer; that alcohol does the same for cancers of the esophagus, mouth, and throat; and that benzene causes leukemia. Indeed, there is now a substantial compendium of chemical compounds and more complex substances known to cause cancer in people or experimental animals, almost always rodents. Complex rules and regulations have been put into effect to minimize exposure to carcinogens or suspect carcinogens in our homes and workplaces. The various bans on smoking in many public places and workplaces in the United States provide an obvious, and important, example.

Many carcinogenic substances occur naturally, whereas others have been made synthetically. The fraction of naturally occurring substances and synthetic substances tested that have proved to be carcinogenic is about the same: that is, nature is no kinder to us in terms of carcinogens than are chemists. The foods that we routinely consume contain many substances known to be carcinogenic. A partial list of such foods is provided in table 24.1. This list is surely interesting and may be surprising but should not be of particular concern. The human body has several ways of protecting us from substances in the environment, including carcinogens in our food. The list does draw attention to an interesting fact. Government regulations and health activists go to great lengths, rightly for the most part, to protect us from synthetic carcinogens but almost no note is taken of those that occur naturally. The major exception is tobacco, without doubt the major carcinogen in the world, whose use is controlled in some ways (in the United States at least) but which remains fully legal. Somehow we have deluded ourselves into believing that chemicals that occur in nature are somehow more benign than those made in the chemistry laboratory. There is little evidence to support such a distinction.

Table 24.1 A partial compilation of common foods and food flavorings known to contain carcinogens.[a]

Food or flavoring	Carcinogen
Sassafras, nutmeg, mace, star anise, cinnamon, black pepper	Safrole[b]
Beets, celery, lettuce, spinach, radishes, rhubarb, mustard greens, kale, turnips, cabbage	Nitrate[c]
Coffee	Many[d]
Meats, charbroiled, smoked, or fried	Many, largely polycyclic aromatic hydrocarbons such as benzo[a]pyrene, and heterocyclic amines
Common mushrooms, false morels, shiitake mushrooms	Hydrazines, including N-methyl-N-formyl hydrazine, gyromitrin, and agaritine
Mustard, horseradish, broccoli, cabbage, arugula	Allyl isothiocyanate
Many herbal teas, comfrey	Pyrrolizidine alkaloids, including intermedine, lycopsamine, and symphytine
Tarragon, basil, fennel	Estragole
Celery, parsnips, parsely	Psoralens
Bread, yogurt, soy sauce, beer, wine	Ethyl carbamate

[a]These compounds have been found to cause tumors in the standard carcinogenicity protocol in rodents. Some are known to be human carcinogens as well. Not all compounds that are carcinogenic in the rodent model are carcinogenic in humans. These data are taken from a report of the American Council on Science and Health (www.acsh.org).

[b]Safrole is abundant in sassafras, no longer approved for use in human foods. It is a very minor constituent of the other flavorings listed here.

[c]Nitrate itself is not a carcinogen. However, nitrates from foods rich in them are acted on by gut microflora and human enzymes to yield nitrite and nitrosamines. The latter are known carcinogens.

[d]Coffee is known to contain at least 1000 different chemicals. Of these, 27 have been tested for carcinogenicity in the rodent model. Nineteen of these 27 have proved to be positive. There is little or no evidence that drinking coffee poses a cancer risk in people.

The observations made on effects of environmental agents on populations have been followed up in the laboratory. For example, if extracts of tobacco or coal tar are painted onto the skin of experimental animals, tumors develop. Exposure of experimental animals to X-rays or radioactive materials induces tumor formation.

The Ames test is useful in identifying potential carcinogens

Demonstrating that a particular chemical or substance is carcinogenic or is suspected to be carcinogenic in people is usually quite difficult. The standard test is to administer as much of the test compound in some suitable vehicle, say olive oil, as the animals will tolerate to both sexes of rats and mice for their lifetime, about 21 months for mice and 24 months for rats. A control group is treated with the vehicle only. As animals die, they are examined for tumors in all major organs. At the end of the study, the remaining animals are sacrificed and similarly examined. Statistical comparison of the treated and control groups is used to search for evidence of carcinogenicity in the test compound. There are a number of possible outcomes: no evidence of tumors, tumors in one organ in one sex of one species only, tumors in one organ in both sexes of both species, tumors in multiple organs in both sexes of both species, and so forth.

These are expensive and time-consuming studies. Beyond that, they do not always yield clear and compelling results. It would clearly be good to have a much simpler test that could be applied early in research and development efforts, even if the answer was not definitive. Bruce Ames devised such a test.

The Ames test actually searches for mutagens, which are molecules that cause mutations in organisms. To identify these, he devised a test using a suitable strain of *Salmonella*. For example, suppose we take a strain of *Salmonella* that is susceptible to the antibiotic streptomycin. Now if we plate a sample of this *Salmonella* onto agar containing streptomycin, nothing will grow. The antibiotic simply kills the bacteria. Now let us treat the *Salmonella* with a test compound. If it is a mutagen, we will expect some fraction of the mutants to become streptomycin-resistant. Those mutants will grow on the streptomycin-containing agar and we can ascertain their number by counting the colonies that grow. The more colonies we observe, the stronger the mutagen. That is basically the Ames test.

Ames was able to demonstrate a good correlation between mutagenicity in his test and carcinogenicity in the rodent model just discussed. Molecules that were positive in one test were generally positive in the other; those that were negative in one were generally negative in the other. Thus, the Ames test proved highly useful. It quickly became the practice to study compounds having promise of some utility in the Ames test. Those that came up positive are almost always discarded. No one wants to spend 2 years and half a million dollars doing a carcinogenicity test on a compound in rodents that is positive in the Ames test.

Now there are two points here: first, the Ames test is useful; second, mutagenicity correlates with carcinogenicity. The second point provides a strong clue as to the molecular origins of cancer. However, the correlation between mutagenicity and carcinogenicity is not perfect. For example, asbestos and alcohol, both known carcinogens, are negative in the Ames test.

Certain viruses are also known to cause cancer

It has been known for nearly a century that viruses can cause cancer. Peyton Rous carried out the pioneering study at the Rockefeller Institute, now the Rockefeller University, in 1911. Here is the experiment. Rous took tissue samples from connective tissue tumors (fibrosarcomas) in chickens, ground them up, and prepared extracts of them. The extracts were passed through the finest filters available at the time, good enough to eliminate all bacteria. When these filtered extracts were injected into chickens, they developed tumors. Clearly, something in the tumor extract retained the capacity to induce new tumors. When those new tumors were, in turn, removed, ground up, filtered, and injected into chickens, tumors again developed. Later work demonstrated that the causal agent is a virus, a particle small enough to pass through the very fine filter that Peyton Rous used. This virus is now known as Rous sarcoma virus, RSV. The importance of this work was not recognized for many years. Peyton Rous finally was rewarded with a share of the Noble Prize in Medicine or Physiology in 1966, more than half a century after his pioneering study.

Following identification of the Rous sarcoma virus, a substantial number of additional tumor-causing (oncogenic) viruses have been identified. We are left with the key question of how to reconcile two observations: on the one hand, chemicals or chemical substances cause cancer, on the other hand, viruses cause cancer. These observations split the oncology community into two camps. As frequently happens in science, neither camp had the full story and the two opposing viewpoints proved to be entirely compatible.

The Rous sarcoma virus proved to be a retrovirus, one in which the genetic material is carried as RNA. A number of oncogenic retroviruses have been identified, though not all retroviruses are oncogenic. All retroviruses have a set of genes in common: *gag, pol,* and *env.* These genes code for proteins that are intimately involved in viral replication. Rous sarcoma virus and other transforming viruses also share a gene termed *src,* for sarcoma. This gene is not present in nontransforming retroviruses. It has been demonstrated that *src* is the gene responsible for transforming activity and the product of this gene, the protein Src, has been identified. It is a protein tyrosine kinase: that is, it catalyzes the ATP-dependent phosphorylation of suitable protein substrates on tyrosine residues, usually one specific tyrosine. *Src* and other genes that have the power to transform cells are known as oncogenes and the proteins for which they code are termed oncoproteins.

A major step forward in the understanding of the origins of transformation of normal cells into cancer cells was made in 1976. This advance came from the laboratories of Harold Varmus and J. Michael Bishop at the University of California, San Francisco. They developed a technique for detecting the *src* gene and set about searching for it. They found it in oncogenic viruses and in mammalian tumor cells. Unexpectedly, they made the remarkable finding that the *src* gene is present in *normal* cells. Indeed, subsequent work revealed that the *src* gene is present in the cells of all vertebrates. The viral gene is frequently identified as v-*src* and the cell gene as c-*src*. The cellular gene became known as a proto-oncogene. This finding strongly suggests that the product of the *src* gene plays some vital role in cell biochemistry and physiology since it has been retained through 600 million years of evolution. Varmus and Bishop shared the 1989 Nobel Prize in Medicine or Physiology for their ground-breaking discovery.

The discovery of proto-oncogenes changed everything

Each cell of the human body carries the potential to become cancerous in the form of its proto-oncogenes. The case of the *src* gene is far from unique. A great many proto-oncogenes have been discovered: some of the best known are *myc, myb, ras, fes, fms, fos,* and *jun* (these names derive from the retrovirus in which they were discovered; e.g., *ras* comes from a rat sarcoma virus and *fes* from a feline sarcoma virus). It follows that each of our cells harbors a number of proto-oncogenes, each of which may become activated and contribute to the formation of a tumor.

The oncogenes of retroviruses are not, basically, viral genes at all. They are genes of mammalian cells that have been stolen by viruses in the process of viral cell infection, activated in some way by the virus, causing the virus to become oncogenic.

It is frequently true that new technologies provide the means to generate new insights into nature. In the 1970s, development of gene transfer technology provided an opportunity to hunt for cellular oncogenes in human tumors. Simply put, gene transfer technology permits the scientist to extract DNA, and therefore genes, from one cell and to insert the genes into another cell. If the genes in the donor cell carry some attribute lacking in the recipient, it may show up as an acquired property of the recipient cell. This provides the means for searching for oncogenes in human or animal tumors.

In 1979, Robert Weinberg at Massachusetts Institute of Technology extracted the DNA from mouse cells that had been transformed by a coal tar carcinogen. When this DNA was injected into a special line of mouse cells, termed 3T3 cells, they were transformed. When DNA from normal cells is injected into mouse 3T3 cells, there is no transformation. This remarkable finding clearly established that animal tumors created by chemical carcinogens, rather than by viruses, possessed oncogenes. These oncogenes must arise by the action of chemical carcinogens on cellular proto-oncogenes. This was quickly followed up with the discovery that human tumors also carry oncogenes. DNA from a number of human tumors proved to have transforming power when injected into mouse 3T3 cells.

Let's pause for a moment and see where we have come to. The central finding in the story thus far is the existence of a family of proto-oncogenes in the DNA of normal mammalian cells. These proto-oncogenes may be converted to active oncogenes through mutagenesis by chemical carcinogens or by radiation. These proto-oncogenes may also be picked up and activated by invading retroviruses, converting them into oncogenes. So we are left with the question of the relationship between the cellular and viral oncogenes.

The technique of gene cloning provided the means to address this question. The *ras* oncogene cloned from a rat sarcoma virus and an oncogene cloned from a human bladder carcinoma proved to be virtually identical. This observation pulled together the threads of the cancer story generated by virologists, on the one hand, and the chemical carcinogenesis community, on the other. As Robert Weinberg puts it:[5]

> So the repertoire of proto-oncogenes discovered by the retrovirologists did have direct relevance to human cancer. The same normal genes that were activated in animals by retrovirus capture and remodeling could serve as targets for mutagenic chemicals in humans. While sitting in place amid the chromosomes of a target cell, these human genes could be altered by mutagenic molecules, remaking them into potent oncogenes.
>
> The plot seemed to be getting much simpler. All vertebrate cells seemed to carry a common set of proto-oncogenes. These genes could become converted into potent cancer-causing genes either by retroviruses or by nonviral mutagens. Proto-oncogenes seemed to represent the ultimate root causes of cancer.

We are clearly making progress in piecing together the story of the molecular basis of cancer. We also have some distance yet to travel along this road. Here is the immediate question that comes to mind at this point in the story. What is the relationship between the proto-oncogenes present in normal cells and the derived oncogene in cancer cells? There is no simple single answer to this question and I will develop several scenarios in what follows. One scenario is provided by the compelling story of the *ras* oncogene.

There is a small but critical difference in the *ras* proto-oncogene and oncogene

It is clear that the cellular proto-oncogene c-*ras* and the viral, or cellular, oncogene v-*ras* must be somehow different. The former has no ability to transform normal cells into cancerous ones while the latter certainly does. The difference in the two genes came from DNA sequencing work. The *ras* gene is about 5000 base pairs long. The two genes are identical throughout this entire stretch of DNA, with the exception of one change: a triplet in c-*ras* that reads GGC is replaced by one in v-*ras* that reads GTC. So the whole story lies in the replacement of one G by a T. The biological consequences of this simple change are profound: normal versus cancerous cells.

GGC codes for glycine; GTC codes for valine. It turns out that the corresponding position in the Ras protein is 12. So replacement of Gly^{12} by Val^{12} in Ras is a root cause of carcinogenesis. As it turns out, replacement of this glycine residue by any other residue in Ras generates an oncoprotein. Clearly, Gly^{12} is critical to the function of this protein, about which more follows below.

Cancer is a disease of multiple mutations

So far, our story has focused on mutated single proto-oncogenes—for example *src* or *ras*—as key entities in the generation of cancer. A little reflection will convince us that the story cannot be that simple.

To begin with, cancer is usually a disease of old age. Typical victims are in their sixties or beyond, although there are obvious exceptions. Now let us suppose that a single mutation in some proto-oncogene were sufficient to cause cancer. Since such a mutation could occur with equal probability at any age, we would expect to see a simple linear increase in cancer with increasing age. Thus, the probability of having some tumor at age 60 years old would be twice as great as at age 30 years old. There is a great body of epidemiological evidence that is contrary to this simple expectation. The chances of dying from colon cancer at age 80 years old are perhaps 20 times greater than the same chance at age 40 years old. This strongly suggests that a succession of mutations may be required and that these will develop independently over the life span of the individual.[6] A final mutation may then trigger the disease.

Mutations in tumor suppressor genes underlie familial susceptibility to cancer development

There is an additional dimension to this story: the concept of tumor suppressor genes. Tumor suppressor genes do exactly what the name implies. They are in some sense the inverse of oncogenes but with a critically important difference. Oncogenes are dominant at the cellular level. That is a sophisticated way of saying that only one of the two cellular copies of an oncogene needs to be mutated to be a causative cancer agent. Mutant oncogenes are rarely manifested in terms of familial susceptibility for cancer development, since such mutations are generally lethal to the embryo. In contrast, tumor suppressor genes are recessive at the cellular level. Thus, you usually need to lose, through mutation, the integrity of both copies of a tumor suppressor gene to abrogate its tumor suppressor function. Mutations in tumor suppressor genes are generally not embryo lethal and, therefore, people may be born with one defective and one intact tumor suppressor gene. Such people will have a familial predisposition for cancer development. Three specific examples follow shortly.

In an individual with two functional copies of a tumor suppressor gene in each cell, the probability that both copies of this gene will be mutated in the same cell is low. That individual is at relatively low risk for cancer development. Now, if we have another individual who inherits a defective tumor suppressor gene, one copy of that gene will be defective in every cell. There is a rather high probability that, sometime over a lifetime, the single functional copy of that gene will be mutated in some cell at some time. That event may lead to cancer. So it is generally mutations in tumor suppressor genes that underlie familial susceptibility to develop cancer at some point in life.

Development of colon cancer is associated with a series of defined mutations

Uniquely among human tumors, the several stages of development of colon cancer have been correlated with specific mutations. This remarkable story is largely the work of Bert Vogelstein at the Johns Hopkins School of Medicine.

Colorectal cancer evolves through a series of well-defined morphological stages. First a polyp forms, then a benign precancerous tumor, followed by adenomas, and finally a malignant carcinoma that may then metastasize. Tissue samples of each stage along the way to carcinoma have been collected and examined for genetic alterations. Specific mutations have been associated with each stage of tumor development. Loss of a tumor suppressor gene is followed by activation of an oncogene, that is followed by successive loss of two more tumor suppressor genes, and so on. Although not all colorectal cancers reveal all these mutations, the correlation of each stage with a specific mutation is remarkable.

There is a disease termed hereditary polyposis. Victims develop a very large number of polyps in their colons, sometimes several thousand. Since the development of a colon polyp is the first step on the road to colon carcinoma, they are at high risk of colon cancer. The associated mutation is in a gene known as the *APC* gene.

The *APC* gene is a tumor suppressor gene. The development of colon polyps requires that both copies of the *APC* gene be defective. Normals have two copies of the intact *APC* gene and so mutations must occur in both copies to permit polyp formation. That is an unlikely combination of events and colon polyps are fairly rare in the population. People with hereditary polyposis inherit one defective gene from one parent and one intact, functional *APC* gene from the other parent. Now it requires just one mutation in the *APC* gene to lead to polyps, a far more likely scenario. Indeed, colon polyps are abundant in these individuals. In a certain sense, people that inherit a defective gene from a parent have a head start on others in terms of getting cancer. Every cell in their body will harbor a defective gene. So a second mutation in any cell has the potential to result in a tumor.

A second example of inherited predisposition to cancer is provided by hereditary retinoblastoma. This is a rare tumor of the retina and, when it develops, it generally develops in one eye only. Nonetheless, in children with hereditary retinoblastoma, the cancer develops early in life and generally in both eyes. This is a striking example of the inherited predisposition to develop a tumor. As in the case of hereditary polyposis, the defect is in a tumor suppressor gene, in this case *RB1*. Here too patients have just one intact gene in each of their cells and a single mutation in that gene may be all that is required to initiate tumor development.

Breast cancer provides a final example. There are two genes known which predispose women to development of breast cancer: *BRCA1* and *BRCA2*. In women with two intact *BRCA1* genes, there is about a 2% chance of developing breast cancer by age 50 years old. For women who have only one intact *BRCA1* gene, that figure rises to 60% by age 50 years old.

Vogelstein and Kinzler have collected a list of the known cancer predisposition genes that can be inherited in mutated form.[7] There are about 50 of these and more are certain to be found. There is a separate set of known cancer predisposition genes that are not inherited but result from some mutation event in somatic cells. There are about 50 of these as well and, again, more are certain to be found.

Remember that we started out this section to establish that cancer is usually a disease of multiple mutations, not just one. We have now cited three strong lines of evidence to support this conclusion. First, the incidence of cancer increases exponentially with age, not linearly. Second, specific mutations have been associated with the successive stages of development of colon cancer. Third, some people inherit a predisposition to develop cancer, reflecting a single mutation, but do not inherit cancer itself.

In sum, the development of tumors generally requires a minimum of two gene mutations and far more often three or more, perhaps as many as 20 in some cases. In general, fewer mutations are required in liquid tumors to cause cancer than in solid tumors. The need for multiple mutations provides a formidable barrier to development of cancer. Were a single mutation sufficient, the family of diseases we call cancer would be far, far more prevalent than it is.

The knowledge of the human genome sequence, coupled with the ability to very rapidly sequence DNA, permits the identification of genetic alterations in cancer in an amazing way. Analysis of 13,023 genes in 11 breast and 11 colorectal cancers revealed that there are, on average, about 11 mutated genes per tumor that contributed to the neoplastic process.[8] Most of these mutations, about 90 all told, were not previously known to be genetically altered in tumors. It is abundantly clear that (a) we have a lot to learn yet about tumorigenesis and (b) we continue to identify targets for diagnostic and therapeutic intervention in cancer. There remains a lot of work to be done but new insights are being generated rapidly and prospects for better days for cancer victims are bright.

By now, we have put together a lot of the cancer story. We know that cancer is a disease originating in a single cell that has overcome the normal limits to its growth and proliferation and that mutations in proto-oncogenes and/or tumor suppressor genes are required for this to happen. Beyond that, we know that multiple such mutations are usually required.

Here are some insights into the functions of tumor suppressor genes and proto-oncogenes

Tumor cells differ from normal cells in one dramatic way: they have lost the susceptibility to normal controls on cell proliferation. It should not surprise us then to learn that tumor suppressor genes and proto-oncogenes fall into one of three key categories. First, some oncogenic mutations directly affect cell proliferation. Second, other oncogenic mutations lead to loss of cell cycle control. Third, still other oncogenic mutations lead to genomic instability.

Since control of cell proliferation is key to development of cancer, we need to pause here for a bit and consider how cell division is regulated. The cell cycle is one of the most tightly regulated events in cell physiology. In its progress through the cell cycle, the chromosomes duplicate, segregate, and cell division occurs. Careful control of this process is required for development and differentiation in multicellular organisms.

This is a precisely choreographed process. Each stage involves complex regulatory mechanisms to ensure that progress through the cell cycle is accomplished one step at a time, is unidirectional, and permits exactly one round of DNA synthesis, resulting in precise duplication of the cellular chromosomes. It is a beautiful, if complex and not yet fully complete, story.

Here are the key players in control of cell proliferation

Let us continue by summarizing those molecular entities involved in cell growth and proliferation. We have (a) growth factors, (b) growth factor receptors, (c) molecules involved in signal transmission from growth factor receptors to the nucleus, and (d) transcription factors in the nucleus that control gene expression. There are oncogenes that are involved with each of these stages.

To understand the function of oncogenes, Vogelstein and Kinzler have come up with a neat automotive analogy:[9] "Oncogenes are mutated in ways that render the gene constitutively active or active under conditions in which the wild-type gene is not .... A mutated oncogene is analogous to a stuck accelerator in an automobile; the car still moves forward even when the driver removes his foot from it." Thus, a mutated oncogene will confer a growth advantage to a cell.

Signaling pathways are central to cell growth

To understand something about how this works, I need to describe a signaling pathway associated with cell growth. There are at least eight of these; I shall describe one—the receptor tyrosine kinase signaling pathway—as an example, omitting a fair amount of detail. It is summarized in figure 24.1.

The story begins with a family of growth factors: epidermal growth factor, EGF; platelet-derived growth factor, PDGF; vascular endothelial growth factor, VEGF; fibroblast growth factor, FGF; among several others. Each of these is a protein. Each of these growth factors possesses specific affinity for its corresponding receptor, confined to the cell surface membrane. For the case considered here, each of these receptors has protein tyrosine kinase activity when associated with its cognate growth factor: that is, when activated by their growth factor, they catalyze the ATP-dependent phosphorylation of proteins, in this case themselves, on tyrosine. So the immediate consequence of association of a growth factor with its receptor is the activation of that receptor by addition of a phosphate group to a specific tyrosine residue. There follows the cascade of activation events shown in figure 24.1, creating activated transcription factors that then find the appropriate sites on genomic DNA. Transcription of growth-promoting genes follows, leading to cell division.

This signaling pathway is complex but the details do not matter. The basic point is that there is an ordered cascade of molecular events, beginning with the association of growth factors with their tyrosine kinase receptors and leading to altered control of the transcription of genes involved with the promotion of cell division and differentiation. There are numerous places in this signaling pathway where there are opportunities for genetic changes to influence cell division and, hence, to facilitate tumor development.

The *sis* oncogene encodes a protein that is a variant of the growth factor PDGF. Consequently, expression of this oncogene stimulates proliferation of cells bearing the PDGF receptor. This is a unique example of an oncogene encoding a growth factor.

A variation on this theme is provided by cases in which mutations in a growth factor receptor simply turn them on in the absence of any effector molecule, including

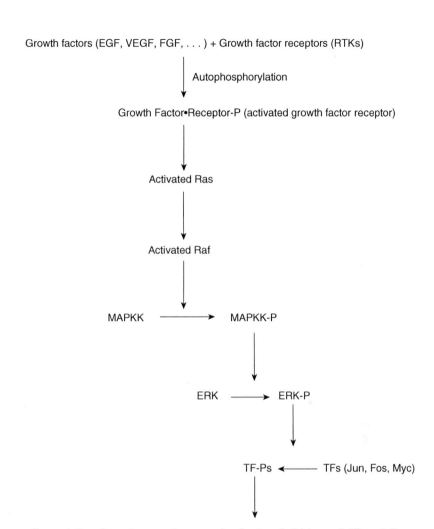

Growth factors (EGF, VEGF, FGF, . . .) + Growth factor receptors (RTKs)

Autophosphorylation

Growth Factor•Receptor-P (activated growth factor receptor)

Activated Ras

Activated Raf

MAPKK ⟶ MAPKK-P

ERK ⟶ ERK-P

TF-Ps ⟵ TFs (Jun, Fos, Myc)

Transcription of growth-promoting genes leading to cell division and differentiation

Figure 24.1 The receptor tyrosine kinase signaling pathway. When growth factors associate with their membrane-bound tyrosine kinase receptors (RTKs), they are activated by phosphorylation on tyrosine. The activated RTKs in turn activate Ras, which then binds to a serine/threonine kinase known as Raf and activates it. Activated Raf then phosphorylates mitogen-activated kinase kinase (MAPKK), activating it. Activated MAPKK then phosphorylates extracellular-signal-related kinase (ERK), activating it. In turn, activated ERK phosphorylates a family of transcription factors (TFs), activating them. The activated transcription factors then interact with appropriate sites on genomic DNA with the result that growth-promoting genes are transcribed. Cell division and differentiation then occur. Phosphorylated proteins are denoted by a P.

the growth factor itself. They are simply "on" all the time. Cell proliferation and transformation result. A good example is provided by the *neu* oncogene. A single point mutation in the gene that codes for the Her2 receptor generates the Neu oncoprotein. The *neu* oncogene is an important player in development of breast cancer.

The protein Ras encoded by the *ras* proto-oncogene is, as we have seen, a signal transmission element. Mutant Ras, in which Gly^{12} is converted to any other amino acid, is always "on" and there is too much transmission of the growth factor signal to transcription factors.

Finally, inappropriate expression of nuclear transcription factors can lead to cell transformation. For example, the products of the *fos* and *myc* proto-oncogenes are transcription factors that regulate the expression of proteins that promote progression through the cell cycle. Levels of the Fos and Myc proteins are tightly regulated in normal cells. Uncontrolled expression of these proteins leads to cell proliferation.

In sum, there are oncogenes that reflect every stage of this signaling pathway: those that are growth factors; mutant growth factor receptors that are "always on"; signal transmission elements that are "always on"; and inappropriate transcription factor expression. In each case, normal controls on the rate of cell division have been overcome.

Now let's think a little more about the converse of oncogenes, the tumor suppressor genes. Before getting any further, I return to the automotive analogy of Vogelstein and Kinzler: "A mutation in a tumor-suppressor gene is analogous to a dysfunctional brake in an automobile; the car does not stop even when the driver attempts to engage it."

One of the most important of the tumor suppressor genes is *Rb*. Mutant *Rb* genes are found in many human tumors, including retinoblastoma (as cited earlier). The Rb protein is a potent inhibitor of cell division. Loss of Rb function will lead to abnormal cell division and, ultimately, to cell transformation.

The product of the *p53* gene plays a key role in protecting and maintaining the integrity of the human genome. Under normal conditions, cells have little of the unstable p53 protein. In contrast, conditions that stress the cell and threaten the genome—UV irradiation, γ-irradiation, and low oxygen—cause p53 to be activated. The consequence is that the cell is arrested in the cell cycle. This gives the cell adequate time to repair whatever damage may have occurred to the genome. Once the damage is repaired, the cell can proceed to complete the cell division process. The progeny cells will inherit the repaired, functional genome. The p53 protein works by acting as a transcription factor and inducing the synthesis of a number of proteins.

It seems reasonable that mutations that render p53 nonfunctional should predispose to cancer. A nonfunctional p53 will permit cells with damaged genomes to bypass the cell cycle checkpoint and generate defective progeny cells. In fact, about 50% of all human tumors have mutations in the *p53* gene, a striking finding.

Finally, let us focus for a moment on those genes that code for the enzymes involved in DNA repair: stability genes. To return to Vogelstein and Kinzler one last time: " Stability genes keep genetic alterations to a minimum, and thus when they are inactivated, mutations in other genes occur at a high rate ... stability genes represent the mechanics and a defective stability gene is akin to an inept mechanic."

Mutations in the genes that code for DNA repair enzymes, and which result in nonfunctional products, should also predispose to cancer. Indeed they do.

Mutations in the PI3K signaling pathway are critically involved in carcinogenesis

The Ras to Raf to Erk signaling pathway depicted in figure 24.1 is important but it is not the most important signaling pathway associated with development of human tumors, particularly those tumors that are most frequently fatal to cancer patients. That honor goes to the phosphatidylinositol-3-kinase, PI3K, signaling pathway. This pathway is complex and I shall not describe it.

Here is evidence for the importance of this signaling pathway in cancer development. About 25% of all breast and colorectal cancers have activating mutations in the gene that encodes PI3K. A third of all lung cancers have the gene for PI3K amplified. Loss of the gene encoding the negative regulator of PI3K (a phosphatase known as PTEN that degrades the product of PI3K action) is nearly as frequent as the loss of *p*53 in human cancers. Activation of PI3K also prevents cell death and promotes cell migration, leading to metastasis. The argument about the central role of the PI3K signaling pathway in human carcinogenesis is compelling.

Perhaps the most optimistic part of this story is that PI3K is a tractable target for drug discovery efforts. Given the central role of this enzyme in cancer development, it will come as no surprise that many pharmaceutical companies have chosen to focus drug discovery and development efforts on it. The immediate fruit of this investment is a family of small molecule inhibitors of PI3K that will enter clinical trials shortly. Among many others, I await the outcome of these clinical trails eagerly.

Insights at the molecular level will lead to better cancer therapy

The story of the origins of cancer related above is remarkable. For the first time in history, we know the basis of cancer at the molecular level. It is certain that we do not know all that there is to know or all that we would like to know. At the same time, the origin of certain cancers is understood in amazing detail. We know what the molecular errors are and how they lead to uncontrolled cellular proliferation. The remarkable insights that have been gained will lead to chemical weapons against cancer that target the underlying problems. Progress will not come easy and it may not come rapidly but it will come. Tomorrow will be a better day for victims of cancer, thanks to insights at the molecular level.

Think for a moment at what we had to know in order to understand the molecular basis of tumor formation. We had to understand genes and gene structure and have the technology to pinpoint defects in specific genes within the human genome. We had to understand the biology and biochemistry of viruses and their genes. We needed to know that certain chemicals could introduce somatic mutations and understand what those mutations were. We needed to know about the enzymes that synthesize and repair DNA. We had to understand about the control of transcription. Understanding the control of the cell cycle was crucial. I could go on and on. Much of what we have developed in the earlier chapters of this book is relevant to understanding the molecular events leading to cancer.

It is important to realize that most of what we had to know to develop an understanding of the molecular basis of cancer came from studies on normal genes, normal enzymes, normal cells, normal tissues, and normal organisms. Once we had an understanding of how things worked normally, we could begin to understand what went wrong. There is an important general lesson here. Everyone should understand it.

Here are some approaches to cancer therapy

Cancer is generally treated in one of three ways: (a) surgery, in an effort to eliminate all tumor cells from the human body; (b) radiation, in an effort to kill all tumor cells; and (c) chemotherapy (drugs), in an effort to kill or prevent the growth of all tumor cells. It is common to follow, or sometimes precede, surgery with either radiation treatment or chemotherapy as well as to use both radiation treatment and chemotherapy alone or as adjuvant therapy for surgery. Beyond that, chemotherapy generally involves regimens that require two or more drugs. Cancer patients frequently receive analgesics to relieve the pain associated with cancer as well as drugs to control the nausea, vomiting, depletion of white cells, and depletion of platelets caused not by the cancer itself but by the drugs used to combat it.

If you think about the content of the above paragraph, you will come to a conclusion that you probably already know: that the treatment of human cancers leaves a lot to be desired. For some pathological states, we have a lot of drugs that work quite well: analgesics and antihypertensives, for example. For others, such as Alzheimer's disease, we have very few drugs and these do not work well. For cancer, we have many drugs and they do not work well. As a class, they are the most toxic drugs employed in medical practice. The use of highly toxic drugs is acceptable since cancer is frequently a life-threatening disease. Extreme measures, including the use of toxic drugs, may be ethically used to preserve life.

Many drugs used in cancer chemotherapy inhibit DNA synthesis or function

The most obvious difference between a normal cell and a cancerous one is that control of cell division has been lost in the latter. Since cell division requires synthesis of DNA, the synthesis, structure, and function of DNA have been frequent targets for discovery of antitumor agents. One of the problems with this approach is that some perfectly normal cells in the human body also turn over rapidly, including the cells in the bone marrow that eventually lead to the blood cells and the cells that line the gut. It is not surprising that the dose-limiting toxicity of many cancer drugs is damage to the bone marrow or the gut.

Let's start with three examples of antitumor drugs that work by the inhibition of DNA synthesis: methotrexate (MTX, Methopterin), 5-fluorouracil (Fluril), and 6-mercaptopurine (Purinethol). These are all old cancer drugs, marketed under numerous trade names, including those indicated above. Although the mechanism of action of each one is different, at the end of the day they all inhibit DNA synthesis.

The functions of DNA—coding for its own synthesis and for the synthesis of mRNA, tRNA, and rRNA—can be disrupted by altering, one way or another, DNA structure. There are two general ways to do this: first, by reacting chemically with DNA and, second, by intercalating between the stacked bases of DNA. Let's have a short look at both possibilities.

One class of cancer drugs that react chemically with DNA to alter its structure is the alkylating agents, some of which are nitrogen mustards. These are rather nasty chemotherapeutic agents. They include busulfan, cyclophosphamide, ifosfamide, and melphelan and several others including a class of compounds known as nitrosoureas. These are old cancer drugs, having been in use for many years, which continue to find wide application in the chemotherapy of cancer.

Cisplatin provides another interesting example:

$$H_3N \diagdown \diagup Cl$$
$$Pt$$
$$H_3N \diagup \diagdown Cl$$

cisplatin

This is an unusual drug in that it contains a metal atom, platinum (Pt) in this case. Cisplatin reacts with DNA to cross-link bases, disrupting normal DNA structure and function. This agent has found broad use in cancer chemotherapy, including efficacy in tumors of the testis, ovary, bladder, head and neck, thyroid, cervix, and endometrium. It is also active against neuroblastoma and osteogenic sarcoma.

Cisplatin was the first example of a platinum-based chemotherapeutic but it has been followed by others: carboplatin, oxoplatin, and cycloplatam. All are variations on the original theme and all find use in the chemotherapy of cancer.

Perhaps a bit more subtle than those agents that react chemically with DNA are those that insert themselves between the stacked bases of the DNA double helix—intercalation. This alters the regular structure of the DNA molecule and may lead, for instance, to inhibition of mRNA synthesis. The structures of the intercalcating agents are generally quite complex and I will spare you the complexity. However, three names may be familiar—dactinomycin (Actinomycin D), daunorubicin (daunomycin), and doxorubicin (Adriamycin)—and intercalation is how they work. All three are natural products and were isolated from the fermentation broths of *Streptomyces* species.

All the agents described so far have one thing in common: they all kill cells, normal as well as tumor ones, but preferentially rapidly dividing tumor cells. They are cytotoxics.

Finally, there are useful antitumor agents that focus on other targets. Among these are paclitaxel (Taxol), which inhibits the deaggregation of microtubules. In addition, the vinca alkaloids vincristine (Oncovin) and vinblastine (Velban) find substantial use in cancer chemotherapy. We noted in an earlier chapter that antiestrogens such as tamoxifen and antiandrogens such as flutamide also find use for control or prevention of specific tumor types. I could extend the list but it seems preferable to conclude this chapter on an upnote: the story of a cancer chemotherapy agent that directly targets the underlying molecular defect in a cancer.

Imatinib caused a revolution in the treatment of chronic myelocytic leukemia

One key exception to the nonspecificity of cancer drugs is *imatinib*, a molecule that is marketed under the trade names Gleevec in the United States and Glivec in Europe. Uniquely, imatinib corrects the causative defect in chronic myelogenous leukemia, usually known more simply as CML. Leukemias are a family of cancers characterized by proliferation of white blood cells. More specifically, CML is a form of leukemia characterized by excessive proliferation of blood cells of the myeloid lineage. So what has gone wrong in patients with CML?

Chromosomes are highly complex structures that house the genetic material (DNA) of eukaryotic cells. For the most part, chromosomes are reasonably stable structures. Once in a while, this stability is lost and chromosomes rearrange.

In the case of CML, there is a reciprocal exchange of material between chromosomes 9 and 22. Part of chromosome 9 is exchanged with part of chromosome 22 so that two hybrid chromosomes result:

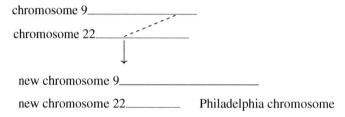

The new chromosome 22, which now has lost part of its original structure and has acquired part of the old chromosome 9, is smaller than the original one and is known as the Philadelphia chromosome, recognizing where it was discovered back in 1960.[10]

Fusing part of chromosome 9 with part of chromosome 22 results in the formation of a new gene on the Philadelphia chromosome, known as the *BCR-ABL* gene. This new gene encodes a new protein, an enzyme in the protein tyrosine kinase class: that is, there is an enzymatic activity present in cells possessing the Philadelphia chromosome that is otherwise absent. Although the details need not concern us, it is the enzymatic activity of this new protein that underlies the uncontrolled proliferation of the myeloid cells in the blood in CML patients. It has been known since 1990 that the creation of the *BCR-ABL* gene is the sole molecular event underlying CML.[11] The point is that the research leading to this insight tells us what to do to treat CML.

Since the enzymatic activity of the protein encoded by the *BCR-ABL* gene is responsible for the cancer, it is logical to search for ways to turn off that activity. The most straightforward way to turn off the activity of an enzyme is to discover an inhibitor of the enzyme. We need a molecule that specifically binds to the enzyme and turns it off. That is precisely what imatinib does; it specifically binds to the tyrosine kinase encoded by the *BCR-ABL* gene and abolishes its activity. In effect, imatinib converts the cancer cells of CML into the functional equivalent of normal cells: normal cells do not have the enzyme and the imatinib-treated CML cells do not have active enzyme. A dead enzyme is a lot like not having the enzyme in the first place.

Imatinib is a genuine breakthrough in cancer chemotherapy. It acts to correct the underlying molecular error in CML, which is rare among cancers in having a single genetic defect that causes the disease. As noted above, the vast majority of cancers have a family of defects (four to six molecular defects are common) that cause the disease. It follows that a drug molecule that corrects one of these defects may be useful, generally as a component of a cocktail of drugs, but cannot by itself correct all that has gone wrong on the pathway to cancer. Nonetheless, molecules that correct molecular errors leading to cancer are highly desirable and are being vigorously sought.

Imatinib is not a perfect drug. A small number of CML patients prove to be resistant to it at the outset and most CML patients eventually become resistant. The common mechanism of resistance development to imatinib involves a mutation in the imatinib-binding pocket of the BCR-ABL protein. This mutation alters the structure of the drug-binding pocket so that effective binding of imatinib can no longer occur. Resistance results. (Incidentally, this finding is what convinced scientists that imatinib really works by hitting the BCR-ABL enzyme.) There are already two additional molecules approved for use in CML that inhibit the tyrosine kinase activity of CML cells and that are effective in imatinib-resistant cells: dasatinib (Sprycel) and nilotinib (Tasigna). Other will surely follow.

Imatinib is a construct of chemists, not nature. It was discovered as one of a large number of molecules synthesized by chemists at Novartis, a large pharmaceutical company, in the search for one that combined all the properties necessary for use in human medicine.[12]

Imatinib is not alone as a targeted cancer drug. Gefitinib (Iressa) is an inhibitor of the receptor tyrosine protein kinase EGFR. It is approved for treatment of non-small-cell lung cancer (though only about 10% of these patients respond to it). Most of these patients have activating mutations in EGFR, arguing strongly that the drug works by hitting the intended target. Finally, a rather nonspecific inhibitor of protein kinases, sorafenib (Nexavar) is approved for treatment of advanced renal cell carcinoma.[13]

In addition to these three targeted small-molecule cancer therapeutics, there are three targeted large molecules approved for various uses. The first among these was trastuzumab (Herceptin), a humanized monoclonal antibody[14] targeted against the HER2 receptor. It is approved for treatment of metastatic breast cancer and has also been shown to be effective as a breast cancer prophylactic. Trastuzumab is an exceptionally important breast cancer agent.

Cetuximab (Erbitux) is a monoclonal antibody against the epidermal growth factor receptor, EGFR. It is approved for treatment of colorectal cancer. The development of Erbitux was much in the news in 2004 as a result of insider stock trading at ImClone in the face of an initial turndown by the FDA. The former head of ImClone, Sam Waksal, is spending a term of 7 years in jail for his role in the affair. Martha Stewart completed a jail term of several months in the same case. Erbitux was subsequently approved by the FDA on the basis of additional clinical information.

Finally, bevacizumab (Avastin), a humanized anti-VEGF monoclonal antibody has also been approved for treatment of colorectal and breast cancer. There are several more targeted anticancer agents in various stages of clinical trials and more will surely be approved and enter clinical practice. Things are looking up for cancer patients.

Key Points

1. A cancer is a clone of cells arising from a single cell. These cells have escaped the usual controls on cell division. Consequently, they have the potential to proliferate and differentiate without control.

2. Tumors may be benign (do not invade surrounding tissues and do not form distant colonies) or malignant (do invade surrounding tissues and have the capacity to form distant colonies—metastases).

3. Some environmental agents are carcinogenic: tobacco, coal tar, asbestos, alcohol, some polycyclic aromatic hydrocarbons, X-rays and other sources of radiation, and many natural products.

4. The Ames test is very useful in identifying potential carcinogens. It uses mutagenicity as a surrogate for carcinogenicity.

5. Some viruses are known to cause cancer; these include both DNA and RNA viruses, retroviruses. These oncogenic viruses possess oncogenes in their genome.

6. Remarkably, viral oncogene relatives are present in normal cells—these are termed proto-oncogenes. Therefore, each cell in the human body carries the potential to become cancerous in the form of its proto-oncogenes, which include *myc, ras, myb, fes, fms, fos,* and *jun.*

7. The action of chemical carcinogens or radiation on cellular proto-oncogenes has the potential to convert them to oncogenes. This is the molecular basis of environmental carcinogenesis.

8. The difference between the normal cellular proto-oncogene, c-*ras,* and the viral oncogene, v-*ras,* lies in a single base pair change among the 5000 base pairs in this gene.

9. Cancer is a disease of multiple mutations.

10. Mutations leading to tumor formation may occur in growth factors, growth factor receptors, molecules along signaling pathways leading to control of gene expression, cell cycle control genes, or DNA repair enzymes. These include the proto-oncogenes and tumor suppressor genes.

11. A predisposition to cancer can be inherited genetically.

12. There are three basic ways to treat cancer: surgery to eliminate all tumor cells; radiation to kill all tumor cells; and chemotherapy to kill all tumor cells or prevent their growth.

13. Many drugs used in cancer chemotherapy inhibit DNA synthesis or compromise its function.

14. Other anticancer agents have distinct mechanisms: paclitaxel and its relatives target microtubules as do the vinca alkaloids. Antiestrogens and antiandrogens target estrogen and androgen receptors.

15. Imatinib was the first anticancer drug that specifically targets a molecular defect in tumor cells. It is a breakthrough drug for chronic myelogenous leukemia (CML). Others have followed and more are yet to come.

25

Chemical communication

Molecular recognition underlies chemical communication, including the phenomena of taste and smell. Pheromones are chemical communicators between individuals, usually members of the same species. Pheromone action has been demonstrated in a number of mammalian species. Several of these affect reproduction. It seems likely that there are human pheromones. The role of these in human behavior is not clear.

I begin the final chapter of this book with a famous quotation from Marcel Proust:[1]

She sent for one of those squat, plump little cakes called "petite madeleines" which look as though they had been molded in the fluted valve of a scallop shell. And soon, mechanically, dispirited after a dreary day with prospect of a dreary morrow, I raised to my lips a spoonful of the tea in which I had soaked a morsel of the cake. No sooner had the warm liquid mixed with the crumbs touched my palate than a shiver ran through me and I stopped, intent upon the extraordinary thing that was happening to me. An exquisite pleasure had invaded my senses, something isolated, detached, with no suggestion of its origin. And at once the vicissitudes of life became indifferent to me, its disasters innocuous, its brevity illusory—this new sensation having had the effect, which love has, of filling me with a precious essence; or rather this essence was not in me, it was me. I had ceased to feel mediocre, contingent, mortal. Whence could it have come to me, this all-powerful joy? I sensed that it was connected with the taste of the tea and cake.

Marcel Proust occupies a distinguished spot in the pantheon of French writers. Born in a suburb of Paris in 1871, he spent his lifetime until his death in 1922 writing. He wrote a great deal. His great work—*A la recherche du temps perdu*—occupies seven very substantial volumes, three of which were published after his death. The quotation from the first volume of that work, *Swann's Way,* which opens this final chapter in our exploration of the molecules of life, is perhaps the most famous passage in all his work. Here Proust is overwhelmed by feelings of joy touched off by a taste of a madeleine dipped in a cup of tea. The episode subsequently evokes a series of childhood memories. Much of his early life is recreated for Proust simply through the repetition of a childhood flavor of tea and cake. This event illustrates the power of chemical communication in human physiology.

Although Proust assigns responsibility for the awakening of his feelings of joy to "the taste of the tea and cake," it is almost certain that the associated aromas contributed as well. The sense of smell, olfaction, is strongly associated with our sense of taste. People who cannot smell—a clinical condition termed anosmia—lose a significant amount of enjoyment of their food. There are many more genes in the human genome that code for odorant receptors than code for tastant receptors. For the moment, we can simply recognize that both receptors for odorants and tastants are involved in the chemical communication of the human senses of smell and taste.

Although most of us have not had an epiphany induced by taste and smell as powerful as that described by Proust, it is surely true that taste and smell influence our physiology, including our frame of mind. Think about your reaction to the aroma of baking bread, roses, lilacs, vanilla, fine wine, saffron, pine trees, cedar, incense, rosemary, cloves, strawberries, a ripe pear, coffee, your favorite perfume, cologne, or after-shave. Think about the taste of a grilled steak, a ripe tomato, dark chocolate, cold beer, hot coffee, iced tea, black raspberries, red onions, green peppers, or blueberries, though smell will contribute to each taste sensation.

Smell and taste matter to us. We spend hundreds of millions of dollars each year in efforts to smell good and avoid smelling bad: perfumes, colognes, after-shaves, antiperspirants, deodorants, room and car fresheners, and the like. My wife and I go to a health resort from time to time, mostly for the exercise opportunities and a better diet than we manage at home. However, there are a lot of ways to spend your time there: aromatherapy is an option. I entered the term "aromatherapy" in Google and got 7,230,000 hits. I chose the AromaWeb site (www.aromaweb.com/essential oils) more or less at random. There I found a substantial list of essential oils intended for emotional well-being, including bergamot, which is good for anger, anxiety, confidence, depression, fatigue, fear, happiness and peace, insecurity, loneliness, and stress; jasmine is recommended for anger, confidence, depression, fatigue, fear, insecurity, and stress; and ylang ylang controls anger, depression, and stress, and induces happiness and peace.

Taste matters as well. We seek out fresh fruits and vegetables for their taste. Restaurants compete to convince potential customers that they have the tastiest beefsteak. Have a look at the shelves in your grocery store: you may find 20 different barbecue sauces and an equal number of hot sauces. Mustard comes as yellow or brown, smooth or grainy, with wine or without, and Dijon or not. There are about as many different cheeses from France as there are days in the year, each with its own particular taste and aroma. We seek out diversity.

We are going to get back to the molecular basis for the human reactions to tastes and smells, examples of chemical communication, shortly. To get there, we need to start with substantially simpler systems of chemical communication and work our way up to more complex ones. As we move forward, we are going to focus on the consequences of small molecules interacting with large ones, molecular recognition, a pervasive theme of this book.

Here are a few words about molecular recognition

At the most basic level, complementarity of shape and surface properties underlies the phenomenon of molecular recognition. This is the most fundamental mechanism of communication effected by molecules. We have noted many examples earlier in this book. Perhaps the most famous is the complementary base-pairing found in DNA: adenine and thymine recognize each other and pair up across the DNA double helix as do cytosine and guanine. This complementary base-pairing underlies the ability of DNA to transmit genetic information from parent cell to progeny cell. Complementary base-pairing also underlies the synthesis of rRNA, mRNA, and tRNA molecules from a DNA template (known as transcription), and is also critical to the fidelity of protein synthesis from an mRNA template, (known as translation). Other key examples of molecular recognition include that of enzymes for their substrates, receptors for their effectors, and antibodies for the antigens that elicited them. Communication among cells depends on mutual recognition of molecules exposed on the cell surfaces. Molecular recognition underlies the ability of virus particles to invade cells. Life depends on molecular recognition.

Earlier, we encountered two examples of chemical communication. First, a small family of hydrocarbons secreted by the female brown algae gamete attracts a free-swimming male gamete. Fertilization follows. Second, bombykol, the sex attractant of the female silkworm moth, is a powerful lure for the male moth. What we need to do now is to expand on these examples. Let's begin with olfaction and taste in mammals. Later, we will move to the other end of the evolutionary scale and start with simple, unicellular organisms (the bacteria) and work our way back up.

The molecular basis of olfaction is understood

In humans and other mammals, the sense of smell begins when we inhale some odorant through our nose. The inhaled air enters the nasal cavity where it encounters a large number of olfactory neurons located in the nasal epithelium associated with bony structures located at the rear of this cavity. These bony structures are known as turbinates. In a human, these turbinates create a surface area of a few square inches. In a medium-size dog, in contrast, the turbinates have a surface area several times larger. It is small wonder that dogs have a more acute sense of smell than we do.

The sense of smell in humans is not limited to detection of those volatile molecules inhaled through the nose, termed orthonasal olfaction. Molecules released at the back of the mouth, particularly in the chewing of food, can make their way up through the nasopharynx to the olfactory epithelium, termed retronasal olfaction.[2] This system is activated when air is exhaled. Orthonasal olfaction is used to detect the scent of flowers and perfumes, food aromas, the presence of skunks, and the like. Retronasal olfaction detects the volatile molecules released from food. It is retronasal olfaction that makes a major olfactory contribution to the taste of food. And it is retronasal olfaction that helped to elicit Proust's profound reaction to a madeleine dipped in tea.

The question that I want to explore here is what happens between inhaling some odorant in the environment or releasing it from food and the recognition of an aroma. Specifically, what are the molecular events that permit you to know, say, that the lilacs are in bloom (without peeking)?

Richard Axel and Linda Buck published the most important scientific paper in the history of the molecular basis of olfaction in 1991 in the molecular biology journal *Cell*.[3] At the time, Axel was a Professor at Columbia University in New York and Buck was an advanced postdoctoral fellow in his laboratory. Linda Buck has gone on to create a distinguished record as an independent scientist, first at Harvard University and later at the Fred Hutchinson Cancer Research Center in Seattle. In the 1991 publication, Axel and Buck reported a large family of genes in the olfactory epithelium of the rat believed to encode receptors for odorants. This finding was a genuine landmark in the history of olfaction. Subsequent research in the area, and in that of taste receptors, moved quickly, particularly spurred by large-scale genome sequencing projects later in the 1990s and since. It turns out that there are about 1200 distinct, functional odorant receptor genes in rats, 900 in mice, 400 in chimpanzees, and 350 in humans.[4] Fish and birds generally have 100–200, roundworms such as *Caenorhabditis elegans* 1000, the *Anopheles* mosquito has 78, and the fruit fly *Drosophila melanogaster* has 62.[5] Fruit odors activate about two-thirds of the odorant receptors in the fruit fly, which should not be too surprising. The genes for odorant receptors constitute the largest gene repertoire identified so far. About 5% of all the genes in rats, 4% of all the genes in mice, and 1.4% of all the genes in humans code for odorant receptor proteins. For their groundbreaking work on the molecular biology of olfaction, Axel and Buck shared the 2004 Nobel Prize in Physiology or Medicine.

The proteins coded for by these genes in vertebrates are G protein-coupled receptors (GPCRs). They are linear chains of amino acids that span the olfactory epithelial membrane in which they are embedded seven times. They act by transmitting the odorant signal to specific olfactory G proteins, G_{olf}.

The olfactory epithelium of mammals contains many types of olfactory neurons, each expressing a specific odorant receptor. Linda Buck has shown that an odorant can activate multiple distinct receptors and that a receptor can be activated by multiple odorants. Thus, there must exist a combinatorial mechanism for odor detection: some sort of pattern recognition.[6] The axons of olfactory neurons converge on glomeruli in the olfactory bulb. There, incoming signals are integrated and the sense of smell is created.

Thus, the process of olfaction in humans begins with more-or-less specific molecular interactions between the small odorant molecules, say the 200–300 in the aroma of coffee, and the 350 macromolecular receptor proteins in the olfactory epithelium. Each odorant binds to some subset of all the receptor proteins and each receptor protein binds to some subset of all the odorants. Each binding interaction causes some change in the properties of the receptor protein that, ultimately, tells us that it is coffee in the air, not roses or incense or the perfume Opium.

Something is known about the events following binding of an odorant to its receptor and generation of an electrical signal to the olfactory bulb. Following activation of a receptor, G_{olf} proteins are activated. These activate the enzyme adenylate cyclase, which catalyzes the synthesis of cAMP, a second messenger that we have encountered several times before. cAMP opens a ligand-gated calcium channel in the epithelial membrane, initiating membrane depolarization, which, in turn, opens a calcium-gated chloride channel completing membrane depolarization. That produces the required electrical signal along the olfactory nerve axon. This is a complex process but, at the end of the day, it works: we can smell the lilacs, jasmine, coffee, postthunderstorm air ....

Like many odors, that for coffee has many individual contributors. Nonetheless, we do not, indeed cannot, identify the individual compounds that contribute to the aroma. All we get is the smell of coffee. With a single odorant, say jasmine, many of us can identify it correctly. But if an aroma contains two components, only about one person in six can identify both correctly. And only about one person in 25 can unravel a mixture containing three components. Nobody gets it right if the aroma contains four or more components. Our brain basically integrates all the inputs and gives us a single sensation, not multiple, individually identifiable sensations.

Beyond that, the sense of olfaction does not depend on the concentration of the odorant: concentration invariance. If you are exposed to jasmine at very low concentration, it smells like jasmine; if the concentration is significantly raised, it still smells like jasmine. Perhaps more to the point is the concentration invariance of complex aromas such as that of coffee. The brain forms a single perception from complex inputs, regardless of the intensity of the signal. Olfaction has this property in common with taste.

From our examples encountered thus far, we know that our chemical senses of smell and taste must be ancient. From bacteria to algae to insects and beyond, through the process of evolution, capacities to find food or mates and avoid predators through sensing of the chemical composition of the environment have been developed. Here we have a small surprise. The odorant receptor genes have now been cloned and sequenced from a number of organisms, including mammals, worms, and flies. There is no sequence homology among the genes for odorant receptors between vertebrates, worms, and insects. Thus, modern odorant receptor genes have not derived from some family of ancestor genes through divergent evolution. Instead, different organisms have found different solutions to the same problem. Beginning from quite distinct starting places, different organisms have come to structurally distinct, functional solutions to the problem of developing useful olfactory systems.

A subset of all odorants is pheromones, about which much more follows later. Basically, pheromones transmit chemical messages among members of the same species. Bombykol is a pheromone for the silkworm moth; the scent of lilacs as perceived by a human being is not. Although the question of human pheromones is difficult, that for many mammals is not. The pheromones in mammals are not detected primarily by the nasal epithelium but by a separate organ known as the vomeronasal organ (VNO), located at the base of the nasal septum. In the mouse, for example, the murine genome encodes another 300 genes that code for pheromone odorant receptors in vomeronasal sensory neurons.

Some natural products have an aroma, and taste, that derives largely from a single compound. For example, oil of wintergreen is about 99% methyl salicylate:

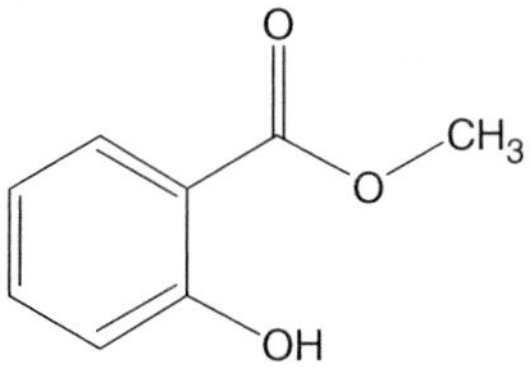

methyl salicylate

Methyl salicylate is a mildly toxic compound and plants that make it may do so as a means of warding off predators. It finds use in human medicine as a constituent of liniments, topical anesthetics.

The principal aroma and flavor agent of Concord grapes, methyl anthranilate, provides a related example:

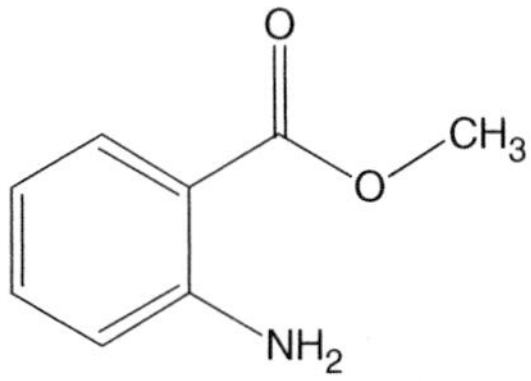

methyl anthranilate

Here we see yet another example of the sensitivity of biological properties to molecular structure: the only difference between methyl salicylate and methyl anthranilate is the replacement of a hydroxy group (OH) by an amino group (NH$_2$). Wintergreen does not smell like grapes. Wintergreen does not taste like grapes. Methyl anthranilate is, in fact, found rather widely in plants: orange flowers owe a part of their aroma to this molecule.

In contrast to the rather simple cases of wintergreen and grape aromas, that for roses is complex. Rose oil contains at least 275 chemical constituents, of which citronellol is the major one. However, two of the minor constituents make the major contribution to rose aroma—β-damascenone and β-ionone:

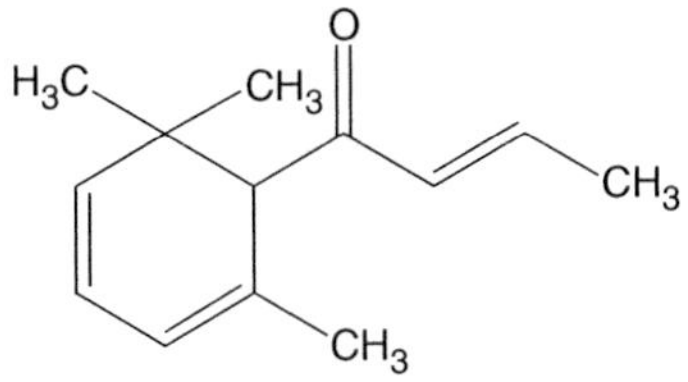

β-damascenone

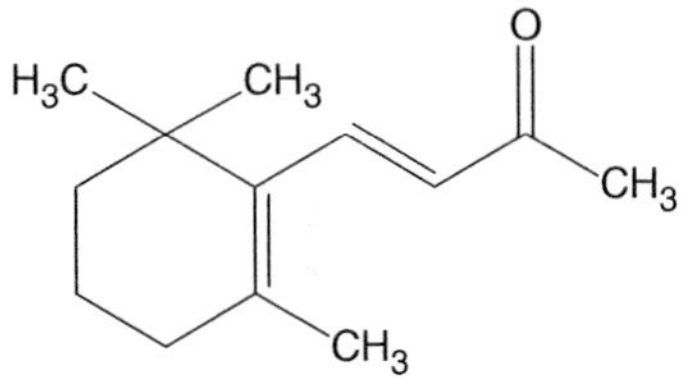

β-ionone

These and related rose aroma constituents find their way into some classes of perfumes: Dior's "Poison" provides one example.

Here is a bit of a complication: there is a lot of individual variation in the sense of human olfaction. Not everything smells the same to everyone. This holds both for the intensity of the perceived smell as well as for its quality: pleasant, floral, skunky, sweaty, or no odor at all. Andreas Keller has recently demonstrated that some significant part of this individual variation in the sense of smell derives from genetic variation in human odorant genes. Specifically, two single nucleotide polymorphisms (SNPs), leading to two amino acid substitutions in an odorant receptor, have dramatic affects on the perception of the odor of androstenone, a steroid derived from testosterone.

Here are some insights into the molecular basis of taste

There are two basic reasons why the sense of taste is important to us. First, things that taste good encourage us to eat and we need to do that in order to get our required calories and nutrients. Second, things that taste bad inhibit us from consuming them. A lot of poisons, for example, taste bad. So does spoiled food.

In humans, there are five primary taste sensations: salty, sour, sweet, bitter, and umami, or savory. The first four of these are familiar to all; umami has been recognized by the Japanese for a long time but has only more recently been recognized in the United States. Umami is the flavor of, for example, beef broth. Salts of certain acids, including those of glutamic and aspartic acids, activate umami receptors. Monosodium glutamate, MSG, widely used as a flavor enhancer, provides a common example. The receptors for these primary tastes are encoded in a few dozen genes in the human genome, contrasted with a few hundred for odorant receptors.

The role of these tastes has been nicely summarized:[7] "Taste is in charge of evaluating the nutritious content of food and preventing the ingestion of toxic substances. Sweet taste permits the identification of energy-rich nutrients, umami allows the recognition of amino acids, salt taste ensures the proper dietary electrolyte balance, and sour and bitter warn against the intake of potentially noxious and/or poisonous chemicals."

In humans, sensors for taste are collected in structures known as taste buds. Your mouth contains about 10,000 taste buds; the majority are located at various sites on your tongue. The remainder are found in the pharynx, epiglottis, and at the entrance to the esophagus. Each taste bud senses and reacts to all five primary taste sensations. I want to dispel an old, untrue but widely held belief that different parts of the tongue are devoted to different tastes. It just isn't so, regardless of what you might have been taught or otherwise have been led to believe.

Here are the bare essentials of taste bud structure. At the top of the taste bud, there is a small pore opening into the bud. Tastants get access to the taste bud molecular machinery through this pore. The body of the taste bud is composed of 30–100 individual neuroepithelial cells, of three types. The key cell type is the sensory cell in whose membrane is embedded the receptors for tastants. The afferent taste nerve axons enter the taste bud from below. They ramify extensively, each axon making synapses with multiple receptor cells.

Tastant receptor structure is complex. The sweet, bitter, and umami receptors are most similar to odorant receptors in terms of mechanism.[8] They too are GPCRs. Binding of an effector molecule, for example, sucrose, to a sweet receptor causes activation of the associated G protein, called gustducin, with activation of an enzyme known as phospholipase C and resultant synthesis of two second messengers: diacylglycerol and inositol-1,4,5-triphosphate. These second messengers ultimately lead to the gating of the taste transduction channel, called TRPM5. We sense sweet.

There is controversy about the nature of the salt receptor. In one theory, Na^+ ions enter the sensory cells through a sodium channel (this is distinct from the sodium channels of nerve and muscle cells), causing membrane depolarization, entry of Ca^{2+}, and generation of an action potential. We sense salty. Acids create the sensation of sour; think about lemon juice or vinegar. In one mechanism, and there may be more than one, the H^+ ions derived from acids block K^+ channels. Blockage of these channels, basically responsible for maintaining the membrane potential, leads to membrane depolarization, Ca^{2+} entry, and so on. We sense sour.

The flavor of the tomato pulls together odor and taste

When you bite into an apple, what you get is flavor. Flavor depends on odor and taste but also on texture, appearance, and mouth feel. Multiple sensory inputs are integrated in our brains to develop the sensation of flavor. We have developed several aspects of the molecular biology of odor and taste, perhaps the major determinants of flavor.

The central question that I want to approach here is the possible relationship between flavor preferences and nutritional value. There are a lot of data to work with. More than 7000 volatile flavor substances have been identified in foods and beverages. The situation may not be quite as complex as this would suggest. While it is true that any single fruit or vegetable may synthesize a few hundred volatile compounds, only a modest subset of these will contribute to its flavor profile. So the task is to sort out what these are, identify their sources, and link, where possible, these sources to nutritional value. Studies with the tomato provide a great example.[9] The bottom line is: "Virtually all of the major tomato volatiles can be linked to compounds providing health benefits to humans."

Of the 400 volatiles detected in the tomato, only 17 have a positive impact on the flavor profile. Two of the most important ones are also key players in the aroma of roses: β-ionone and β-damascenone. Another player is methyl salicylate, a compound we previously encountered in oil of wintergreen. Some of the most important flavor elements are present in very small concentrations but can be perceived by us at these extremely small concentrations.

Here are the key points. First, the most abundant volatiles in tomatoes are derived from catabolism of essential fatty acids. Linoleic acid is the precursor for hexanal and linolenic acid is that for *cis*-3-hexenal, *cis*-3-hexenol, and *trans*-2-hexenal. All of these are important flavor elements in the tomato. A healthy diet for people requires both of these fatty acids. Note that these volatiles are also important flavor components in apple, sweet cherry, olive, bay leaf, and tea.

Second, the essential amino acids leucine, isoleucine, and phenylalanine are precursors for several additional tomato flavor elements. Here, too, these flavor elements are important flavor constituents in other fruits, including strawberries and apples. They are also found in breads, cheeses, wine, and beer.

Finally, carotenoids are the metabolic precursors for three more of the flavor volatiles of the tomato. Although the role of carotenoids, light-harvesting pigments in plants, in human nutrition is the subject of debate, these compounds are antioxidants and β-carotene is the principal source of the visual pigments of the eye.

Taken together, these findings strongly suggest that the flavor attractiveness of the tomato, and other fruits and vegetables, is strongly correlated with nutritional value to people. Seen in the light of a good deal of related research, there seems to be an important connection between sensory perception, flavor preferences, and health benefits in foods. Note that this connection may be lost in processed foods: I am talking here about what nature provided for our tables not what General Mills or McDonalds does.

Finally, the research just noted on the tomato has been done on two varieties: a wild variety isolated in Peru and a commercial cultivated tomato. It should come as little surprise that the wild variety contained two to seven times greater concentrations of the flavor volatiles than did the variety created for commercially attractive properties: yield, color, shape, and disease resistance.

Bacteria can do more than you think they can

Bacteria are surely among the simplest of the living constructs of nature. Single-celled organisms, surrounded by a membrane and cell wall, about the size of a mitochondrion of a human cell, for which they are the forerunners, and lacking a nucleus or other intracellular membranous structures, they nevertheless get on quite well. Bacteria exhibit some remarkable and, to my mind at least, quite unexpected properties that depend on chemical communication among bacterial cells. Let's consider one compelling example: quorum sensing.

Quorum sensing is a mechanism that permits a population of bacteria to assess their population density and, as a direct consequence, to alter their behavior as a population.[10] Put simply, quorum sensing permits a bacterial population, consisting entirely of single-celled organisms, to function in some sense as a multicellular organism. How do these simple organisms manage to carry out this unexpected task?

Let's begin with the most common and simplest case: communication among members of the same bacterial genus and species. In this case, quorum sensing essentially involves counting the number of like bacteria in the population.

The fundamental molecular mechanism underlying quorum sensing in bacteria is the interaction of secreted small molecules known as autoinducers with their receptors, known as sensor proteins. The logic is simple. The higher the density of bacteria secreting the autoinducers, the higher the concentration of the autoinducers in the medium and the more frequently will a sensor protein detect them. The sensor proteins basically determine the population density by counting the number of autoinducer molecules arriving per unit of time. Activated sensor proteins alter gene expression in the bacterium, with resulting behavioral changes.[11]

To make the underlying ideas more concrete, let's consider a specific example: quorum sensing in the Gram-negative bacterium *Vibrio fischeri*. *V. fischeri* and the Hawaiian squid *Euprymna scolopes* live together to their mutual benefit. The bacteria colonize the light organ of the squid where they encounter an abundance of nutrients and can proliferate to numbers not attainable in seawater. At high density, *V. fischeri* are bioluminescent: that is, they give off visible light, just as do fireflies, and in much the same way. The squid uses the illumination provided by the bacteria as a type of camouflage to avoid predators.

There are redundant mechanisms to ensure that communication among *V. fischeri* is faithful: a bacterial enzyme catalyzes the synthesis of only one autoinducer and the bacterial receptor responds only to that autoinducer. So members of a *V. fischeri* population can speak to one another with clarity even in the presence of one or more other bacterial species that also make autoinducers and sensor molecules.

As the concentration of autoinducers increases, the fraction of sensors bound by the autoinducer molecules will increase, activating them. At a critical concentration of autoinducer, the activated sensors bind to specific sites on the bacterial DNA and alter gene function. In this specific case, the synthesis of the enzyme required for light production, luciferase, is induced. Light happens.

A very important case is the production of virulence factors, structures that are responsible for disease in humans and other species. Take *Escherichia coli* for example. Most strains of this bacterium are quite harmless and, indeed, we have *E. coli* as a normal member of our gut microbiota. However, equipped with a set of virulence factors, we can get the *E. coli* strain known as H157:O7, which can cause a fatal infection.

If we think about it for a bit, quorum sensing as a mechanism regulating the production of virulence factors makes good sense for the bacterial population. If the bacterial population is too low, secretion of virulence factors would seem to be of little use: too few bacteria secreting too few virulence factor molecules to serve whatever purpose these serve for the bacteria. It would seem better to wait until the bacterial forces have massed, then release the virulence factors and have some real impact.

The chemistry of quorum sensing may have very practical applications. Quorum sensing is peculiar to bacteria, though it may have some real connections to signaling in mammalian cells. Finding molecules that antagonize quorum sensing in pathogenic bacteria has the potential to prevent expression of virulence factors. Such molecules may find use as novel antibacterials.

Pheromones are chemical communicators between individuals

We have seen three example of chemical communication among organisms: quorum sensing in bacteria, the attraction of a male gamete to a female one in brown algae, and the attraction of the male silkworm moth to the female one in *Bombyx mori*. We are going to continue along this track for the remainder of this chapter, culminating in an examination of the question of the existence and importance of human pheromones. Before we get there we need to lay some groundwork: what are pheromones and what kinds of roles do they serve?

Karlson and Luscher first introduced the term "pheromone" in 1959 in a publication about chemical communication in insects in the scientific journal *Nature*.[12] A pheromone was viewed as a substance that is secreted or excreted into the environment by one individual that elicits some behavioral, developmental, or endocrine response when received by another individual of the same species. The behavioral change in the male silkworm moth in response to bombykol secreted into the environment by the female of the same species fits this mold, for example.

As frequently happens in science, simplicity gives way to complexity when our horizons are broadened by further research. There are now several definitions of "pheromone" available in the relevant literature, each more elegant than the other. However, we do not have to worry much about the arcane issues that amuse and challenge researchers in the field. For our purposes, we can pretty much do with the original definition, expanded to include a classification system.

Pheromones are divided into four classes.[13] The first is *releaser* pheromones. These generate immediate and, for the most part, behavioral responses. The silkworm moth and brown algae molecules fall into this class. In both cases, the pheromone elicits an immediate behavioral response in the recipient: movement toward a female silkworm moth by the male moth in one case and movement of a male gamete toward a female gamete in the second. Sex attractants are typical releaser pheromones.

The second class is *primer* pheromones that generate longer-term physiological or endocrine responses. There are many examples of primer pheromones. Honeybee queen mandibular pheromone controls the age at which worker honeybees get busy looking for food outside the nest. The mechanism of action of this pheromone has been worked out. High levels of queen mandibular pheromone lower those of juvenile hormone, which in turn, delays worker bee maturation. Chemical signals from adults of one sex may speed the onset of puberty in the other sex. Female chemical signals may induce surges of testosterone in recipient males. Toward the end of this chapter, we will encounter apparent examples of primer pheromones in humans that influence the menstrual cycle in women. All these cases meet the definition of a primer pheromone.

Signaler pheromones act to provide information to the recipient but do not necessarily elicit a response. These are perhaps the most numerous pheromones. For example, animals using only chemical signals can determine the sex of the sending animal and the status of the reproductive cycle in females. At the extreme, chemical signals from a sender may serve to identify the sender. This chemical signal is known as the odor-print of the individual. It is determined by the nature of the gene family known as the major histocompatibility complex (MHC). Here, too, we shall encounter examples of apparent human signaler pheromones later in this chapter.

Finally, we have the subtlest category: the *modulator* pheromones. These pheromones change "stimulus sensitivity, salience, and sensorimotor integration" in the recipient. These pheromones may determine how a recipient organism will respond to a signal in a specific context. This adds another level of complexity to pheromone action.

Many examples of pheromone action are collected in two important books: Tristram Wyatt's *Pheromones and Animal Behavior*[14] and William Agosta's *Chemical Communication*.[15] The examples that follow here are taken from these books. They may be consulted both for additional details and further examples of pheromones.

Pheromones are everywhere

After several years in the ocean, pheromones guide marine salmon back to the river of their birth, where they will spawn and die. In the forest, dying trees emit an odor, a pheromone, which attracts bark beetles. The examples that one could cite are numerous. So let's have a look at a few, beginning with rather simple organisms—though not retreating again to bacteria—and moving toward greater complexity.

Moving a step up the evolutionary ladder from bacteria, we encounter *Allomyces*, a water mold. This mold occupies a variety of moist niches in nature: on plant or animal debris along the edges of ponds, in ditches, or other wet sites. Sexual reproduction in *Allomyces* provides an interesting story of a releaser pheromone. Here is what happens.

Male and female gametes in this water mold are contained in structures called gametangia. When covered with water, the gametes are released from these structures into the environment where, equipped with flagella, they can swim around hunting for each other. The whole point here is to get a female gamete to meet up with a male gamete. Given such an encounter, the two gametes can fuse to create a zygote, the first cell in a new generation of *Allomyces*. That is the *Allomyces* version of sex, and when it happens life goes on for the water mold. It is not a random process: A few minutes before taking leave of the gametangium, the female gamete begins to release sirenin, a rather complex molecule in the terpene family of molecules.

Let's consider the female gamete as a point source of sirenin. Once released into the water, molecules of sirenin will begin to diffuse away from their source. The further they diffuse away from their source, the greater the volume of water occupied and, consequently, the lower the concentration of sirenin. Concentration gradients are good things to have since they can be followed. A blind person can find a lilac bush by following his or her nose: one just walks in the direction where the lilac aroma is getting stronger: that is, up the concentration gradient. It is also pretty easy to walk away from a skunk: you walk down the concentration gradient in a direction where the smell is getting weaker. The male *Allomyces* gamete has a receptor for sirenin. This permits him to move up the concentration gradient pretty smoothly until he is quite close to the female gamete where he fumbles around randomly until contact is made. This works really well: fertilization in the water mold is close to 100% effective.

The male gamete samples the environment for sirenin repeatedly at definite time intervals. If the concentration of sirenin is found to increase on repeated sampling, then the male gamete is swimming up the gradient. If it is decreasing, then the male gamete is headed in the wrong direction and a corrective turn is required to get him headed in the right way.

Sirenin is a really potent molecule. It effectively attracts male gametes at a concentration of 24 picograms per milliliter (0.000000000024 grams/ml). The potency of sirenin posed a difficult problem for scientists interested in isolating it and determining its molecular structure. The female gamete need not make much of this stuff to be effective and, in fact, doesn't. It follows that huge quantities of solutions containing female gametes had to be processed to get enough material for identification. Once the structure was in hand, chemists synthesized sirenin and many other molecules very closely related to it. Only sirenin works as a pheromone for the male gamete. Even the smallest molecular changes result in an indifferent response in the male gamete. So we have yet another example of the sensitive dependence of a biological property on molecular structure.

Not all releaser pheromones are sex attractants. One alternative example of a releaser pheromone is the sending out of an alarm message, warning other members of the species that some threat is at hand. The California sea anemone *Anthopleura elegantissima* provides a neat example of alarm signaling.

Sea anemones are coelenterates, a group that includes jellyfish and corals. The basic structure of a sea anemone is quite simple: it has a closed base that permits the organism to anchor itself to a rock. Coming up from this base is a cylindrical column forming, at the top, a mouth that the anemone can open and close. Surrounding the mouth is a ring of tentacles that wave about in the water, seeking out particles of food and directing them into the mouth. That is about all there is to a sea anemone structurally.

Anthopleura elegantissima is a small, common sea anemone. It is a bit less than 1 inch across normally. However, it can do some interesting chemistry. If one is wounded, it emits an alarm pheromone as a warning to neighboring members of the same species. When this alarm pheromone hits its receptor, the recipient anemone reacts with a series of quick flexes of its basal tentacles, followed by withdrawing the tentacles into the mouth that then snaps shut. This takes a bit less than 3 seconds overall.

The alarm pheromone from this sea anemone, anthopleurine, has been isolated and characterized. Like most pheromones, anthopleurine is highly potent. It is an effective alarm pheromone at a concentration of 74 picograms per milliliter (0.000000000074 grams/ml).

So far in our discussion, life has been simple in one respect: each action of a pheromone has been the consequence of one or two, or a few in the case of the brown algae, molecules in one species. In nature, things can be more complex. One molecule can serve as a pheromone for multiple species: benzaldehyde serves as a trail pheromone for bees, *Trigona* genus; a defense pheromone for ants, *Veromessor* genus; and as a male sex pheromone in the moth, *Pseudaletia* genus.

A more striking example is provided by *cis*-7-dodecenyl-1-acetate:

cis-7-dodecenyl-1-acetate

This long-chain unsaturated ester serves as one component of a multicomponent female sex pheromone for 140 species of moths as well as the Asian elephant. This is unlikely to cause any real confusion in nature. On the one hand, the 140 species of moths keep out of each other's way and out of the elephant's way by varying the nature and relative concentrations of other components in the multicomponent pheromone. On the other hand, the male elephant is not likely to detect the extremely small amounts of this pheromone released by female moths. Chemistry aside, there are other obvious difficulties in moths mating with elephants in the wild.

It should not surprise us that quite distinct species use some of the same molecules as pheromones. Remember that all extant life had a single common origin and that multiple biochemical pathways are shared by basically all living organisms. Thus, using much the same set of biochemical tools, multiple organisms have come to common solutions to common problems. This is one more example of the unity of life on Earth.

Ants, emergence, and pheromones are linked

Ants are a remarkably successful group of social insects. They are almost everywhere: only Iceland, Greenland, Polynesia, and Antarctica lack ants. There are 10,000 known species of these insects. Their engineering and social coordination is amazing, particularly given that no one is in charge. Steven Johnson has employed the ants as a model of emergence—properties of a system not possessed by any member of that system. Johnson notes:[16]

> They think locally and act locally, but their collective action produces global behavior. Take the relationship between foraging and colony size. Harvester ants constantly adjust the number of ants actively foraging for food, based on a number of variables: overall colony size (and thus mouths needed to be fed); amount of food stored in the nest; amount of food available in the surrounding area; even the presence of other colonies in the near vicinity. No individual ant can assess any of these variables on her own .... The perceptual world of an ant, in other words, is limited to the street level. There are no bird's eye views of the colony, no ways to perceive the overall system—and indeed, no cognitive apparatus that could make sense of such a view. "Seeing the whole" is both a perceptual and conceptual impossibility for any member of the ant species.

Nonetheless, the ants get on extremely well. It is clear that doing so requires an elaborate system of communication among them. There is simply no way of achieving highly organized behavior if no one knows what anyone else is doing. Communication is required.

E. O. Wilson and his collaborators made an intensive and revealing study of the fire ant *Solenopsis invicta* in the 1960s.[17] Wilson determined that communication among fire ants is limited to 10 signals, nine of which are based on pheromones. The 10th is tactile communication directly between ants. As Johnson puts it: "Among other things, these semiochemicals code for task-recognition ('I'm on foraging duty.'); trail attraction ('There's food over here.'); alarm behavior ('Run away!'); and necrophoric behavior ('Let's get rid of these dead comrades.')."

The power of pheromones to act as agents of emergent behavior is perhaps nowhere better exemplified than by the ants. The termites, another highly successful group of social insects, provides a comparably compelling story. However, let's move on to communication by pheromones in mammals.

Pheromones are important in the lives of mammals

Mice are a favorite experimental animal for scientists interested in mammalian pheromones:[18] they are easy to breed; they have well-defined genetic backgrounds; and they rely on pheromones in a number of ways. My former colleague at Indiana University, Milos Novotny, did the first definitive work on mammalian pheromones in the house mouse. He has identified a number of primer pheromones in the urine of male mice that accelerate puberty and induce estrus in female mice.[19]

These effects have names associated with their discoverers: the induction of estrus in adult females by males is known as the Whitten effect; the acceleration of puberty in immature females by males is known as the Vanderburgh effect. Perhaps you can work these terms into the conversation at your next cocktail party. People may be impressed.

These effects are hardly limited to mice: they have been observed in Norway rats, hamsters, rabbits, voles, goats, sheep, pigs, cattle, and nonhuman primates.

The list of pheromone controls on reproductive behavior can be extended: female pheromones inhibit estrus in adult females and delay puberty in juvenile females, Lee-Boot effect; male pheromones can block pregnancy, Bruce effect; pheromones from dominant females can suppress reproduction in subordinate females; and so on. Have a look at Wyatt's book for an abundance of examples.

In some cases, commercial use has been made of reproductive pheromonal effects. For example, spraying sows with synthetic 5α-androstenone, a steroid sold under the trade name Boar Mate, helps to ready the animal for mating with a boar. 5α-Androstenone is one of two steroids in the saliva of the boar that primes the sow for mating.

It is not always clear how pheromone signals are detected in mammals. Most vertebrates, mice for example, have a VNO in addition to the main olfactory system. The VNO has two separate families of olfactory receptors: V1r, 137 functional receptors in mice; V2r, 60 functional receptors in mice. The genes for these are only distantly related to those for the main olfactory receptors, suggesting that these systems evolved independently. As a general rule, it is the VNO and not the olfactory epithelium that is responsible for detecting pheromone molecules. However, it has been demonstrated that mice whose VNO has been surgically removed can discriminate MHC-determined odor types. This finding clearly implicates the main olfactory system in the detection of pheromones.

Are there human pheromones?

The question of the existence and role, if any, of human pheromones has a long, interesting, and incomplete history.[20] Here is the beginning of the story.

As an undergraduate at Wellesley College, Martha McClintock noted that her female dormmates tended to synchronize their menstrual cycles over time. When she pointed this out to a group of scientists at a pheromone conference, she was challenged to study the phenomenon scientifically if she wanted to be believed. She took up the challenge, recruited 135 women from her dorm, and collected the data on dates of onset of menstrual cycles. The data strongly supported her informal observations and she wrote up the results of her study as her senior thesis.

A year later as a graduate student in the Department of Psychology at Harvard University, E.O. Wilson, a noted biologist and authority on communication among ants, encouraged her to write up her study for publication. She did so and her breakthrough publication appeared in the scientific journal *Nature* in 1971.[21] It provided the first evidence for human pheromones. Martha McClintock was 23 years old at the time: an auspicious start for a young scientist. Incidentally, her class at Wellesley College included Hillary Rodham Clinton and this story appears in somewhat more detail in Miriam Horn's book: *Rebels in White Gloves: Coming of Age with Hillary's Class—Wellesley '69.*[22] Martha McClintock has gone on to a distinguished scientific career at the University of Chicago.

It is difficult to be confident that McClintock's findings on the synchronization of menstrual cycles through social interaction reflects the action of a pheromone or family of pheromones. There are other possible explanations, though the action of a human pheromone is perhaps the most plausible one. As a general rule, experimental evidence of pheromone activity has been followed up by isolation, characterization, and chemical synthesis of the pheromone. Once it can be shown that the chemically synthesized molecule has the same chemical structure and biological properties as the isolated material, one has proof of the nature of the pheromone. That work has not been done for the volatile agent or agents that may be responsible for menstrual cycle synchronization in women. Isolating and characterizing pheromones is almost always a tough task and doing so for potential human pheromones provides special hurdles (see below). Put succinctly, people are difficult experimental animals.

Despite the difficulties, there is substantial additional reason to believe in human pheromones, although their role in shaping our lives is far from clear. The evidence just summarized from the 1971 McClintock study of menstrual synchrony in women provides an example of a primer pheromone at work in humans. Others have replicated McClintock's pioneering study. A 1980 study suggesting that underarm secretions might mediate menstrual synchrony,[23] has been followed up by a more detailed study, also by McClintock, together with her coworker Kathleen Stern.[24] Here it is.

Nine healthy women agreed to wear cotton pads under their arms for 8 hours a day at various points in their menstrual cycles. These pads were then cut into sections and treated with alcohol in an effort to dissolve volatile substances. The alcohol extracts were then wiped under the noses of 20 other women every day for a month. These women reported that they could smell nothing except the alcohol.

The results: 68% of the women treated with extracts from the follicular phase of the menstrual cycle shortened their own current cycle by an average of 1.7 days. A different 68% of the women treated with extracts from the ovarian phase of the menstrual cycle lengthened their own current cycle by an average of 1.4 days. Finally, underarm extracts taken following ovulation, in the luteal phase, had no effect on the timing of other women's periods. This is a more robust result than that reported in 1971, since both lengthening and shortening of ovulation cycles were observed, which strongly argues against the results being a coincidental synchronization of menstrual cycles. However, in this case too, no pheromone has been isolated and characterized.

A related study strongly suggests that contact with males, as well as with females, may have an effect on menstrual cycle kinetics.[25] Specifically, wiping extracts of male underarm secretions on the upper lip of normally cycling women in the early to mid phases of their menstrual cycle three times in 6 hours significantly advanced the next pulse of luteinizing hormone, LH. Although this does not directly demonstrate an effect on the menstrual cycle, it is strongly suggestive of one, given the role of LH in controlling it. All these examples are evidence for primer pheromones in humans.

There is also evidence for signal pheromones in humans. For example, soon after birth, infants can recognize their mothers by smell alone. The converse is also true: mothers can identify their children by smell alone. Finally, newborn babies can recognize the smell of a lactating woman.

The evidence for modulator pheromones in humans is perhaps less convincing. Here is the evidence. First, the odors of some people viewing a film designed to elicit fear are different from those of the same people viewing a film designed to elicit laughter. Although the difference can be detected by other humans, it is not clear what the consequences are for the emotional state of the odor detectors. Second, exposure of women to extracts of male underarm secretions tends to make them more relaxed, less tense. This study was carried out in a hospital setting, not ideal for the purpose of establishing a modulator pheromone effect. Nonetheless, there are the beginnings of a story to suggest that human modulator pheromones may exist and have some effect on the emotional state of the recipients. Doubtless, this story will become more fully complete as additional, careful studies come to fruition.

In contrast to the evidence for primers, signalers, and modulators, there is no decent evidence to suggest that there are human releaser pheromones. That is not to argue that there are none but to state that there is no evidence for them at present. Nonetheless, products purported to be human releaser pheromones—specifically sex attractants—are widely available on the Internet. They go by such suggestive names as Scent of Eros, The Edge, Alter-ego, and Pheromone Additive. Many of these products contain either androstenone or androstenol, steroids of unknown influence on the human emotional state.

Earlier, I noted that pheromone chemistry is usually enormously challenging, in part because the potency of these molecules means that they are frequently present in very small amounts. Life is particularly difficult in the case of human pheromones. One aspect of this work is efforts to characterize the volatile organic compounds (VOCs) of human skin. Such molecules have the potential to be chemosignals among humans. Here is a list of reasons why such work is so demanding.

First, human skin possesses four different glands capable of emitting VOCs: sebaceous, eccrine (sweat), apocrine, and apoeccrine glands. Although the chemical activity of these glands is not fully understood, each is, in principle at least, capable of emitting its own spectrum of VOCs. Second, the distribution of these glands is anatomically distinct. So it matters where on the body one collects the VOCs.

Third, the VOCs of each type of gland may well depend on the hormonal status of the individual. Thus, the VOCs may depend, for example, on the stage of a woman in her estrous cycle. Fourth, the VOCs are emitted in extremely low concentrations. Finally, the VOCs are complex mixtures and it is never clear which components of these mixtures are important chemosignals. Deconvoluting these mixtures is a challenge. Despite the challenges (or perhaps because of them—chemists like challenges) progress is made.

The field of human pheromones is an active area of research into human behavior.[26] Perhaps we should be less aggressive in our use of colognes, perfumes, after-shaves, and the like that may dull our ability to sense what nature has put in place for us. Time will tell.

Key Points

1. Taste and smell are two important, and linked, human senses.

2. Life depends on molecular recognition: specific base-pairing in nucleic acids; recognition of substrates by enzymes; of effectors by receptors; of antigens by antibodies.

3. Molecular recognition underlies chemical communication, including the phenomena of taste and smell.

4. The human genome includes about 350 genes that encode receptors for odors. These are expressed in the olfactory epithelium. Some other species contain even more odorant genes: rats have about 1200.

5. In the olfactory epithelium, each olfactory neuron expresses a specific odorant receptor. Each may be activated by multiple odorants.

6. Olfactory receptors are G protein-coupled receptors; cAMP is the key second messenger.

7. The sense of taste and olfaction are concentration-independent.

8. There are five primary taste sensations: salty, sour, sweet, bitter, and umami (or savory). The receptors for these tastes are encoded in a few dozen genes in the human genome. These are expressed in taste buds.

9. There are substantial data to suggest that the flavor attractiveness of natural foods is strongly correlated with nutritional value to people.

10. Chemical communication among bacteria underlies the phenomenon of quorum sensing.

11. Pheromones are chemical communicators between individuals (usually members of the same species).

12. There are four classes of pheromones. Releaser pheromones elicit immediate, behavioral responses. Primer pheromones elicit longer-term physiological or endocrine responses. Signaler pheromones act to provide information to the recipient. Modulator pheromones determine how a recipient organism will respond to a signal in a specific context.

13. Pheromones are everywhere in nature.

14. The highly organized, emergent social behavior of ants reflects the action of a number of ant pheromones.

15. Pheromone action has been demonstrated in a number of mammalian species, particularly in mice. Several of these affect reproduction.

16. It seems likely that there are human pheromones. The role of these in human behavior is not clear.

Appendix

Some examples of explicit and condensed molecular structures

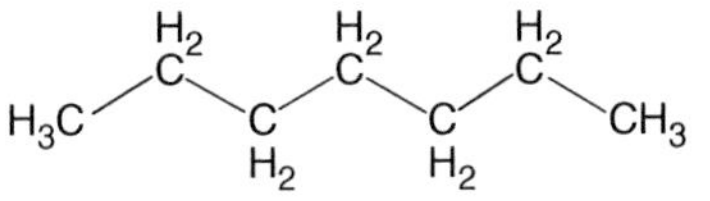

heptane

heptane

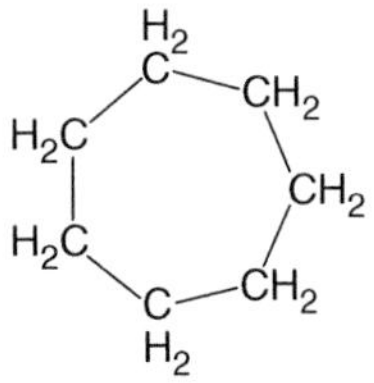

cycloheptane

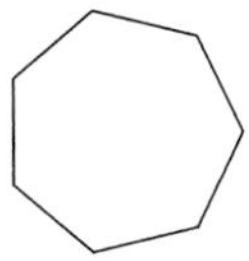

cycloheptane

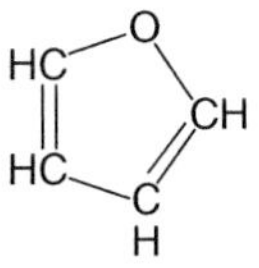

furan

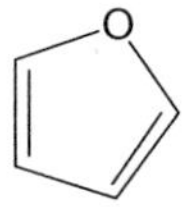

furan

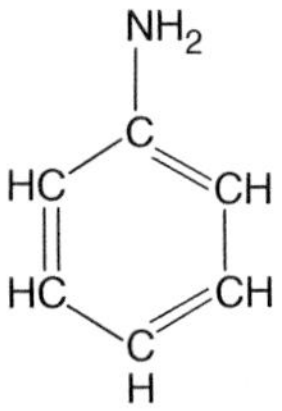

aniline

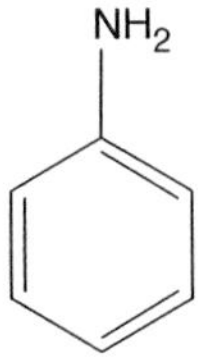

aniline

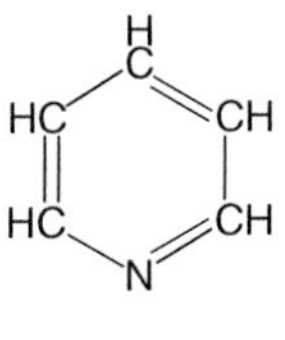

pyridine

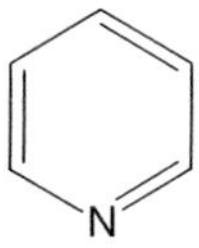

pyridine

Notes

Chapter 1

1. As far as I know, the noted biologist, E. O. Wilson coined the term "biophilic." See: E. O. Wilson, *The Diversity of Life*, Norton, New York, 1992.

2. A millimeter is one one-thousandth of a meter. A meter is a little longer than a yard—about 39.37 inches. So a millimeter is about 1/25th of an inch. These organisms are, therefore, about one thousand times shorter than this value.

3. The key role of size in life has been eloquently described: J. T. Bonner, *Why Size Matters: From Bacteria to Blue Whales*, Princeton University Press, Princeton, N.J., 2006.

4. There are two recent and excellent references to the diversity of deep sea life: C. Nouvian, *The Deep: The Extraordinary Creatures of the Abyss*, University of Chicago Press, Chicago, Ill., 2007; T. Koslow, *The Silent Deep: The Discovery, Ecology and Conservation of the Deep Sea*, University of Chicago Press/University of New South Wales Press, 2007.

5. See: E. O. Wilson, *The Diversity of Life*, Norton, New York, 1992 for a thorough discussion of the possible number of living species on Earth.

6. See: S. M. Stanley, *Extinction*, Scientific American Library, Freeman, New York, 1987. This short book provides a useful and informed discussion of the history of mass extinctions of living species.

7. Nick Lane has published a detailed picture of the Permian extinction: see *Nature* **448:** 122 (2007). He estimates that this extinction killed off up to 96% of all species then alive.

8. D. E. Erwin, The mother of mass extinctions *Sci Am* **275(1):** July 1996.

9. The figures for rate of extinction are taken from E. O. Wilson, *The Diversity of Life*, Norton, New York, 1992.

10. See E. O. Wilson, *The Creation: An Appeal to Save Life on Earth*, Norton, New York, 2006.

11. *Physicians Desk Reference*, 62nd edn. Thomson PDR, Montvale, NJ, 2008; this publication is updated annually: www.PDR.net.

12. A. H. Knoll, *Life on a Young Planet: The First Three Billion Years of Evolution on Earth*, Princeton University Press, Princeton, N.J., 2003, p 25.

13. For a detailed story of the Archaea, see: T. Friend, *The Third Domain. The Untold Story of the Archaea and the Future of Biotechnology*, Joseph Henry Press, Washington, DC, 2007.

14. Abstracted from a longer essay written by J. C. Rettie: *But a Watch in the Night: A Scientific Fable;* originally published in *The Land*, Vol. VII, No. 3, Fall, 1948.

15. For a conceptual history of the development of life on Earth, see: C. de Duve, *Singularities: Landmarks on the Pathways of Life*, Cambridge University Press, New York, 2005; C. de Duve, *Life Evolving: Molecules, Mind, and Meaning*, Oxford University Press, New York, 2002; C. de Duve, *Vital Dust: Life as a Cosmic Imperative*, Basic Books, New York, 1995.

16. The seminal study of Miller is: S. Miller, *Science* **117**: 528–529 (1953).

17. Lateral transfer of genetic information refers to information transfer from one cellular structure to another or from one organism to another, other than transfer from parent to progeny. Lateral transfer of genetic information still occurs but at a much smaller rate.

18. The quotation is taken from: C. Woese, *Proc Natl Acad Sci USA* **95**: 6854 (1998).

Chapter 2

1. The book referred to is: G. Brown, *The Energy of Life: The Science of What Makes Our Minds and Bodies Work*, The Free Press, New York, 2000.

2. Here are two excellent books on the phenomenon of emergence: S. Johnson, *Emergence: The Connected Lives of Ants, Brains, Cities, and Software*, Scribner, New York, 2001; and J. H. Holland, *Emergence: From Chaos to Order*, Addison Wesley, Reading, Mass., 1998.

3. Several metabolic pathways are described in later chapters. Basically, these are sequences of chemical reactions leading from some molecular starting point to some distinct end point. Photosynthesis and respiration provide two examples.

4. While sunlight is the overwhelmingly important source of energy to sustain life, it is not the only one. For example, some exotic organisms living far underground manage to capture energy from decay of radioactive elements and use this to sustain themselves.

5. The science of thermodynamics deals with energy and its changes. For a useful summary of thermodynamics see: P. Atkins, *Four Laws that Drive the Universe*, Oxford University Press, Oxford, 2007.

6. Quantitatively, the relationship between E and H is given by $H = E + pV$, in which p is pressure and V is volume. If we talk about changes accompanying some process in which p is constant, we can write: $\Delta H = \Delta E + p\Delta V$. The last term is a measure of work done. Suppose that we consider a system composed of a candle. If we light the candle, two things will happen. First, energy will be generated in the form of heat; that is a measure of ΔH. Second, the hot gases will expand against the pressure of the atmosphere and that takes work: $p\Delta V$. It turns out that, in this case, ΔH is a great deal larger than $p\Delta V$ so that ΔH and ΔE are approximately equal. The underlying meaning, however, is distinct.

7. Note that there are plants that do not have roots. The aerial branches of plants can move and consume cellular energy in the process.

8. Ernst Mayr, *This Is Biology: The Science of the Living World*, Harvard University Press, Cambridge, Mass., 1997. This book provides a highly thoughtful discussion of the science of living organisms by an eminent biologist.

9. For a detailed but readable discussion of the Second Law of Thermodynamics, see P. W. Atkins, *The Second Law*, Scientific American Library, Freeman, New York, 1986.

10. Here are a few additional statements of the Second Law of Thermodynamics:

1 "Every system which is left to itself will, on average, change toward a condition of maximum probability." (G. N. Lewis)

2 "The state of maximum entropy is the most stable state for an isolated system." (Enrico Fermi)

3 "A transformation whose only final result is to transform into work heat extracted from a source which is at the same temperature throughout is impossible." (Max Planck)

4 "Gain in information is loss in entropy." (G. N. Lewis)
5 "Entropy is time's arrow." (A. Eddington)

11. E. Schrödinger, *What is Life?* Cambridge University Press, London, 1944. This short book by one of the twentieth centuries, leading physicists provides an unusual view of the nature of life. Not all the ideas put forth in this early book are correct, however.

Chapter 3

1. The definition of chemistry comes from: *The American Heritage Dictionary of the English Language*, 3rd edn, Houghton Miflin, New York, 1993.

2. The molecular mass has units termed daltons (Da), after the famous English chemist John Dalton (1766–1848). For example, the molecular mass of insulin might be said to be about 6000 daltons. However, this designation does not add much to understanding the concept of molecular size and we shall delete the dalton unit in all that follows.

3. The number of atoms of each type in a molecule is denoted by a subscript; if the number is one, no subscript is used. Thus, CH_4 tells us that there is one carbon atom and four hydrogen atoms in a molecule of methane. CO_2 tells us that there is one carbon atom and two oxygen atoms in a molecule of carbon dioxide.

4. The formulas for getting back and forth between the Fahrenheit and Celsius temperature scales are: $°F = 9/5°C + 32$ and $°C = 5/9(°F − 32)$. These temperature scales are named for their inventors, Daniel Gabriel Fahrenheit and Anders Celsius.

5. J. Shreeve, *The Genome War: How Craig Venter Tried to Capture the Code of Life and Save the World*, Alfred A. Knopf, New York, 2004.

6. The genome of an organism is captured in the DNA of the nucleus of each cell. Mature red cells lack a nucleus (though immature red cells have a nucleus) and do not have, in consequence, a copy of the genome.

Chapter 4

1. The proton weighs 1.673×10^{-27} grams. Its charge is 1.602×10^{-19} coulombs. The neutron weighs 1.675×10^{-27} grams.

2. The nature and chemistry of the elements have been described in considerable detail in a number of works, including: P. Strathern, *Mendeleyev's Dream: The Quest for the Elements*, St. Martin's Press, New York, 2001; J. E. Brady and J. R. Holum, *Descriptive Chemistry of the Elements*, Wiley, New York, 1996; John Emsley, *Nature's Building Blocks: An A-Z Guide to the Elements*, Oxford University Press, New York, 2001.

3. J. Buckingham, *Chasing the Molecule*, Sutton Publishing, Phoenix Mill, Stroud, Glos., UK, 2004.

4. For a detailed look at hydrogen: J. S. Rigden, *Hydrogen: The Essential Element*, Harvard University Press, Cambridge, Mass., 2002.

5. Hydrogen is converted to helium in the sun "little by little' in terms of the total amount of hydrogen present. In terms of the actual rate of conversion—about 5 billion tons per second—hydrogen is converted to helium in the sun at a phenomenal rate.

6. Unhappily, there is another, older convention for designating chiral molecules: here the labels are D and L. The *R/S* system is more systematic and more generally useful but old usages die hard. We shall encounter D and L when we talk about amino acids in chapter 10.

7. For a detailed look at the history and uses of thalidomide, see: T. D. Stephens and R. Brynner, *Dark Remedy: The Impact of Thalidomide and its Revival as a Vital Medicine*, Perseus Books, New York, 2001.

Chapter 5

1. In the interest of full disclosure, it is true that the halogens, particularly bromine and iodine, do make molecules and ions in which they form more than one chemical bond.

However, the vast majority of all molecules in which halogens appear have exactly one chemical bond to each halogen atom.

2. The basic nature of the carbon–carbon double bond is, in fact, quite distinct from that for carbon–carbon single bonds. However, this is a complex matter and, at any event, this need not concern us.

3. A note about nomenclature: the "2" in 2-butene designates the carbon atom at one end of the double bond, numbering the carbon chain from the end that assigns the position of the double bond the lowest number. For example, there is also 1-butene, $CH_2{=}CH{-}CH_2{-}CH_3$. 1-Butene has the double bond at the end of the molecule while 2-butene has it in the middle. To provide one more example, 3-hexene is $CH_3{-}CH_2{-}CH{=}CH{-}CH_2{-}CH_3$.

4. For more details and a wealth of additional examples, see: T. D. Wyatt, *Pheromones and Animal Behavior: Communication by Smell and Taste*, Cambridge University Press, Cambridge, UK, 2003.

Chapter 6

1. Modern agriculture is practiced as intensive monoculture of annuals: that is, fields are planted yearly with a single crop. In Nebraska or Iowa, for example, you can see endless fields of corn. In Kansas, you are more likely to see endless fields of wheat. It does not have to be this way: for years there has been a movement to make perennial polyculture competitive with annual monoculture. The goal is to be able to plant legumes along with other perennial crops as a source of nitrogen and to avoid the need to replant each year. This movement is worth following.

2. D. Charles, *Master Mind: The Rise and Fall of Fritz Haber, The Nobel Laureate Who Launched the Age of Chemical Warfare*, HarperCollins, New York, 2005.

3. Things are a little more complicated than implied by this statement. Molecules that donate protons and molecules that accept them are formally known as Brönsted acids and bases. There is a more general definition of acid and base—Lewis acids and bases—in which a Lewis acid is any molecule having the ability to accept a pair of electrons and a Lewis base is any molecule having the ability to donate a pair of electrons to a Lewis acid. The Brönsted concept will suffice for all our purposes.

4. N. Lane, *Oxygen: The Molecule that Made the World*, Oxford University Press, New York, 2002.

5. L. J. Henderson, *The Fitness of the Environment: An Inquiry into the Biological Significance of the Properties of Matter*, Macmillan, New York, 1913.

6. The heat of vaporization of ammonia and the heat of fusion of ammonia are also abnormally high, though not as abnormally high as that of water. The explanation of this fact is that ammonia also forms hydrogen bonds with itself.

Chapter 7

1. For the original research identifying sweat-component receptors in mosquitoes, see: E. A. Hallem, A. N. Fox, L. J. Zwiebel, and J. R. Carlson, *Nature* **427:** 212 (2004).

2. T. Eisner, *For the Love of Insects*, Harvard University Press, Cambridge, Mass., 2003.

3. The statement that the aroma of apples is due to methyl butyrate is too simple. Life is more complex: a number of chemical compounds contribute to apple aroma. However, methyl butyrate makes a major contribution. If you were to smell this compound, you would recognize the aroma of apples.

4. N. Lane, *Power, Sex, Suicide: Mitochondria and the Meaning of Life*, Oxford University Press, New York, 2005.

Chapter 8

1. The book referred to is: P. Levi, *The Periodic Table*, Schocken Books, New York, 1984.

2. The quotation comes from: J. Emsley, *The Thirteenth Element: The Sordid Tale of Murder, Fire, and Phosphorus*, Wiley, New York, 2000.

3. The interest of alchemists in urine was driven by a search for a yellow pigment known as India Yellow, which is produced in the urine of cows fed with mango leaves. This interest, then, led to the discovery of elemental phosphorus.

4. The quotation is taken from: J. Emsley, *Nature's Building Blocks: An A-Z Guide to the Elements*, Oxford University Press, New York, 2001, p 312.

5. Although hydrogen sulfide is a noxious and toxic compound, recent evidence indicates that it is synthesized in the human body and probably acts as a neurotransmitter. Its lifetime in the body is short, a few seconds.

6. Methylmercaptan is also known as methanethiol. The designations "mercaptan" and "thiol" are giveaways for a molecule containing an —SH group. "Thia" is a giveaway that the molecule contains a sulfur atom. For example, thiadiazole is a molecule containing a sulfur atom (thia) and two nitrogen atoms (diazole; azole being a giveaway for a nitrogen atom).

7. The acetate esters of the skunk thiols cited in the text are really thioesters. The acetyl group is attached to the sulfur atom: CH_3—CO—S—R. These are similar to the oxygen esters encountered in chapter 7, except that a sulfur atom replaces one of the oxygen atoms in the ester group.

8. For a review of the therapeutic uses of garlic constituents, see: K. C. Agarwal, *Med Res Rev* **16:** 111–124 (1996).

9. For a review of the chemistry of garlic, see: E. Block, *Sci Am* **252:** 114–119 (1985).

10. There are several mutations in the CFTR protein that render it nonfunctional. However, one specific mutation underlies about 70% of the cases.

Chapter 9

1. Urease is a very unusual enzyme in the sense that it contains an atom of nickel that is required for activity. Many enzymes contain metal ions but very few have nickel as that metal. Incidentally, urease is not an easy enzyme to crystallize. I tried it as a graduate student many years ago and developed high regard for the skills of James Sumner.

2. Wendell Stanley also gained a share of the 1946 Nobel Prize in Chemistry based on his work in providing the first pure preparations of a virus.

3. Although most catalysts in living systems are enzymes, they are not the only ones. Ribozymes are catalytic RNA molecules. These also play important roles in the metabolism of living organisms, including specifically the synthesis of proteins.

4. There is a huge literature on enzymes and how they work. Two of the best treatments are: A. Fersht, *Structure and Mechanism in Protein Science: A Guide to Enzyme Catalysis and Protein Folding*, W. H. Freeman, New York, 1999; and P. A. Frey and A. D. Hegeman, *Enzymatic Reaction Mechanisms*, Oxford University Press, New York, 2007.

5. The huge factors by which enzymes increase reaction rates have been established quantitatively by Richard Wolfenden and coworkers. See: R. Wolfenden, and M. J. Snider, *Acc Chem Res* **34:** 938–945 (2001).

6. The basic problem with the lock and key analogy is that locks and keys are generally rigid structures while proteins and their substrates have substantial freedom of movement: that is, proteins and their substrates have a number of conformations available to them in space and they change conformation a few billion times per second. So, our biological locks and keys are wobbly.

7. Molecular recognition captures the sense of two molecules interacting strongly and specifically, recognizing each other in the same sense as a key and its lock might be said to recognize each other. Molecular recognition depends on several factors. Notable among them is a complementarity of shapes. Here the analogy of a key and a lock seems apt: if the shapes are not complementary, the two do not fit together. In the case of molecular recognition, more than complementary shapes is required. It is also the case that electronic properties on the

molecular surfaces must be complementary. Thus, a region of high electron density on one partner might pair up with a region of low electron density on the other. A hydrophobic patch on one molecule might pair up with the hydrophobic patch on the other. A hydrogen bond donor might find a hydrogen bond acceptor; a positive charge might find a negative charge, and so on.

8. There is an immense literature on enzyme inhibitors and their role in clinical medicine. A number of examples are cited in later chapters of this book.

9. A good general introduction to immunology and the human immune system is: P. Parham, *The Immune System*, Garland, New York, 2000.

10. See M. Bliss, *The Discovery of Insulin: 25th Anniversary Edition*, University of Chicago Press, Chicago, Ill., 2007.

11. Human growth hormone has certain antiaging effects, though it is not clear that administration of growth hormone to the elderly is safe. Growth hormone is also employed, as are other agents, by some athletes in efforts to improve their performance. It is not clear that this is safe and it is certainly not legal.

12. The number of distinct cell types in the human body varies somewhat from textbook to textbook. For our purposes, the differences in classification that underlie the differences in number do not matter. However, you should not be surprised if you see figures somewhat different from 200 from time to time.

Chapter 10

1. In the more rigorous system of identifying enantiomers, L-alanine is known as *S*-alanine and D-alanine is known as *R*-alanine.

2. Biochemists are marvelous at inventing abbreviations, perhaps because many of the molecules of biochemistry are quite complex and have correspondingly complex and clumsy names. So abbreviation seems like the right way to go. A capable biochemist can speak for quite a while in a way quite unintelligible to those not in the know about the abbreviations. Physicians and lawyers do the same thing but without the abbreviations.

3. Here is an important early reference to requirements for essential amino acids: W. C. Rose, R. L. Wixom, H. B. Lockhart, and G. F. Lambert, *J Biol Chem* **7**: 987 (1955).

4. The techniques of genetic engineering have made it possible to create high-lysine corn.

5. The sequence is unique for insulin from a specific species but insulins from different species have slightly different primary structures.

6. Note that we encountered disulfide bonds in chapter 4 when we described certain compounds that are formed in cut cloves of garlic.

7. The primary structure for ribonuclease: D. G. Smyth, W. H. Stein, and S. Moore, *J Biol Chem* **238**: 227 (1963).

8. Protein primary structure databases include the following: ExPASy Molecular Biology Server (Swiss-Prot): expasy.ch/; Protein Information resources (PIR): pir.georgetown.edu; Protein Research Foundation (PRF): prf.or.jp/en/os.html.

Chapter 11

1. Dickerson and Geis collaborated on a protein chemistry book that contains many of the drawings and paintings of Geis: R. E. Dickerson and I. Geis, *The Structure and Action of Proteins*, Benjamin, Menlo Park, Calif., 1969.

2. The work on the myoglobin structure has been summarized.: J. C. Kendrew, *Science* **139**: 1259 (1963).

3. Here is a summary reference to the pioneering work on the structure of hemoglobin: M. F. Perutz, *Sci Am* December 1978.

4. Christian Anfinsen did the crucial experiment that showed that primary structure determines tertiary structure for proteins: C. J. Epstein, R. F. Goldberger, and C. B. Anfinsen, *Cold Spring Harbor Symp Quant Biol* **28**: 439 (1963). This experiment is not always as simple

as the discussion here would suggest. Getting the proper refolding to occur can be problematic. In some living organisms, there is a family of proteins that help this process along. Nonetheless, the principle arrived at here is correct.

5. These compositions for hemoglobins are taken from: R. Hoffmann, *The Same and Not the Same*, Columbia University Press, New York, 1995.

Chapter 12

1. Different people count different cell types differently. Thus, it is possible that you will encounter statements that assert that there are more or fewer than 200 different cell types in the human body.

2. DNA is frequently said to be a self-replicating molecule. This is nonsense. DNA contains the information required for its faithful replication but the process is complicated and involves a number of enzymes, as described later.

3. The early history of the foundations of modern molecular biology has been recounted many times. In my view, the best source is: H. F. Judson, *The Eighth Day of Creation*, Simon and Schuster, New York, 1979.

4. The demonstration that DNA is the substance of genes is reported in: O. T. Avery, C. M. MacLeod, and M. McCarty, *J Exp Med* **79:** 137–158 (1944). Cited on page 39 of H. F. Judson, *The Eighth Day of Creation*, Simon and Schuster, New York, 1979.

5. The argument begun here in which DNA is revealed as information has been developed in more detail and more elegantly in: R. Pollack, *Signs of Life: The Language and Meanings of DNA*, Houghton Miffin, New York, 1994. This book also provides a wealth of insights into how genetics works.

6. The actual codon for an amino acid is the base complement of the "word" in DNA. So the codon corresponding to GGT is CCU (see below and table 12.1).

7. The codon in mRNA for Phe is the complement to AAA: that is, UUU.

8. Table 12.1. The genetic code presented in this table is very nearly universal. There are isolated exceptions in the genome of mitochondria, which is described later in this chapter. Beyond that, the genetic code has been expanded to include codons for two unusual amino acids that occur in a modest number of proteins. These amino acids are selenomethionine, in which an atom of selenium replaces the sulfur atom of methionine, and pyrrolysine, a cyclized form of lysine. For details, see: A. Ambrogelly, S. Palioura, and D. Söll, *Nat Chem Biol* **3:** 29–35 (2007).

9. The double helix structure for DNA was first described in: J. D. Watson and F. H. C. Crick, *Nature* **171:** 737–738 (1953). This is the single most famous publication in all of molecular biology.

10. The story of the discovery of the structure of DNA, and all that preceded it and followed, has been told many times. In addition to the Judson book cited above, of particular interest are the following books: J. D. Watson, *The Double Helix: A Personal Account of the Discovery of the Structure of DNA*, Atheneum Press, New York, 1968; J. D. Watson and J. Tooze, *The DNA Story: A Documentary History of Gene Cloning*, Freeman, San Francisco, Calif., 1981; R. Cook-Deegan, *The Gene Wars*, Norton, New York, 1994; P. Berg and M. Singer, *Dealing with Genes: The Language of Heredity*, University Science Books, Mill Valley, Calif., 1992.

11. Chargaff's rules were first defined in: E. Chargaff, *Experientia* **6:** 201–209 (1950).

12. The Meselson–Stahl experiment is described in: M. Meselson and F. W. Stahl, *Proc Natl Acad Sci USA* **44:** 671–682 (1958).

13. Kornberg shared the Nobel Prize that year with Severo Ochoa.

14. The pioneering research in RNAi is described in: A. Fire, et al., *Nature* **391:** 806–811 (1998).

15. For a summary of RNAi structure and function, see for example: *Science* **298:** 2296–2297 (2002); D. M. Dykxhoorn, C. D. Novina, and P. A. Sharp, *Nat Rev: RNAi Collection* December 2003, pp 7–17.

Chapter 13

1. Dr. Richard Schweet later joined the faculty at the University of Kentucky in Lexington. His highly promising career in biochemistry ended way too soon in a fatal commuter airplane crash on his way to Louisville. He was an excellent mentor and I remain indebted to him for his help and attention during my undergraduate years at Caltech.

2. The central dogma of molecular biology was first defined by Francis Crick. See: R. Olby, 'Francis Crick, DNA, and the Central Dogma'. *Daedalus* Fall 1970, pp 970–986.

3. The structure of the ribosome is described in: N. Ban, P. Nissen, J. Hansen, P. B. Moore, and T. A. Steitz, *Science* **289:** 905–920 (2000); A. Yonath, *Annu Rev Biophys Biomol Struct* **31:** 257–273 (2001).

Chapter 14

1. The unraveling of the first RNA sequence: R. W. Holley, J. Apgar, G. A. Everett, et al, *Science* **147:** 1462–1465 (1965).

2. The two articles revealing the human genome are published in: *Nature* **409:** 15 February, 2001, and *Science* **291:** 16 February, 2001.

3. The race to get the human genome has been detailed: N. Wade, *Life Script: How the Human Genome Discoveries Will Transform Medicine and Enhance Your Health*, Simon and Schuster, New York, 2001.

4. The first publications on the Neanderthal genome are: J. P. Noonan, G. Coop, S. Rudaravalli, et al., *Science* **314:** 1113–1118 (2006); R. E. Green, J. Krause, S. E. Ptale, et al., *Nature* **444:** 330–336 (2006).

5. For a detailed account of mitochondria, including its genome, see N. Lane, *Power, Sex, Suicide: Mitochondria and the Meaning of Life*, Oxford University Press, New York, 2005.

6. For a more complete listing of genomic sizes, see: www.genomics.com/statistics.php.

7. The elegant early work on homeotic genes has been summarized: C. Nüsslein-Volhard, *Harvey Lectures* **86:** 128–148 (1992); C. Nüsslein-Volhard, *JAMA* **266:** 1848–49 (1991).

8. The Nobel Prize that year was shared with Edward Lewis.

9. See: M. Ridley, *Genome: The Autobiography of a Species in 23 Chapters*, HarperCollins, New York, 2000.

Chapter 15

1. In addition to RDIs and RDAs, there is a set of values termed the dietary reference intakes (DRIs). These are the most recent set of dietary recommendations established by the Food and Nutrition Board of the Institute of Medicine, 1997–2001. DRIs vary significantly from the current RDIs and may provide the basis for updating the RDIs in the future (see www.nal.usda.gov/fnic/etext/000105.html).

2. Upper limit values are from the Council for Responsible Nutrition.

3. Values of RDI and RDA for vitamin A are given as retinol activity equivalents (RAE) where 1 microgram of retinol equals 1 RAE, 12 micrograms of β-carotene equals 1 RAE, and so forth for all provitamins A.

4. Retinal, an aldehyde, is reactive with amino groups of proteins. Retinol, an alcohol, is not. The conversion of retinol to retinal is, therefore, critical.

5. P. J. Hilts, *Rx for Survival: Why We Must Rise to the Global Health Challenge*, The Penguin Press, New York, 2005.

6. *Hox* genes or their equivalent are widely distributed in nature, from fruit flies to humans.

7. For a detailed story of *Hox* genes and development, see: S. B. Carroll, *Endless Forms Most Beautiful: The New Science of Evo Devo*, Norton, New York, 2005.

8. For a more complete story of the vitamin C/scurvy issue, see pages 214–218 in the book: L. Bergreen, *Over the Edge of the World*, Morrow, New York, 2003 in which the circumnavigation of the world by Magellan is recounted.

9. Though getting adequate sunlight is beneficial in terms of vitamin D chemistry, there is also good reason to avoid sunburn, a well-known risk factor for development of melanoma.

10. For a detailed review of vitamin D, with emphasis on vitamin D deficiency, see: M. F. Holick, *New Engl Med J* **357**: 266–281 (2007).

11. Just as FAD occurs in oxidized and reduced forms, so does FMN. The reduced form is known as $FMNH_2$. FMN is a coenzyme for some biochemical reactions but is significantly less important in this regard than is FAD. Since the chemistry involved is identical in the two cases, I have chosen to focus on FAD.

12. NMN is basically half of the NAD^+ molecule: nicotinamide ribose phosphate. $NADP^+$ is NAD^+ bearing a phosphate group at $C3'$ of the ribose group attached to the adenine. The redox chemistry is the same in all three forms of the coenzymes. NAD^+ is the form most frequently employed for biochemical oxidation reactions in catabolism and $NADP^+$ (in its reduced form NADPH) is the form usually employed for biochemical reduction reactions in anabolism. NMN is employed infrequently.

13. The quotation comes from the following website: http://history.nih.gov/goldberger/docs/south.htm.

Chapter 16

1. Solubility is a complex issue, since solubility of crystalline solids, for example, reflects a balance between the attractive forces in the crystal and those in the dissolved ions.

Chapter 17

1. Oxidation is a nontrivial concept. Formally, oxidation is the removal of one or more electrons from some chemical species. Less formally, you can get a good idea of oxidation state by simply counting the number of bonds from a carbon atom to oxygen or other heteroatom (e.g., nitrogen or sulfur). Let's see how this works, beginning with methane. In methane, there are no chemical bonds to oxygen. So the carbon atom in methane is at its lowest oxidation state. In methanol, we have one chemical bond to oxygen so that the carbon atom in methanol is at a higher oxidation state than that in methane. In formaldehyde, $H_2C{=}O$, there are two chemical bonds to oxygen; in formic acid, HCOOH, there are three, and finally, in carbon dioxide, CO_2, there are four. It follows that carbon dioxide is at the highest oxidation state, followed by formic acid, formaldehyde, methanol, and methane.

2. Formally, one says that the Gibbs free energy of ATP hydrolysis under physiological conditions is large and negative. That is, the value of ΔG for the hydrolysis of ATP under physiological conditions is near -12 kcal/mole.

3. The de Duve quotation is taken from: C. de Duve, *Singularities: Landmarks on the Pathways of Life*, Cambridge University Press, New York, 2005.

4. Kinases are enzymes that employ ATP to add a phosphate group to some substrate. There are a great many kinases encoded in the human genome, some 520 of them. This testifies to their importance in human metabolism.

5. Note that there are actually more than two beta adrenergic receptors. A beta-3 has been definitely identified and there may be a beta-4 as well.

6. For a more detailed examination of the role of mitochondria in cellular energetics, see: G. Brown, *The Energy of Life: The Science of What Makes Our Minds and Bodies Work*, The Free Press, New York, 2000.

Chapter 18

1. For a detailed description of orlistat and its approved uses (or for any other prescription medication approved for sale in the United States), see *The Physicians Desk Reference*.

2. For a general summary of the issue of eating and weight gain, see: Steven Shapin, Eat and Run, *The New Yorker*, January 2006, pp 76–82.

3. For a discussion of the role of leptin in weight control, see: J. M. Friedman, *Nat Med* **10**: 563–569 (2004).

4. See: A. B. Keys, *Seven Counties: A Multivariate Analysis of Death and Coronary Heart Disease*, Harvard University Press, Cambridge, Mass., 1980.

5. For some details and references to original literature linking *trans* fatty acids and coronary heart disease, see: www.hsph.harvard.edu/reviews/transfats.html.

6. For the original work showing the effect of dietary *trans* fatty acids on LDL and HDL cholesterol levels, see: R. P. M. Mensink and M. B. Katan, *N Engl J Med* **323:** 439–445 (1990).

7. Although having 60 trade names under which acetaminophen is marketed around the world is exceptional, different trade names for the same molecule in different countries or areas is very common. Indeed it is the rule. A single *generic* name for each molecule marketed as a drug for human medicine is employed worldwide but multiple trade names are generally employed. For example, the blood pressure-lowering molecule enalapril is marketed by Merck as Enap, Enapren, Innovace, Lotrial, Olivin, Pres, Renitec, Renivace, Vasotec, and Xanef.

8. Compare this situation with that for normal and sickle cell hemoglobins. The two COX enzymes are both fully functional and differ by a single conservative amino acid replacement. In contrast, normal hemoglobin is fully functional but sickle cell hemoglobin is not and these differ by a single nonconservative amino acid replacement (see chapter 11).

Chapter 19

1. Actually, we are not quite done here. If we want to push the argument all the way to the end, we would have to include the ATP molecules generated by the catabolism of the glycerol part of the tripalmitin molecule. Doing that is a bit beyond the scope of what we are up to here, but suffice it to say that it would push the total yield of ATP from one molecule of tripalmitin to more than 400 molecules.

2. For a detailed summary of what is known about resveratrol, see: J. A. Baur and D. A. Sinclair, *Nat Rev Drug Discov* **5:** 493–506 (2006).

3. H. Y. Cohen, C. Miller, K. J. Bitterman, et al., *Science* **305:** 390–392 (2004).

4. See: J. M. Dhabi, H. J. Kim, P. L. Mote, R. J. Beaver, and S. R. Spindler, *Proc Natl Acad Sci USA* **101:** 5524–5529 (2004).

5. For those sophisticated in the art, Sir2 is a NAD-dependent histone deacetylase.

Chapter 20

1. The usual reference is to plasma cholesterol, rather than blood cholesterol. Plasma is simply blood from which the blood cells (red cells, white cells, and platelets) have been removed. Since most clinical chemistry measurements are made on plasma rather than blood, they are rightly reported as plasma values. If the plasma is permitted to clot and the clot removed, the resulting liquid is known as serum.

2. The AHA Dietary Guidelines can be found in: R. M. Krauss, R. H. Eckel, B. Howard, et al., *Circulation* **102:** 2284–2289 (2000).

3. Achieving really significant reductions in plasma cholesterol levels through diet alone generally requires that one limit dietary cholesterol to 200 mg per day or less. This is about fivefold less than the typical US diet contains. The typical Japanese diet prior to the influence of Western dietary habits contained about 200 mg of cholesterol per day.

4. When the role of HMGR inhibitors in human medicine became clear, the generic designation "statin" came into general use.

5. The nature of the tumors and the conditions under which they are observed remain clouded. I was involved in work with lovastatin at Merck at the time and we were rightly concerned about the Sankyo findings, given the close structural relationship between mevastatin and lovastatin. Rumors suggested that the problem was with gastrointestinal tumors in dogs. Efforts to obtain details from Sankyo proved unrewarding, even at the CEO to CEO (or shogun to shogun) level. This problem was not seen with lovastatin or other statins in long-term animal trials.

6. For a summary of the story leading to lovastatin and simvastatin, see: J. A. Tobert, *Nat Rev Drug Discov* **2:** 517–526 (2003).

7. For clinical studies demonstrating that statins are effective in the various populations mentioned, see: Scandinavian Simvastatin Survival Study: T. R. Pedersen A. G. Olsson, O. Faergerman, et al., *Circulation* **97**: 1453–1460 (1998); WOSCOPS Study, see: J. Shepherd et al., *N Engl J Med* **333**: 1301–1307 (1995); AFCAPS/TexCAPS Study, see: J. R. Downs et al., *JAMA* **279**: 1615–1622 (1998); CARDS Study, see: P. S. Sever et al., *Lancet* **361**: 1149–1158 (2003).

8. Like all drugs, the statins have their adverse effects. Hepatotoxicity is occasionally seen and potentially serious muscle problems arise rarely. See your doctor.

9. The fact that we all begin life basically as females can be appreciated in several ways. For example, if the testes are removed from an embryonic male fetus, the individual is born female. Males lacking functional androgen receptors provide a second example: these individuals are phenotypically female, a phenomenon known as testicular feminization. It follows that embryonic mammals are targeted to become females unless androgens intervene in the development and differentiation process.

10. The issues around HRT for men have been detailed: H. Pearson, *Nature* **431**: 500–501 (2004); E. L. Rhoden and A. Morgentaler, *N Engl J Med* **35**: 482–492 (2004).

11. A male pseudohermaphrodite is the central subject of the Pulitzer Prize winning novel *Middlesex* by Jeffrey Eugenides.

12. The earliest study linking HRT to heart attacks in women having heart disease is: S. Hulley et al., *JAMA* **280**: 605–613 (1998).

13. The results of the WHI (Women's Health Initiative) Estrogen Plus Progesterone clinical trial revealing increased incidence of breast cancer, heart disease, stroke, and dementia are provided in: J. E. Rossouw et al., *JAMA* **288**: 321–333 (2002).

14. Results of the clinical studies of HRT using estrogen alone have been detailed in the following three publications: G. L. Anderson et al., *JAMA* **291**: 1701–1712 (2004); J. Hsia et al., *Arch Intern Med* **166**: 357–365 (2006); M. L. Stefanick et al., *JAMA* **295**: 1647–1657 (2006).

15. The complex, tissue-specific actions of SERMs have been described: B. L. Riggs and L. C. Hartmann, *N Engl J Med* **348**: 618–629 (2003).

Chapter 21

1. I have found the following several references useful in gaining understanding of the nervous system. The most general of these is a standard textbook: M. F. Bear, B. W. Connors, and M. A. Paradiso, *Neuroscience: Exploring the Brain*, 3rd edn, Lippincott Williams and Wilkins, Philadelphia, 2007. More specific references include: S. H. Barondes, *Molecules and Mental Illness*, Scientific American Library, New York, 1993; T. Stone and G. Darlington, *Pills, Potions, Poisons: How Drugs Work*, Oxford University Press, New York, 2000; E. R. Kandel and L. R. Squire, Neuroscience: breaking down scientific barriers to the study of brain and mind, *Science* **290**: 1113–1120 (2000).

2. E. R. Kandel, *In Search of Memory: The Emergence of a New Science of Mind*, Norton, New York, 2006.

3. pH is a measure of the acidity or basicity (alkalinity) of a water-based preparation. Mathematically, pH is defined as the negative logarithm of the hydrogen ion, H^+, concentration: $pH = -\log(H^+)$. Preparations that have a high concentration of hydrogen ions, say a solution of sulfuric acid in water, have low values of pH and are said to be acidic. Those that have very low concentrations of hydrogen ion, such as a solution of sodium hydroxide in water (lye), have high values of pH and are said to be basic or alkaline.

4. Ions or molecules flowing down their concentration gradients is one aspect of a very general statement known as the Second Law of Thermodynamics. The Second Law is a mathematical statement to the effect that all real processes increase the disorder, captured in a quantity known as entropy, of the universe. Entropy is a measure of disorder or randomness and may be thought of as negative information.

5. In specialized cells of the auditory system, there are also receptors that are gated by pressure changes.

Chapter 22

1. For revealing stories of people dealing with depression, see: A. Solomon, *The Noonday Demon: An Atlas of Depression;* K. R. Jamison, *Unquiet Mind: A Memoir of Moods and Madness*; K. R. Jamison, *Night Falls Fast: Understanding Suicide*; P. D. Kramer, *Listening to Prozac*; W. Styron, *Darkness Visible: A Memoir of Madness.*

2. Establishing the efficacy of a drug for depression is challenging. One reason is the huge placebo effect in clinical trials for depression. Typically, about 40% of trial participants will report significant improvement on placebo. Since no antidepressant works for everyone, there is a rather narrow window for showing efficacy. In many cases, the distinction between a successful test molecule and a placebo is that the test molecule works in 60% of trial participants versus 40% for the placebo. In addition, measuring efficacy is not trivial. If you want to know if something works for high blood pressure, there is a simple, quantitative measurement available. Nothing equivalent is available for measuring antidepressant efficacy.

3. For background information about schizophrenia, see: R. Freedman, *N Engl J Med* **349:** 1738–1749 (2003).

4. S. Nasar, *A Beautiful Mind*, Simon and Schuster, New York, 1998.

5. For the clinical study comparing perphenazine to second-generation antipsychotics, see: J. A. Lieberman et al., *N Engl J Med* **353:** 1209–1223 (2005).

6. O. Sacks, *Awakenings*, Simon and Schuster, New York, 1987; several subsequent editions have appeared.

7. For more information about drugs that affect the human nervous system, see: C. Regan, *Intoxicating Minds: How Drugs Work*, Columbia University Press, New York, 2001; N. C. Andreasen, *Brave New Brain: Conquering Mental Illness in the Era of the Genome*; Oxford University Press, New York, 2001.

8. A ganglion is a structure formed by the cell bodies of a collection of neurons.

9. The neural circuits shown have been redrawn from circuits provided in: E. Kandel, *In Search of Memory: The Emergence of a New Science of Mind*, Norton, New York, 2006.

Chapter 23

1. Antibiotics are also employed to prevent disease though this accounts for only a small fraction of total antibiotic use.

2. The state of public health measures in the United States and beyond leaves a lot to be desired. See: L. Garrett, *Betrayal of Trust: The Collapse of Global Public Health*, Hyperion, New York, 2000.

3. C. Walsh, *Antibiotics: Actions, Origins, Resistance*, ASM Press, Washington, DC, 2003.

4. R. Rappuoli, *Nat Med* **10:** 1177–1185 (2004). The article provides an excellent summary of progress of the fight against infectious diseases over time and the challenges that remain.

5. For information on the role of infectious diseases in the New World prior to the arrival of Europeans, see: C. C. Mann, *1491: New Revelations of the Americas Before Columbus*, Knopf, New York, 2005.

6. For the results of the Sargasso Sea DNA sequencing experiment, see: J. C. Venter et al., *Science* **304:** 66–74 (2004).

7. The market potential for vaccines is usually limited compared to those for drugs. Vaccines may give a lifetime of disease prevention after one or a few treatments. Drugs are generally given repeatedly over time and the therapy may be repeated as often as the patient gets the disease.

8. The study of virulence factors in *Salmonella* is reported in: D. Becker, M. Selbach, C. Rollenhagen, et al., *Nature* **440:** 303–307 (2006).

9. For an enlightening look at early efforts at disease therapy, see: D. L. Cowen and W. H. Helfand, *Pharmacy*, Abrams, New York, 1988. This unique book provides multiple examples of advertisements for early agents asserted to have useful medicinal properties.

10. The results of clinical trials of *Echinacea* for the common cold are provided in: R. B. Turner, R. Bauer, K. Woelkart, T. C. Hulsey, J. D. Gangemi, *N Engl J Med* **353:** 341–348 (2005); W. Sampson, *N Engl J Med* **353:** 337–339 (2005).

11. The results of clinical trials of saw palmetto in prostatism are provided in: S. Bent, C. Kane, K. Shinohara, et al., *N Engl J Med* **354:** 557–566 (2006); R. S. DiPaola and R. A. Morton, *N Engl J Med* **354:** 632–634 (2006).

12. For the story of the discovery of penicillin, see: W. Howard Hughes, *Alexander Fleming and Penicillin*, Crane Russak, 1977; R. Bud, *Penicillin: Triumph and Tragedy*, Oxford University Press, Oxford, UK, 2007.

Chapter 24

1. R. A.Weinberg, *One Renegade Cell: How Cancer Begins*, Basic Books, New York, 1998.

2. Among the currently FDA approved drugs for colorectal cancer, cetuximab and bevacizumab are large molecules (antibodies), the products of the biotechnology industry. The others are small molecules.

3. Although tumors begin as clones of single cells, subsequent mutations generally result in tumors that are genetically heterogeneous, one of the problems for tumor therapy.

4. For some early history of efforts to understand the origins of cancer, see: R. A. Weinberg, *Racing to the Beginning of the Road: The Search for the Origin of Cancer*, Harmony Books, New York, 1996.

5. The quotation is taken from: Robert A. Weinberg, *One Renegade Cell: How Cancer Begins*, Basic Books, New York, 1998.

6. There are exceptions to the multiple mutations origin of cancer. There are a few rather rare cancers that appear to be the consequence of a single mutation. Much more generally, two or more mutations are required. However, I discuss in detail later a prominent exception and a breakthrough in cancer chemotherapy.

7. For a list of cancer predisposition genes: see B. Vogelstein and K. W. Kinzler, *Nat Med* **10:** 789–799 (2004).

8. For the analysis of mutated genes in breast and colorectal cancer, see T. Sjöblom, S. Jones, L. D. Wood, et al., *Science* **314:** 268–274 (2006).

9. The quote and the related ones below come from: B. Vogelstein and K. W. Kinzler, *Nat Med* **10:** 789–799 (2004).

10. The discovery of the Philadelphia chromosome is reported in: P. C. Nowell and D. A. Hungerford, *Science* **132:** 1497 (1960).

11. The relationship between the *BCR-ABL* gene and CML is reported in: G. Q. Daley, R. A. Van Etten, and D. Baltimore, *Science* **247:** 824–830 (1990).

12. For the story of the discovery of imatinib, see: R. Capdeville et al., *Nat Rev Drug Discov* **1:** 493–502 (2002).

13. For background on sorafenib, see: S. Willhem, et al., *Nat Rev Drug Discov* **5:** 835 (2006).

14. For background information on monoclonal antibodies and targeted oncology drugs, see: A. Gschwind, O. M. Fischer, and A. Ullrich, *Nat Rev Cancer* **4:** 361–70 (2004).

Chapter 25

1. Proust's great work is titled *A la recherche du temps perdu* in the original French and is also known as *Remembrance of Things Past* or *In Search of Lost Time* in English translations.

2. For a thoughtful analysis of orthonasal and retronasal olfaction, see: G. M. Shepherd, *Nature* **444:** 316–321 (2006).

3. The pioneering molecular biology work in olfaction is: L. Buck and R. Axel, *Cell* **65:** 175–187 (1991).

4. Actually the human genome contains about 600 odorant receptor genes. However, about 250 of these are pseudogenes that have no function. They are remnants of evolution.

5. For odorant receptors in the fruit fly and other insects, see. C. I. Bargmann, *Nature* **444:** 295–300 (2006).

6. Buck's work that established a pattern recognition mechanism for olfaction is: L. B. Buck, *Cell* **116**(Suppl): S117–119 (2004).

7. The quotation is from: J. Chandrashekar, M. A. Hoon, N. J. Ryba, and C. S. Zuker, *Nature* **444:** 288–294 (2006).

8. Receptors for taste have been described in: J. Chandrashekar, M. A. Hoon, N. J. Ryba, and C. S. Zuker, *Nature* **444:** 288–294 (2006). For the original work on the bitter taste receptors see: K. L. Mueller et al., *Nature* **434:** 225–229 (2005); for the original work on the sour taste receptors, see: A. L. Huang et al., *Nature* **442:** 934–938 (2006).

9. The story of tomato tastants and nutrition is found in: S. A. Goff and H. J. Klee, *Science* **311:** 815–819 (2006).

10. See the excellent summaries of quorum sensing by Bonnie Bassler and her colleagues at Princeton University: A. Camilli and B. L. Bassler, *Science* **311:** 1113 (2006); C. M. Waters and B. L. Bassler, *Annu Rev Cell Dev Biol* **21:** 319–346 (2005); J. M. Henke and B. L. Bassler, *Trends Cell Biol* **14:** 648–656 (2004).

11. The molecular steps that occur between autoinducer and sensor protein interaction and alteration of gene expression vary from case to case and some have been worked out in substantial detail.

12. For the original use of the term pheromone, see: P. Karlson and M. Luscher, *Nature* **183:** 55 (1959).

13. For additional insight into the classification of pheromones, see: M. K. McClintock, Human pheromones: primers, releasers, signalers, or modulators? in *Reproduction in Context: Social and Environmental Influences on Reproduction*, K. Wallen and J. E. Schneider, eds, MIT Press, Cambridge, Mass., 2000.

14. T. D. Wyatt, *Pheromones and Animal Behaviour: Communication by Smell and Taste*, Cambridge University Press, Cambridge, UK 2003.

15. W. C. Agosta, *Chemical Communication: The Language of Pheromones*, Scientific American Library, New York, 1992.

16. S. Johnson: *Emergence: The Connected Lives of Ants, Brains, Cities, and Software*, Scribner, New York, 2001.

17. Wilson's pioneering work on communication among ants is detailed in: B. Hölldobler and E. O. Wilson, *The Ants*, Harvard University Press, Cambridge, Mass., 1991. See also: B. Hölldobler and E. O. Wilson, *Journey to the Ants: A Story of Scientific Exploration*, Harvard University Press, Cambridge, Mass., 1998.

18. The earliest definitive reports of mammalian pheromones are reported in: M. Novotny et al., *Proc Natl Acad Sci USA* **82:** 2059–2061 (1985); B. Jemiolo, S. Harvey, and M. Novotny, *Proc Natl Acad Sci USA* **83:** 4576–4579 (1986).

19. Isolation and characterization of house mouse pheromones are reported in: M. V. Novotny, B. Jemiolo, D. Wiesler, et al., *Chem Biol* **6:** 377–383 (1999).

20. A useful review of the science of human pheromones is: C. J. Wysocki, K. Yamazaki, M. Curran, L. M. Wysocki, and G. K. Beauchamp, *Horm Behav* **46:** 241 (2004).

21. The pioneering study of human pheromones is: M. K. McClintock, *Nature* **291:** 244 (1971).

22. M. Horn, *Rebels in White Gloves: Coming of Age with Hillary's Class—Wellesley '69*, Random House, New York, 1999.

23. For the 1980 study, see: M. J. Russell, G. M. Switz, and K. Thompson, *Pharmacol Biochem Behav* **13:** 737–738 (1980).

24. For the follow-up study of effects of underarm secretions on timing of the menstrual cycle, see: K. Stern and M. K. McClintock, *Nature* **392:** 177–179 (1998).

25. For the effect of male secretions on the menstrual cycle, see: G. Preti, C. J. Wysocki, K. T. Barnhart, S. J. Sondheimer, and J. J. Leyden, *Biol Reprod* **68:** 2107–2113 (2003).

26. For a summary and references to the original publications concerning human pheromones, see: C. J. Wysocki and G. Preti, *Anat Rec A Discov Mol Cell Evol Biol* **281:** 1201–1211 (2004).

Glossary

A

Acetylcholine: a neurotransmitter acting at synapses and the neuromuscular junction.

Acetylcholinesterase: the enzyme that inactivates acetylcholine.

Acid: a molecule or ion that has the capacity to donate a proton to some acceptor.

Action potential: a fluctuation in membrane potential caused by opening and closing of voltage-gated ion channels.

Adenylate cyclase: the enzyme that catalyzes the formation of cyclic AMP (cAMP).

Adiponectin: a polypeptide hormone produced and released from adipose tissue that acts to increase insulin sensitivity.

Adipose tissue: fat; composed of adipocytes, fat-filled cells.

Adrenaline: a potent adrenergic agonist; also known as epinephrine.

Adrenergic receptor: one of several classes of membrane-localized receptors that recognize adrenaline and noradrenaline as agonists.

Afferent pathway: a neural pathway leading from the periphery to the central nervous system.

AIDS: acquired immunodeficiency disease.

Alcohols: a class of organic molecules possessing at least one hydroxyl group, —OH.

Alkanes: hydrocarbons containing only carbon–carbon single bonds.

Alkenes: hydrocarbons containing at least one carbon–carbon double bond.

Alkynes: hydrocarbons containing at least one carbon–carbon triple bond.

Allotropes: alternative forms of elements.

Alpha helix: one form of secondary structure in proteins in which the polypeptide chain forms a helix having 3.6 amino acid residues per turn.

Ames test: a simple and widely used experimental test for mutagenicity.

Amine: a class of organic molecules possessing at least one amino group, $-NH_2$, or substituted amino group, $-NR_2$.

Amino acid: a compound possessing both amino and carboxylic acid functional groups.

Aminoglycoside: a structurally complex antibacterial that works as bacterial protein synthesis inhibitor.

Ammonia: the hydride of nitrogen: NH_3.

Ammonium ion: a cationic species derived from an amine by the addition of a proton or alkyl group: e.g., ammonium ion itself, NH_4^+.

Amphetamine: a stimulant drug that decreases appetite.

Anabolic steroid: a growth-promoting steroid.

Anabolism: biosynthesis; the creation of more complex products from simpler substrates.

Androgen: a male sex hormone.

Androgen antagonist: a molecule that antagonizes the action of androgens at the androgen receptor.

Anion: an atom or group of atoms bearing a negative charge.

Anosmia: the clinical condition of having lost the sense of smell.

Antibacterial: a general term for a drug that kills or inhibits the growth of bacteria.

Antibiotic: a general term for a drug that inhibits the growth or kills pathogens of humans and animals.

Antibody: a protein produced by plasma cells as part of the humoral immune response that has the ability to recognize and bind to the antigen that elicited it.

Anticodon: a sequence of three bases on tRNA that is complementary to a codon on mRNA.

Antigen: a molecule capable of eliciting the formation of antibodies by plasma cells.

Apoptosis: programmed cell death, as opposed to necrosis.

Archaea: one of the two prokaryotic great domains. The other is the Eubacteria.

Aromatase: the enzyme the catalyzes a key reaction in the conversion of androgens to estrogens.

Aromatic molecule: a molecule based on the cyclic hydrocarbon benzene.

Ascorbic acid: a coenzyme for prolyl hydroxylase; the preventive and cure for scurvy; also known as vitamin C.

Atom: the smallest representative particle of an element.

Atomic mass unit: a unit based on a value of exactly 12 atomic mass units for the mass of the isotope of carbon that has exactly six protons and six neutrons in the nucleus: ^{12}C.

ATP: a molecule that serves as the energy currency of the cell.

Autoimmune disease: a disease in which the immune system attacks normal tissues (i.e., fails to recognize "self").

Autonomic nervous system: a system of central and peripheral nerves responsible for involuntary movements.

Avogadro's number: the number of ^{12}C atoms in exactly 12 grams of ^{12}C. This value is approximately 6.022×10^{23}.

Axon: an outgrowth (neurite) of a neuron that conducts action potentials away from the cell body.

B

Basal metabolic rate: the rate of energy consumption when at rest.

B cell: a type of white blood cell that develops in the blood marrow and can be induced to differentiate into an antibody-secreting plasma cell.

B-cell receptor: a protein receptor on the surface of B cells that has the ability to bind antigens.

Benign tumor: a localized tumor that lacks the capacity to invade surrounding tissues or to establish distant colonies.

Benzene: the iconic aromatic molecule: C_6H_6.

Beriberi: a thiamine deficiency disease.

Beta blocker: a compound that blocks β adrenergic receptors.

Beta-lactam antibiotic: one of several classes of antibacterials possessing the β-lactam nucleus.

Beta sheet: a form of secondary structure in proteins.

Bile acids: derivatives of cholesterol that facilitate the digestion of fats and oils.

Biodiversity: the full range of living organisms at some point in time.

Biophilia: the love of life.

Biosynthesis: a sequence of chemical transformations leading from some starting molecule to one or more product molecules.

Blood–brain barrier: a barrier created by tight junctions among cells of the brain capillaries that limits the movement of substances in the blood into the brain.

Body mass index (BMI): defined as weight/height^2 (kg/m^2); a useful index to determine one's deviation from ideal body weight.

Bruce effect: blockage of pregnancy by male pheromones.

C

Calcitonin: a peptide hormone secreted by the thyroid gland that lowers plasma calcium levels by increasing the deposition of bone.

cAMP: cyclic AMP; a second messenger that regulates the activity of protein kinase A, among that of many other enzymes.

Carbapenem: an antibacterial based on a carbapenem skeleton.

Carbohydrate: a general term for a class of molecules of composition $C_nH_{2n}O_n$, including the simple sugars (monosaccharides), oligosaccharides, and polysaccharides.

Carbon: an element in which the nucleus contains six protons.

Carbon dioxide: CO_2; an important greenhouse gas.

Carbon monoxide: the simplest oxide of carbon: CO.

Carboxylic acids: the class of organic compounds containing the carboxylic acid function: COOH. Loss of a proton creates carboxylates: COO^-.

Carcinogen: a chemical that has the capacity to induce tumor formation.

Carotene: one of a family of 40-carbon molecules that are effective as vitamin A.

Catabolism: degradation; the conversion of more complex compounds to simpler ones.

Catalyst: a molecule capable of accelerating a chemical reaction without itself being changed.

Cation: an atom or group of atoms bearing a positive charge.

Cell: the unit of structure and function of all living organisms.

Central nervous system: the brain and spinal cord.

Cephalosporin: one of an important class of antibacterials based on the cephem skeleton.

Cerebellum: a brain structure attached to the brainstem that is important for the control of movement.

Cerebrum: the largest structure in the brain.

CFTR (cystic fibrosis transmembrane conductance regulator): a membrane protein that functions as a chloride channel.

Chargaff's rules: for all normal DNAs, $A = T$, $G = C$ and $A + G = C + T$.

Checkpoint: a point in the cell cycle where the cell basically takes stock of the integrity of its genome and makes any corrections necessary prior to proceeding to divide.

Chemiosmotic hypothesis: the concept that electron transport along the electron transport chain of mitochondria is coupled to the creation of a proton concentration gradient across the inner mitochondrial membrane.

Chemotaxis: movement in response to a gradient in the concentration of some substance or substances.

Chemotherapy: the treatment of cancer with chemical drugs.

Chirality: a property of a molecule that exihibits handedness.

Chlorine: an element in which the nucleus contains 17 protons.

Cholecystokinin: a gut-derived polypeptide hormone released in response to a food load that decreases food intake.

Cholesterol: the steroid precursor of all steroid hormones in humans and a critical component of biological membranes.

Chromosome: a gene-carrying structure consisting of a single, long molecule of DNA, or RNA in retroviruses, and associated molecules.

Citric acid cycle: a central metabolic pathway responsible for converting acetate to carbon dioxide and water with the generation of chemical energy.

Clone: a line of genetically identical cells or individuals.

Codon: a triplet of bases in DNA or mRNA that specifies a specific amino acid in proteins or a termination signal.

Coenzyme: a small molecule essential for the activity of one or more enzymes.

Coenzyme A: a derivative of pantothenic acid required for many reactions in human metabolism.

Collagen: a strong, insoluble protein that forms the organic component of bone.

Compound: a substance formed from more than one element.

Conditioning: a type of learning in which an unpleasant stimulus becomes associated with one that would ordinarily elicit no response.

Conformations: rapidly interconverting forms of the same molecule involving different dispositions of its atoms in space.

Constitutional isomers: isomers differing in atom-to-atom connectivity.

Corpus callosum: a brain structure consisting of axons connecting the two cerebral hemispheres.

Covalent bond: a chemical bond formed by the sharing of electrons between atoms.

CREB-1 and CREB-2: cAMP response element binding proteins.

Cyclooxygenase (COX): PGH synthase; the first enzyme in the generation of the eicosanoids from polyunsaturated fatty acids.

Cystic fibrosis: a hereditary disease in which there is a mutation in the CFTR protein.

Cytotoxic: one of several classes of cell-killing drugs commonly employed in cancer chemotherapy.

D

Dendrite: a neurite that receives signals from other neurons.

Dendritic spine: a small sac of membrane that protrudes from the dendrites of some cells and receives synaptic input.

Diabetes: a metabolic disease characterized by hyperglycemia.

Diastereomers: stereoisomers not related as mirror images.

Disulfide bonds: covalent chemical bonds formed between two sulfur atoms: $-S-S-$.

DNA: deoxyribonucleic acid. DNA is one of two forms of nucleic acids; the repository of genetic information in most living organisms.

DNA gyrase: a bacterial enzyme that catalyzes the unwinding of DNA; the molecular target of the quinolone antibiotics.

DNA polymerases: a family of enzymes that are involved in the replication of DNA.

DNA replication: the enzyme-catalyzed process in which one molecule of DNA is converted to two molecules of DNA, each identical to the first.

Dopa (L-dopa): a metabolic precursor of dopamine useful in the treatment of parkinsonism.

Dopa decarboxylase: the enzyme that catalyzes the conversion of L-dopa to dopamine.

Dopamine antagonist: an antagonist at one or more of the family of dopamine receptors.

Dysplasia: precancerous changes in cells or tissues.

E

Efferent pathway: a neural pathway leading from the central nervous system to the periphery.

Eicosanoids: the prostaglandins, prostacylins, thromboxanes, and leukotrienes.

Electron: a subatomic particle carrying a unit negative charge.

Electron transport chain: a series of multienzyme complexes organized in the inner layer of the mitochondrial membrane that catalyze the transport of electrons from reduced coenzymes to molecular oxygen coupled to the synthesis of ATP.

Element: a substance that cannot be separated into simpler substances by chemical means.

Emergent property: any property that cannot be explained through a reductionist approach.

Enantiomers: nonsuperimposable mirror images.

Enthalpy: a quantity, H, very closely related to energy, E.

Entropy: a quantity, S, that describes the randomness of a system in terms of its individual elements.

Enzyme: protein catalyst.

Enzyme inhibitors: molecules that bind to and inhibit the action of an enzyme. Many drugs in clinical use are enzyme inhibitors.

Essential amino acid: one of the family of amino acids that the human body cannot make or cannot make in adequate amounts to sustain optimal health.

Essential fatty acid: one of a family of polyunsaturated fatty acids that are essential dietary requirements.

Esters: organic molecules formed by the union of a carboxylic acid and an alcohol. They relevant functional group is —COOR.

Estrogen: a female sex hormone.

Estrogen antagonist: a molecule that antagonizes the action of estrogens at the estrogen receptor.

Ethanol: the iconic alcohol: CH_3CH_2OH.

Eubacteria: one of the two prokaryotic great domains of life. The other is the Archaea.

Eukarya: the domain of organisms whose cells have a membrane-enclosed nucleus and membrane-enclosed intracellular organelles; the third great domain of living organisms.

Exon: a coding sequence in pre-mRNA.

Exoskeleton: a hard encasement on the surface of an animal that provides for maintenance of structure and for protection.

F

Fat: a triglyceride that is solid at room temperature.

Fatty acid: any long-chain carboxylic acid; a key building block of more complex lipids.

Feedback inhibition: inhibition of some metabolic pathway (or other process) by a product of that pathway (or process).

First Law of Thermodynamics: energy is conserved; quantitatively, $\Delta E = q - w$, where q is heat added to the system and w is work done by the system.

Folic acid: the vitamin precursor to a family of coenzymes involved with the metabolism of one-carbon fragments.

Free energy: defined as $G = H - TS$; changes in G are a measure of the useful work that can be done by a system.

Free radicals: reactive oxidizing agents bearing an unpaired electron.

Fructose: a monosaccharide (hexose) found in high concentrations in honey; a constituent of sucrose, common table sugar.

G

Gene: the entire nucleic acid sequence that is necessary for the synthesis of a functional polypeptide or RNA molecule.

Genetic code: the relationship between triplets of bases in mRNA and the corresponding amino acids in proteins (or a termination signal) for which the triplet codes.

Genome: the complete complement of an organism's genes.

Ghrelin: a gut-derived peptide hormone that acts to increase food intake.

Glia: a family of cell types found in the central nervous system other than neurons.

Glucagon: a polypeptide hormone that signals the liver to produce glucose.

Glucocorticoid: one of a class of steroids having potent anti-inflammatory activity.

Glucose: the iconic monosaccharide, a hexose.

Glutamic acid: an amino acid commonly found in proteins and the most important excitatory neurotransmitter in the human central nervous system.

Glycerol: a three-carbon molecule possessing one hydroxyl group attached to each carbon atom; the core building block of the triglycerides.

Glycerophospholipid: a complex lipid based on glycerol, containing two fatty acids, and a polar headgroup; an important constituent of biological membranes.

Glycine: an amino acid commonly found in proteins and the most important inhibitory neurotransmitter in the human central nervous system.

Glycogen: a polysaccharide of glucose; an important energy store in muscle and liver.

Glycolipid: a molecule in which sugars and lipids are linked together.

Glycolysis: the metabolic pathway leading from glucose to pyruvate and lactate.

Glycoprotein: a protein linked to carbohydrate groups.

Goiter: enlargement of the thyroid gland as a consequence of inadequate dietary iodine.

Greenhouse gas: any atmospheric gas that has the capability of re-reflecting light reflected from the surface of the Earth back to the Earth, thus warming it.

Growth factor: a protein that must be present in the extracellular environment of a cell or tissue for the growth and normal development of sensitive cell types.

Growth hormone: a protein hormone that promotes growth in the young.

H

Habituation: a type of learning that leads to decreased responses to repeated stimulation.

Halogen: one of a specific group of elements, comprising fluorine, chlorine, bromine, and iodine.

HDL (high-density lipoprotein): a protein–lipid complex that carries cholesterol away from the tissues; "good cholesterol."

Heat of fusion: the quantity of heat required to convert 1 gram of a solid to the liquid at the melting point of the solid at a pressure of 1 atm.

Heat of vaporization: the quantity of heat required to convert one gram of a liquid to the gas at the boiling point of the liquid at a pressure of 1 atm.

Hemoglobin: the protein of red blood cells responsible for transporting oxygen from the lungs to the tissues.

Heterocycle: a cyclic organic molecule containing at least one noncarbon atom in the ring, typically, oxygen, nitrogen, or sulfur.

Hexose: a monosaccharide containing six carbon atoms.

Homeotic gene: a gene responsible for encoding a part of the body plan.

Hormone: a general term for the many classes of circulating chemical signals in multicellular organisms.

HRT (hormone replacement therapy): administration of estrogens to women or androgens to men who, due to menopause or age, have decreased levels of these plasma steroids.

Hydrocarbon: a compound formed only from the elements carbon and hydrogen.

Hydrogen: an element with only one proton in the nucleus.

Hydrogen bond: a weak chemical bond formed by sharing a hydrogen atom between two atoms chosen from oxygen, nitrogen, and fluorine.

Hydrogen peroxide: a hydride of oxygen: H_2O_2. It is an oxidizing agent.

Hydrophilicity: the property of being water-loving; most carbohydrates are hydrophilic.

Hydrophobicity: the property of being water-hating; fats, oils, and waxes are hydrophobic.

Hydroxyapatite: the mineral form of bone: $Ca_5(PO_4)_3(OH)$.

I

Insulin: a polypeptide pancreatic hormone involved in the regulation of blood glucose levels.

Integral membrane protein: a protein molecule that forms an essential part of the structure of biological membranes.

Intercalation: the insertion of a molecule between adjacent base pairs in the double helix of DNA.

Interleukin: one of a family of proteins secreted by white blood cells that have a spectrum of regulatory effects on other white blood cells.

Intermediate filament: a protein component of the cytoskeleton that includes filaments larger than the microfilaments and smaller than the microtubules.

Intron: a noncoding sequence in pre-mRNA.

Ion channel: a specialized protein structure that spans biological membranes and, when open, permits the passage of specific ions.

Ionic bond: a chemical bond formed by the transfer of one or more electrons from a donor to an acceptor.

Isomerism: the general term referring to the existence of isomers and their relationship to each other.

Isotope: atoms of the same element containing different numbers of neutrons.

L

Lateral gene transfer: the transfer of genetic material from one organism to another that is not its offspring.

LDL (low-density lipoprotein): the primary carrier of cholesterol in the blood; a positive risk factor for cardiovascular disease; "bad cholesterol."

Lee-Boot effect: inhibition of estrus in adult females or delay of puberty in juvenile females by female pheromones.

Leptin: a polypeptide hormone produced by and released from adipose tissue; its levels are inversely correlated with body weight; a homeostatic hormone.

Leukemia: a cancer of the blood or bone marrow characterized by abnormal proliferation of leukocytes.

Lymphoma: a cancer of the lymphatic system characterized by abnormal proliferation of lymphocytes.

M

Macrolide: a structurally complex antibacterial that acts as a bacterial protein synthesis inhibitor.

Malignant tumor: a tumor having the capability of invading surrounding tissues or establishing distant colonies.

MAO (monoamine oxidase): two related enzymes that inactivate neurotransmitters such as serotonin, dopamine, and norepinephrine.

Metabolic pathway: a sequence of enzyme-catalyzed reactions beginning with some defined substrates and ending with some defined products.

Metabolism: the subset of all potential chemical processes in a living organism actually expressed at some point in time.

Metabolite: a compound found on metabolic pathways.

Metalloprotein: a protein containing one or more bound metal ions.

Meter: a unit of measurement of length in the metric system. One meter equals approximately 39.37 inches in the English system.

Methane: the simplest hydrocarbon, CH_4.

Microtubule: a hollow rod formed from the protein tubulin. Microtubules form part of the cytoskeleton of cells as well as cilia and flagella.

Millimeter: one one-thousandth of a meter (10^{-3} meters).

Mineralocorticoid: a steroid that influences salt and water balance in the human body.

Miscibility: the property of two compounds forming solutions in all relative proportions.

Mitochondria: membrane-enclosed subcellular organelles that possess a small genome and are principally concerned with energy generation.

Mitosis: the process of nuclear division in eukaryotic cells.

Modulator pheromone: a pheromone that changes the sensitivity to a stimulus in the recipient.

Molecular mass: the mass of the collection of atoms that comprise the molecule.

Molecule: a construct formed by linking atoms together with chemical bonds.

Monobactam: a molecule based on a minimal β-lactam skeleton.

Monosaccharide: a simple sugar; a building block for polymeric, complex carbohydrates.

mRNA: messenger RNA.

Muscarinic acetylcholine receptors: a class of acetylcholine receptors for which muscarine is the iconic agonist.

Mutagen: a chemical that has the capacity to cause mutations.

Myoglobin: an oxygen-storing protein found in muscle.

N

Natural gas: an abundant gas collected in millions of tons annually that is largely methane in composition.

Natural product: a substance produced by living organisms.

Neoplastic: abnormal growth in a tissue or organ.

Neuromuscular junction: a chemical synapse between a spinal motor neuron axon and a skeletal muscle fiber.

Neuron: the information-processing cell of the nervous system.

Neurotransmitter: a substance that communicates a chemical signal across a synapse.

Neutron: a nuclear subatomic particle carrying no electrical charge.

Niacin: the generic name for nicotinic acid and nicotinamide; precursors for the coenzymes NAD^+ and $NADP^+$.

Nicotinic acetylcholine receptor: one of a class of acetylcholine receptors for which nicotine is the iconic agonist.

Nitric oxide: the simplest oxide of nitrogen, NO.

Nitrogen mustard: one of a family of reactive molecules that covalently modify DNA.

Noradrenaline: a potent adrenergic agonist; a neurotransmitter; also known as norepinephrine.

NSAIDs (nonsteroidal anti-inflammatory drugs): inhibitors of cyclooxygenase, including aspirin, ibuprofen, and ketoprofen, among many others.

Nuclear hormone receptor: one of a large class of cytoplasmic receptors that, when activated by binding to their ligands, migrate to the nucleus and act as transcription factors.

Nucleoside: a molecule in which a sugar is linked to a nitrogenous base.

Nucleotide: a molecule in which a sugar is linked to a nitrogenous base and phosphate: base–sugar–phosphate.

O

Obesity: a body mass index of 30 or greater.

Odorant receptor: one of a class of proteins encoded by several hundred genes that are receptors for odorant molecules.

Oil: a triglyeride that is liquid at room temperature.

Olfaction: the sense of smell.

Olfactory bulb: a structure in the vertebrate forebrain involved in the sense of smell.

Olfactory epithelium: the anatomical site for olfactory receptors.

Oncogene: a gene found in viruses or as part of a normal genome that is involved in eliciting cellular transformation.

Open system: a system that exchanges matter and energy with the environment.

Opioid: one of a class of potent analgesics; morphine provides a prominent example.

Osteoblast: a cell that forms bone.

Osteoclast: a cell that mobilizes bone.

Osteoporosis: a disease in which calcium is lost from bones, leading to weak, porous bones and an increased potential for fractures.

Oxidation: the removal of one or more electrons from some chemical species, molecule or ion.

Oxygen: the element having six electrons in the nucleus.

Ozone: a less common form of elemental oxygen: O_3.

P

Pancreatic lipase: a gut enzyme that initiates the digestion of dietary fats.

Pantothenic acid: a vitamin that is a precursor for coenzyme A.

Parathyroid hormone: a peptide hormone that acts to mobilize calcium from bone.

Pellagra: a dietary deficiency disease of niacin.

Penam: a molecule based on the penam mucleus (e.g., certain penicillins).

Penem: a molecule based on the penem nucleus (e.g., certain penicillins).

Penicillin: one of a family of important antibacterials based on the penam or penem skeleton.

Pentose: a monosaccharide containing five carbon atoms.

Peptide bond: the amide bond formed between two amino acids in proteins.

Peptide YY (PYY): a gut-derived peptide hormone that acts to decrease food intake.

Peripheral nervous system: the nervous system excluding the brain and spinal column.

Pheromone: a molecule employed for communication among different members of the same species.

Phosphates: chemical compounds derived from phosphoric acid.

Phosphatidic acid: the structural backbone of the glycerophospholipids; two molecules of fatty acids are esterified to a molecule of glyceryl phosphate.

Photosynthesis: a process in which radiant energy from the sun is employed to convert carbon dioxide and water into complex carbohydrates.

Phylogenetic tree: a tree-like structure showing the relationship among diverse organisms.

Plasma cell: an antibody-secreting cell of the blood.

Plasma membrane: the membrane that surrounds and encloses the cell.

Polymer: a large molecule consisting of many repeating units.

Polynucleotide: a polymer formed from nucleotide subunits; a synonym for nucleic acid.

Polypeptide: a synonym for protein.

Preproprotein: a proprotein synthesized with the N-terminal signal peptide targeting it to a membrane site.

Preprotein: a protein synthesized with the N-terminal signal peptide targeting it to a membrane site.

Primary structure: the structure of a protein defined by the sequence of amino acids along the polypeptide backbone.

Primer pheromone: a pheromone generating long-term physiological or endocrine responses.

Progenote: a very early form of life from which more complex living forms evolved.

Progestin: a steroid that is essential for the implantation of a fertilized egg.

Promoter: a specific nucleotide sequence in DNA that binds RNA polymerase and establishes the starting place for transcription.

Proprotein: a protein that is synthesized in an inactive form.

Prosthetic group: a nonprotein component of the protein.

Protein: a polymer formed from a family of 20 common L-α-amino acids.

Protein kinase A: an enzyme that catalyzes the ATP-dependent addition of phosphate to other proteins; subject to regulation by cAMP.

Proton: a nuclear subatomic particle carrying a unit of positive electrical charge.

Proto-oncogene: a normal cellular gene corresponding to an oncogene.

Pseudogene: a derivative of a coding gene that has been mutated over time with loss of function.

Purine: a heterocycle, the core of which consists of two fused rings, one six-membered and one five-membered, containing nitrogen atoms at positions 1, 3, 7, and 9.

Pyrimidine: a heterocycle, the core of which consists of a single six-membered ring containing nitrogen atoms at positions 1 and 3.

Q

Quaternary structure: the protein structure that results from the assembly of two or more independent polypeptide chains.

Quinolone antibiotic: one of a family of antibacterials based on the quinolone skeleton.

Quorum sensing: the ability of bacteria to coordinate their behavior through response to signaling molecules.

R

Receptor: a general term for a protein that is the recipient of chemical signals.

Recommended Daily Allowance (RDA): an older standard for the FDA-recommended daily intake of a vitamin.

Reduction: the addition of electrons to a chemical substance.

Reference Daily Intake (RDI): the current standard for the FDA-recommended daily intake of a vitamin.

Releaser pheromone: a pheromone generating immediate, behavioral responses.

Respiration: a process in which complex carbohydrates are consumed with oxygen to generate carbon dioxide and water and useful chemical energy in the form of ATP.

Resting potential: the membrane potential maintained by a cell when it is not generating action potentials.

Resveratrol: a natural product polyphenol found in red wine and elsewhere that has the property of extending the life spans of several organisms.

Retinal: a 20-carbon molecule critically involved in the visual process.

Retinoic acid: a 20-carbon carboxylic acid involved in development and differentiation; a potent teratogen.

Retinol: a 20-carbon molecule having vitamin A activity.

Riboflavin: also known as vitamin B_2; the precursor to the coenzymes flavin adenine mononucleotide, FMN, and flavin adenine dinucleotide, FAD.

Ribosome: a cell organelle consisting of RNA and proteins organized into two subunits; the site of protein synthesis.

Ribozyme: a catalytic RNA molecule.

RNA, ribonucleic acid: a polynucleotide in which the sugar is ribose and the four nitrogenous bases are adenine, guanine, cytosine, and uracil.

RNA interference (RNAi): a mechanism for control of gene expression through the targeting of mRNA molecules by small-interfering RNAs (siRNA).

RNA polymerases: enzymes that catalyze the transcription of DNA into RNA.

rRNA: ribosomal RNA; that form of RNA complexed with proteins in the ribosome.

S

Secondary structure: those protein structures that result from hydrogen bond formation between the amino and carbonyl groups of peptide bonds. The most important ones are the alpha helix and beta sheets.

Second Law of Thermodynamics: natural processes are accompanied by an increase in the entropy of the universe.

Semiochemical: a signaling molecule.

Sensitization: a type of learning leading to an intensified response to all stimuli.

Sickle cell anemia: a genetic disease that compromises the ability of red blood cells to deliver oxygen to the tissues.

Signaler pheromone: a pheromone that acts to provide information to a recipient.

Signaling molecule: a general term for a molecule that carries chemical signals in living organisms.

Signaling pathway: a communication channel that begins with a molecule that carries a message and ends with the regulation of some physiological process.

Signal peptide: an N-terminal amino acid sequence that targets preproteins to membrane sites.

Somatic nervous system: the part of the peripheral nervous system that innervates skin, joints, and muscles; responsible for the control of voluntary movements.

Spliceosome: a protein complex that catalyzes the splicing out of introns in pre-mRNA.

SSRI (selective serotonin reuptake inhibitor): one of a class of drugs useful in the treatment of depression.

Statin: an inhibitor of the enzyme HMGCoA reductase (HMGR).

Stereoisomers: isomers of identical constitution but differing in the arrangement of their atoms in space. Subclasses include constitutional isomers, enantiomers, and diastereomers.

Steroid: one of a class of biologically active molecules built on a specific molecular framework consisting of four fused rings.

Sucrose: simple table sugar; a disaccharide containing glucose and fructose.

Sugar: a common term employed to refer to sucrose specifically and mono- or disaccharides generally.

Sulfa drug: an antibacterial that acts by inhibiting bacterial folic acid synthesis.

Sulfonamide: a synonym for sulfa drug.

Superoxide: an oxygen-centered free radical: $O_2 \bullet^-$.

Suppressor gene: a gene that acts to prevent inappropriate cell division.

Synapse: a region where a neuron transmits information to another cell, neuron or muscle cell.

Synaptic vesicle: a membrane-enclosed vesicle found near synapses and housing neurotransmitters.

T

Targeted cancer drug: a cancer drug that targets a specific defect in cancer cells, as opposed to cytotoxics.

Taste bud: a structure in the mouth housing receptors for tastants.

T cell: one of a class of white blood cells that are responsible for cell-mediated immunity.

Teleonomic: goal-directed or outcome-directed.

Teratogen: a molecule that causes birth defects.

Tertiary structure: that structure of proteins that identifies the position in three-dimensional space of each atom in the protein.

Tetracycline: one of a class of antibacterials based on a tetracyclic skeleton and that act as a bacterial protein synthesis inhibitors.

Tetrahedron: a solid figure having four triangular faces. If the four faces are identical, the tetrahedron is said to be regular.

Thyroid hormone: an iodine-containing amino acid that affects the action of the thyroid gland.

Thyroxine: one of the iodine-containing amino acids that act as a thyroid hormone.

Transcription: the synthesis of RNA from a DNA template.

Transcription factor: a regulatory protein that binds to DNA and influences gene expression.

Trans fats: fats containing unsaturated fatty acids that have the *trans* geometry at their carbon–carbon double bonds.

Transformation: the conversion of normal cells into neoplastic cells.

Translation: the synthesis of proteins from mRNA templates.

Triglyceride: a triester of fatty acids with glycerol; they form the fats and oils.

tRNA: transfer RNA.

Type II hypercholesterolemia: a genetic disease in which the level of hepatic LDL receptors is compromised.

U

Upper limit (UL): the maximum daily intake of a vitamin not associated with adverse effects.

V

Vanderburgh effect: the acceleration of puberty in immature females by males.

Virulence factor: a protein responsible for the ability of bacteria to cause disease.

Virus: in the simplest cases, an infectious aggregate of nucleic acid and protein, lacking both structures and metabolic machinery usually found in cells.

Vitamin: a general term for a required human nutrient.

Vitamin A: a family of molecules having vitamin A activity, such as retinol and several carotenes.

Vitamin B_1: thiamine; the precursor to the coenzyme thiamine pyrophosphate.

Vitamin B_2: riboflavin.

Vitamin B_6: a family of molecules having vitamin B_6 activity; these include pyridoxal, pyridoxine, and pyridoxamine; precursors to the coenzyme pyridoxal phosphate.

Vitamin B_{12}: a cobalt-containing molecule that is the precursor for two coenzymes.

Vitamin C: ascorbic acid.

Vitamin D: a family of molecules that prevent the diseases rickets and osteomalacia; cholecalciferol and ergocalciferol are two important examples.

Vomeronasal organ: a structure possessing odorant receptors for pheromones.

W

Whitten effect: the induction of estrus in adult females by males.

X

Xerophthalmia: night blindness, occasioned by a deficiency of vitamin A.

Index